Thermodynamics of Crystalline States

Minoru Fujimoto

Thermodynamics of Crystalline States

Second Edition

Springer

Minoru Fujimoto
Department of Physics
University of Guelph
Guelph, ON NIG 2W1
Canada

ISBN 978-1-4939-0238-5 ISBN 978-1-4614-5085-6 (eBook)
DOI 10.1007/978-1-4614-5085-6
Springer New York Heidelberg Dordrecht London

Printed on acid-free paper

Springer is part of Springer Science+Business Media (www.springer.com)

Preface

Originated from steam engines, in thermodynamics the energy exchange is formulated traditionally for homogeneous states of materials. Applying to condensed matters however, laws of thermodynamics have to be used with respect to the structural detail. Following Kirkwood's *Chemical Thermodynamics*, this book is written for lattice dynamics in crystalline states under the laws of thermodynamics. It is noted that lattice symmetry remains implicit in thermodynamic functions of crystalline states, if assumed as homogeneous. In contrast, deformed crystals with disrupted symmetry are not stable and inhomogeneous, exhibiting *mesoscopic* properties, for which the lattice dynamics should be redefined under thermodynamic principles. I have selected the topics of structural changes, magnetic crystals, and superconducting transitions in this book to discuss these basic thermodynamic processes in crystalline states.

Born and Huang have laid ground for thermodynamics of crystalline states in their book *Dynamical Theory of Crystal Lattices*. However, they considered that order–disorder phenomena was independent from the lattice dynamics, and so excluded it from their book. On the other hand, today new evidence indicates that the problem should be treated otherwise; in fact, the lattice plays a vital role in ordering processes. Accordingly, I was motivated to write about the physics of crystal lattice in the light of Born-Huang's principles, which constitutes my primary objective in this book.

In modern experiments, mesoscopic objects in crystals can be investigated within the timescale of observation, yielding results that appear somewhat unusual, if compared with macroscopic experiments. It is important that thermodynamic relations in mesoscopic states should be described with respect to the timescale of observation. Also significant is that mesoscopic quantities in crystals are driven by internal interactions in nonlinear character. I have therefore discussed thermodynamic quantities with regard to the timescale of observation, including elementary accounts of nonlinear physics to deal with long-range correlations. For the convenience of readers who are not particularly familiar with nonlinear dynamics, Appendix is attached for some useful formula of elliptic functions.

I have rewritten this second edition for advanced students in physics and material science, assuming basic knowledge of traditional thermodynamics and quantum theories. This book is designed for serving as a useful textbook for classroom and seminar discussions. I hope that it will stimulate advanced studies of condensed matter.

I should mention with my sincere appreciation that I have benefited for my writing from numerous discussions with and comments from my colleagues and students. Finally, I thank my wife Haruko for her continuous support and encouragements.

Mississauga, Ontario Minoru Fujimoto

Contents

1 Introduction 1
- 1.1 Crystalline Phases 1
- 1.2 Structural Changes 4
- 1.3 New Concepts 7
  - 1.3.1 Modulated Phases and Renormalized Coordinates 7
  - 1.3.2 Thermodynamic Changes and Nonlinear Dynamics 8
  - 1.3.3 Fields within Thermodynamic Boundaries 8
  - 1.3.4 Adiabatic Potentials 8
- 1.4 Sampling Modulated States 9
- Exercise 1 9
- References 10

2 Phonons 11
- 2.1 Normal Modes in a Simple Crystal 11
- 2.2 Quantized Normal Modes 14
- 2.3 Phonon Field and Momentum 16
- 2.4 Thermal Equilibrium 18
- 2.5 Specific Heat of a Monatomic Crystal 20
- 2.6 Approximate Models 22
  - 2.6.1 Einstein's Model 22
  - 2.6.2 Debye's Model 23
- 2.7 Phonon Statistics Part 1 25
- 2.8 Compressibility of a Crystal 28
- Exercise 2 30
- References 31

3 Order Variables and the Adiabatic Potential 33
- 3.1 A One-dimensional Ionic Chain 33
- 3.2 Displacive Variables 35
- 3.3 Born-Oppenheimer's Approximation 39

3.4 Equilibrium Crystals and the Bloch Theorem 45
3.4.1 Reciprocal Lattice and Renormalized Coordinates 45
3.4.2 The Bloch Theorem 47
3.4.3 Lattice Symmetry and the Brillouin Zone 49
3.5 Phonon Scatterings 51
3.5.1 Phonon Scatterings in Crystals 51
3.5.2 The Born-Huang Relaxation and Soft Modes 52
Exercise 3 53
References 54

4 Mean-Field Theories of Binary Ordering 55
4.1 Probabilities in Binary Alloys 55
4.2 The Bragg-Williams Theory 57
4.3 Becker's Interpretation 60
4.4 Ferromagnetic Order 62
4.5 Ferromagnetic Transition in Applied Magnetic Field 64
Exercise 4 66
References 66

5 Pseudospin Clusters 67
5.1 Pseudospins for Binary Displacements 67
5.2 A Tunneling Model 70
5.3 Pseudospin Correlations 71
5.4 Pseudospin Clusters in Crystals 73
5.4.1 Classical Pseudospins in Crystals 73
5.4.2 Clusters of Pseudospins 74
5.5 Examples of Pseudospin Clusters 78
5.5.1 Cubic to Tetragonal Transition in $SrTiO_3$ 78
5.5.2 Monoclinic Crystals of Tris-Sarcosine Calcium Chloride (TSCC) 80
Exercise 5 82
References 82

6 Critical Fluctuations 83
6.1 The Landau Theory of Binary Transitions 83
6.2 Internal Pinning of Adiabatic Fluctuations 87
6.3 Critical Anomalies 88
6.4 Observing Anomalies 90
6.5 Extrinsic Pinning 92
6.5.1 Pinning by Point Defects 93
6.5.2 Pinning by an Electric Field 93
6.5.3 Surface Pinning 94
Exercise 6 94
References 95

7 Pseudospin Correlations 97
7.1 Propagation of a Collective Pseudospin Mode 97
7.2 Transverse Components and the Cnoidal Potential 101
7.3 The Lifshitz Condition for Incommensurability 103
7.4 Pseudopotentials 105
Exercise 7 108
References 109

8 Soliton Theory of Long-Range Order 111
8.1 A Longitudinal Dispersive Mode of Collective Pseudospins 111
8.2 The Korteweg–deVries Equation 115
8.3 Solutions of the Korteweg–deVries Equation 117
8.4 Cnoidal Theorem and the Eckart Potential 120
8.5 Condensate Pinning by the Eckart Potentials 122
8.6 Born–Huang's Transitions 126
8.7 Topological Mapping of Mesoscopic Fields 129
Exercise 8 132
References 135

9 Soft Modes 137
9.1 The Lyddane–Sachs–Teller Relation 137
9.2 Soft Lattice Modes of Condensates 141
9.2.1 The Lattice Response to Collective Pseudospins 141
9.2.2 Temperature Dependence of Soft-Mode Frequencies 144
9.2.3 Cochran's Model 148
9.3 Symmetry Change at T_c 151
Exercise 9 153
References 154

10 Experimental Studies on Critical Fluctuations 155
10.1 Diffuse X-Ray Diffraction 155
10.2 Neutron Inelastic Scatterings 160
10.3 Light Scattering Experiments 164
10.3.1 Brillouin Scatterings 165
10.3.2 Raman Scatterings 169
10.4 Magnetic Resonance 171
10.4.1 Principles of Nuclear Magnetic Resonance and Relaxation 173
10.4.2 Paramagnetic Resonance of Impurity Probes 175
10.4.3 The Spin Hamiltonian 176
10.4.4 Hyperfine Interactions 180
10.5 Magnetic Resonance in Modulated Crystals 180
10.6 Examples of Transition Anomalies 183
References 187

11 Magnetic Crystals 189
11.1 Microscopic Magnetic Moments in Crystals 189
11.2 Paramagnetism 193
11.3 Spin–Spin Correlations 194
11.4 Spin Clusters and Magnetic Symmetry 198
11.5 Magnetic Weiss Field 200
11.6 Spin Waves 203
11.7 Magnetic Anisotropy 204
11.8 Antiferromagnetic and Ferrimagnetic States 208
11.9 Fluctuations in Ferromagnetic and Antiferromagnetic States 210
11.9.1 Ferromagnetic Resonance 211
11.9.2 Antiferromagnetic Resonance 212
Exercise 11 214
References 214

12 Phonon and Electron Statistics in Metals 215
12.1 Phonon Statistics Part 2 215
12.2 Conduction Electrons in Metals 218
12.2.1 The Pauli Principle 218
12.2.2 The Coulomb Interaction 220
12.2.3 The Bloch Theorem in Equilibrium Crystals 221
12.3 Many-Electron Systems 222
12.4 Fermi–Dirac Statistics for Conduction Electrons 226
Exercise 12 226
References 227

13 Superconducting Metals 229
13.1 Superconducting States 229
13.1.1 Near-Zero Electrical Resistance 229
13.1.2 The Meissner Effect 231
13.1.3 Thermodynamic Equilibrium Between Normal and Superconducting Phases 233
13.2 Long-Range Order in Superconducting States 235
13.3 Electromagnetic Properties of Superconductors 237
13.4 The Ginzburg–Landau Equation and the Coherence Length 244
Exercise 13 247
References 248

14 Theories of Superconducting Transitions 249
14.1 The Fröhlich's Condensate 249
14.2 The Cooper Pair 252
14.3 Critical Anomalies and the Superconducting Ground State 255
14.3.1 Critical Anomalies and a Gap in a Superconducting Energy Spectrum 255

14.3.2 Order Variables in Superconducting States 257
14.3.3 BCS Ground States 262
14.3.4 Superconducting States at Finite Temperatures 266
Exercise 14 267
References 268

Appendix 269

Index 271

Chapter 1
Introduction

The role played by the lattice for symmetry changes in a crystal is a basic subject for discussion in this book. On the other hand, as related to *lattice excitations*, the concept of *order variables* needs to be established in thermodynamics of crystalline states. Normally inactive at elevated temperatures, order variables at lattice sites should be responsive to a symmetry change at a critical temperature. Exhibiting a *singular* behavior, an *adiabatic potential* should also emerge with order variables in finite magnitude with lowering temperature. Well documented experimentally, we should analyze their properties under thermodynamic principles.

In reviewing thermodynamic principles, the present thermodynamics of crystalline states is overviewed in this chapter, emphasizing dynamic roles played by the lattice structure.

1.1 Crystalline Phases

Originated from early studies on steam engines, thermodynamics today is a well-established discipline of physics for thermal properties of matter. Specified by a uniform density, a thermodynamic state of a crystal can be described by state functions of temperature and pressure of the surroundings. In contrast, characterized by symmetry of the lattice structure, properties of a *single crystal* are attributed to masses and other physical properties of constituents at lattice points, while symmetry per se cannot be responsible for the physical properties. We realize that structural changes cannot be described by uniform state functions, unless these functions are defined as related with the structural detail. The structural transition is generally *discontinuous* at a critical temperature T_c, below which even a chemically pure crystal becomes *heterogeneous*, composed of substructures in smaller volume, for example, *domains* in thermal equilibrium. Nevertheless, an external force or field can transform domains from one type to another while

M. Fujimoto, *Thermodynamics of Crystalline States*,
DOI 10.1007/978-1-4614-5085-6_1, © Springer Science+Business Media New York 2013

maintaining structural stability. In addition, there are inhomogeneous crystals among various types, which are signified by a *sinusoidal* density, representing *modulated* structure. Following Kirkwood [1] in *Chemical Thermodynamics* we can consider state functions of continuous internal variables in crystalline states in equilibrium with the surroundings.

In *Dynamical Theory of Crystal Lattices*, Born and Huang [2] proposed the principle for lattice stability, laying the foundation of thermodynamics of crystalline states. However, their principle was not quite verified with experimental results, which were not fully analyzed at that time. Today, in spite of many supporting results, their theory is still not considered as sufficiently substantiated.

A crystalline state is very different from a gaseous state of a large number of *free* independent particles. A crystal structure packed with identical constituents is characterized by distinct symmetry; in contrast, fast molecular motion prevails in gaseous states. The *internal energy* of a gas is primarily kinetic energies of constituent particles in free motion, whereas for a crystal, it is intermolecular potential energies in the packed structure. The structural transformation in crystals is a dynamic process, whereas a gas is normally in a single phase, except in condensation process. In addition, a gas confined to a container has a finite volume, whereas external conditions determine a crystal volume. Simplifying by a mathematical conjecture, we can say that surfaces have a little contribution to bulk properties of a sizable crystal. However, a small volume change is inevitable in practical crystals under structural changes.

In thermodynamics, a crystal must always be in contact with the surroundings. Assuming no chemical activity, surfaces are in *physical contact* with the surroundings, exchanging *heat* between them. Joule demonstrated that heat is nothing but energy, although it is of unique type microscopically. Boltzmann considered that heat originates from randomly distributed microscopic energies in the surroundings. On the other hand, excitations in a crystal are primarily lattice vibrations at a given temperature, which are quantum mechanically expressed as *phonons* in random motion. The phonon spectrum is virtually continuous; the *internal energy* of lattice is given by a statistical average of distributed phonon energies. In this context, the lattice symmetry is implicit in thermodynamic functions, as specified by a uniform density in a constant volume.

Equilibrium between two bodies in thermal contact, as characterized by a common *temperature* T, signifies that there is no net flow of heat across the surfaces. For a combined system of a crystal and its surroundings, the total fluctuating heat is always *dissipative*, so that ΔQ is expressed as negative, that is, $\Delta Q \leq 0$, implying that any thermal process is fundamentally *irreversible* in nature, as stated by the *second-law* of thermodynamics. Depending on the nature of heat dissipation, such ΔQ cannot generally be expressed by a function of temperature and pressure; namely, the *heat quantity* Q is not a state function. Instead, a function S, defined by $\frac{\Delta Q}{T} = \Delta S$, can be employed as a state function if the *integrating denominator* T can be found for an infinitesimal S to be a *total differential*, representing a reversible contact between the crystal and surroundings. Clausius generalized this argument in integral form $\oint_C \frac{Q}{T} \leq 0$ or $\oint_C \mathrm{d}S \leq 0$ around a closed curve C in a *phase diagram*, where the equality sign indicates an idealized case of *reversible* process. Thermal equilibrium can thus be determined by *maximizing* the state function S, which is known as the *entropy*.

Boltzmann interpreted the entropy S in terms of *thermodynamic probability* $g(T)$ at constant pressure to represent a thermal average of randomly distributed microscopic energies in the surroundings, for which he wrote the relation

$$S = \mathrm{k_B} \ln g(T), \tag{1.1}$$

where $\mathrm{k_B}$ is the Boltzmann constant. For crystalline states in equilibrium at temperature T, such a probability is a valid concept if random phonon energies are responsible for the statistical description under a constant volume V. However, if a small volume change ΔV is considered at constant T and p, a process at a constant crystal volume V is not an acceptable assumption for a crystal under constant p. Hence, Boltzmann's statistical theory is valid only if V is constant; otherwise, the dynamical system in a crystal is statistically *non-ergodic*.

In thermodynamics, physical properties of a crystal can be represented by the internal energy $U(p,T)$, which is varied not only by heat ΔQ but also by external work W, as stated by *the first law of thermodynamics*. Namely,

$$\Delta U = \Delta Q + W. \tag{1.2}$$

A mechanical work W can change the volume V, where a structural modification may be induced, for example, by an electric or magnetic field, straining crystals by their forces X. Considering that such variables σ_m respond to applied X, the work W can be expressed as $W = -\sum_m \sigma_m X$. It is noted that such X can be associated with internal interactions with σ_n at different sites n, signifying *correlations* between internal variables σ_m and σ_n. Such internal correlations can occur *adiabatically* in a crystal, independent of temperature. However, they are weakly temperature dependent in practice, via interactions with phonons. In contrast, the heat energy Q depends only on the temperature of the surroundings, flowing in and out of a crystal at a constant V.

Correlated σ_m is generally in collective motion at long wavelength, propagating through an excited lattice. Such motion in a crystal should occur when driven by an *adiabatic potential* ΔU_m at the lattice point m, as will be discussed in Chap. 7. In a *modulated structure*, the propagation can be detected in a standing wave whose period is not necessarily the same as in the lattice. On the other hand, if the correlation energy in short range is expressed in a form $-J_{mn}\sigma_m\sigma_n$, where J_{mn} is a function of the distance between σ_m and σ_n, we can write

$$-\sum_n J_{mn}\sigma_m\sigma_n = -\sigma_m X_m \tag{1.3}$$

which allows us to define the *internal field* $X_m = \sum_n J_{mn}\sigma_n$ at a site m. Averaging X_m and σ_m over all lattice sites, we can define the *mean-field average* $\langle X_m \rangle$ and $\langle \sigma_m \rangle$, respectively. For a nonzero $\langle \sigma_m \rangle$, the expression (1.3) can be interpreted as work performed by a field $\langle X_m \rangle$ on the $\langle \sigma_m \rangle$in mean-field accuracy. On the other hand, by writing (1.3) as $-\langle \sigma_m \rangle X_{\text{int}}$, X_{int} can represent another kind of internal field, defined in better accuracy than the mean-field average $\langle X_m \rangle$. In this case, attributed to the adiabatic potential ΔU_m, we can write $X_{\text{int}} = -\nabla_m(\Delta U_m)$, corresponding to the *Weiss field* that was originally proposed for a magnetic crystal.

In the presence of an external field X, the *effective field* can be defined as $X + X_{\text{int}} = X'$, considering X' as if applied externally. Such a field X' is usually temperature dependent, as contributed by temperature-dependent X_{int}; hence, we can write $X' = X'(T)$ under a constant p. The macroscopic energy relation can therefore be expressed as

$$\Delta U \leq \Delta Q - \Delta(\sigma X') \quad \text{or} \quad \Delta U - T\Delta S + \Delta(\sigma X') \leq 0.$$

Writing the Gibbs potential as $G = U - TS + \sigma X'$, we can obtain the inequality

$$\Delta G \leq 0, \tag{1.4}$$

indicating that the minimum of G determines the equilibrium at constant T and p.

In practice, *order variables* σ_m should be identified experimentally; thereby, equilibrium can be determined from the Gibbs function at given T and p. On the other hand, the mean-field average $\langle \sigma_m \rangle = \sigma$, called *order parameter*, can be utilized for simplifying description. In this case, the Gibbs function can be written as $G(p,T,\sigma)$, where the order parameter σ is defined for the range $0 \leq \sigma \leq 1$, where 0 and 1 are designated to perfect *disorder* and *order*, respectively. In *spontaneously* ordered crystals, the internal field X_{int} can be related to spatially distributed σ_m. For a ferromagnetic crystal, Weiss (1907) postulated that $X_{\text{int}} \propto$ (*magnetization*), which is traditionally called *Weiss' molecular field* in a ferroelectric crystal.

1.2 Structural Changes

The structure of an equilibrium crystal is stable, which is however transformable from one stable state to another by applying an external field. In addition, we consider that the order variables σ_m emerging at T_c in finite magnitude are in collective motion, leading to spontaneous *phase transitions*. Although assumed primarily as independent of the hosting lattice, the correlated order variables modulate translational lattice symmetry. Interpreting this by Newton's action–reaction principle, correlated lattice displacements $\boldsymbol{u}_m$ should interact with correlated vectors σ_m. Born and Huang [2] proposed their principle for the lattice to maintain stability with minimum strains, which can be evaluated by minimizing the Gibbs potential. Such a minimizing process can nevertheless be observed as thermal relaxation.

Regarding a transition from a phase 1 to another phase 2 at a critical point (p_c, T_c), the Gibbs potential may be written as $G(p_c, T_c;\ \sigma_1, \sigma_2)$, where σ_1 and σ_2 represent these phases 1 and 2 which coexist during the transition. Although somewhat imprecise, we write the whole Gibbs potential under an applied field X as

$$G_{\mathrm{trans}} = G(p_c, T_c; \sigma_1, \sigma_2) - (\sigma_1 + \sigma_2)X.$$

In the noncritical region below T_c, we consider two Gibbs potentials G_1 and G_2, coexisting at p and T, which are expressed as

$$G_1 = G(p, T) - \sigma_1 X$$

and

$$G_2 = G(p, T) - \sigma_2 X,$$

respectively.

For a *binary transition* characterized by inversion $\sigma_1 = -\sigma_2$, we consider fluctuations $\Delta\sigma_1 = -\Delta\sigma_2$, and hence, $\sum_{i=1,2}\left(\frac{\partial G_{\mathrm{trans}}}{\partial \sigma_i}\right)_{p_c,T_c}\Delta\sigma_i = 0$. Therefore, expanding with respect to $\Delta\sigma_i$, the Gibbs potential can be expressed as

$$\Delta G_{\mathrm{trans}} = \frac{1}{2}\sum_{i=1,2}\left(\frac{\partial^2 G_{\mathrm{trans}}}{\partial \sigma_i \partial \sigma_j}\right)_{p_c,T_c}\Delta\sigma_i\Delta\sigma_j + \ldots. \tag{1.5}$$

Neglecting higher than second-order terms for small $\Delta\sigma_i$, a nonvanishing $\Delta G_{\mathrm{trans}}$ gives a discontinuity due to the leading term of the second-order in (1.5). On the other hand, if such an inversion does not apply, $\Delta G_{\mathrm{trans}}$ is dominated by the finite first-order derivative terms; hence, the transition is called the *first-order*, according to the *Ehrenfest's classification*. In the binary case, $\Delta G_{\mathrm{trans}}$ is dominated by the second-order terms, and hence, the transition is called the *second-order*. It is noted from (1.5) that both the correlations $\Delta\sigma_i\Delta\sigma_j$ and the second-order derivatives should not vanish for the leading term to come into effect; nonzero correlations are essential, as substantiated by *transition anomalies* observed for the specific heat. Such fluctuations $\Delta\sigma_i$ should be quantum mechanical in nature, being responsible for the threshold anomalies, while σ_i behave like a *classical vector* at and below T_c.

The transformation below T_c is signified by the difference $\Delta G_{1,2} = G_2 - G_1$, which is finite at $X \neq 0$, however zero if $X = 0$. Accordingly,

$$\Delta G_{1,2} = (\sigma_2 - \sigma_1)X. \tag{1.6}$$

For $X = 0$ and $X \neq 0$, the transformations are the second- and the first-order, respectively. In thermodynamics, both the Gibbs potential and order parameter are *extensive variables* by definition, as they are proportional to the corresponding volume. Denoting domain volumes by V_1 and V_2, the total volume is $V_1 + V_2 = V$, which can be assumed as invariant under a constant p. If so, (1.6) can be determined by either the volume ratio V_1/V_2 or order-parameter ratio σ_1/σ_2; both can be changed by applying X (see Chap. 4).

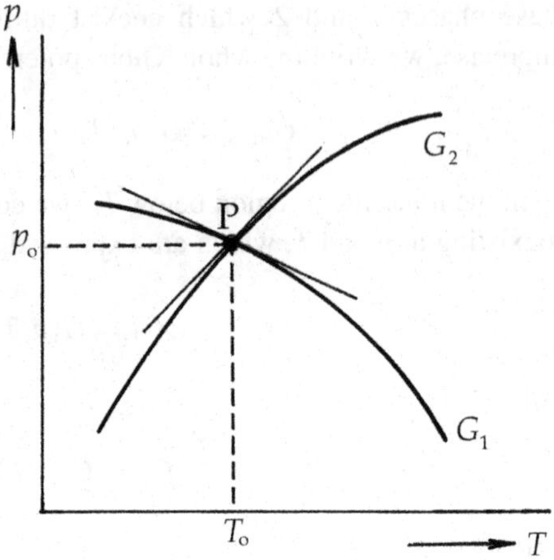

Fig. 1.1 Phase equilibrium in the p–T diagram. Two curves of Gibbs functions. $G_1(p, T)$ and $G_2(p, T)$ cross at a point $P(p_o, T_o)$, representing thermal equilibrium between phase 1 and phase 2. If the transition is discontinuous at P, such Gibbs functions cannot specify the equilibrium sufficiently at P, requiring another variable such as σ.

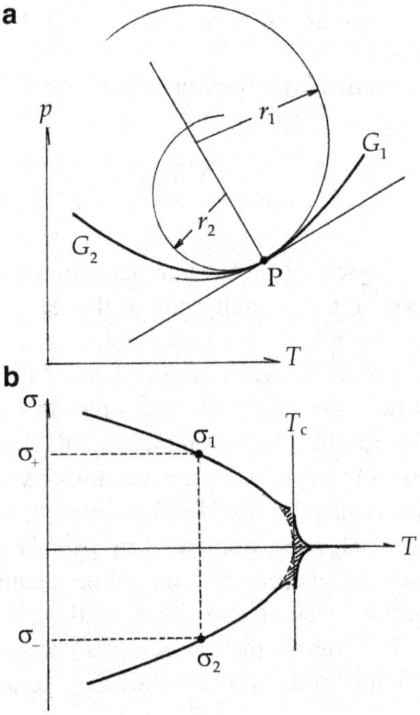

Fig. 1.2 (**a**) The second-order phase transition is characterized by two functions $G_1$and G_2 that have a common tangent and different curvatures at P in the p–T diagram. (**b**) Equilibrium between binary domains specified by σ_1 and σ_2, which are separated phases at temperatures below the critical temperature T_c. Transition anomalies near T_c are shown schematically by the *shaded area*.

Figure 1.1 shows a phase diagram, where two Gibbs functions of phases G_1 and G_2 are schematically plotted in a p–T plane. The crossing point $P(p_o, T_o)$ indicates that a transition occurs between G_1 and G_2 at $P(p_o, T_o)$. The transition is generally first-order, because of discontinuity in the perpendicular direction to p–T plane, although implicit in the two-dimensional p–T diagram; there is a finite change of curvature at P as related to ΔG_{trans}. Nevertheless, the discontinuity vanishes, if we ignore fluctuations. Figure 1.2(a) illustrates a continuous binary transition $G_1 \rightarrow G_2$ at T_c, assuming that there are no fluctuations. In the diagram σ–T in Fig. 1.2(b), a continuous transition is sketched by solid lines, which are broadened by fluctuations in the vicinity of T_c. Below T_c, the transformation $\sigma_1 \leftrightarrow \sigma_2$ can take place in the first order, if performed by an external X. It is noted that such a process under a constant p condition is not strictly adiabatic, because X_{int} is temperature dependent in practical crystals.

1.3 New Concepts

In this book, we discuss thermodynamics of crystalline states under the laws of thermodynamics, for which new concepts, such as *renormalized coordinates*, *solitons*, and *fields*, are added to express collective properties of order variables in periodic structure. These are listed in the following.

1.3.1 Modulated Phases and Renormalized Coordinates

We describe fluctuations in crystals as arising from quantum-mechanical space–time uncertainties for order variables σ_i that represent classical vectors at all sites of a periodic lattice. Their correlated motion described as propagation in the *continuum field* suited particularly to low excitations of lattice under critical conditions. Associated with sinusoidal lattice displacements $\boldsymbol{u}_i$, the related variables σ_i are signified by their displacements and expressed as a sinusoidal field $\sigma_i = \sigma_o \exp i(\boldsymbol{k}.\boldsymbol{r}_i - \omega t_i + \varphi_i)$, accompanying phase uncertainties $\Delta\varphi_i$.

Surfaces are significant for thermal properties of a crystal; however, by ignoring them, we can idealize a crystal as infinite periodic structure, where phases of order variables σ_i fluctuate by $\Delta\varphi_i$ at a given T. Such sinusoidal fluctuations are detectable if their *standing waves* are *pinned* by the modulated lattice, visualizing *mesoscopic* fluctuations. (See Chap. 8 for the detail.) The order variable $\sigma(\boldsymbol{r}, t) = \sigma(\phi)$ is continuous in space–time or phase $\phi = \boldsymbol{k}.\boldsymbol{r} - \omega t$, for which one full angular period $0 \le \phi \le 2\pi$ is sufficient instead of a repeated expression. Hence, such a continuous variable $\sigma(\phi)$ can be utilized for expressing thermodynamic functions in a mesoscopic state.

Expressed in the form $\sigma = \sigma_o f(\phi)$ incorporated with periodic boundary conditions, σ_o is a finite amplitude and $\phi = \boldsymbol{k}.\boldsymbol{r} - \omega t$ in the range $0 \le \phi \le 2\pi$. The *spatial phase* $\phi_s = \boldsymbol{k}.\boldsymbol{r}$ in standing waves visualizes distributed mesoscopic phases in the angular range $0 \le \phi_s \le 2\pi$ in a whole crystal without referring to lattice sites. Noted that this definition is similar to a *renormalization group* of order variables in statistical mechanics [3], we shall call ϕ_s and$(\boldsymbol{r}, t)$ *renormalized* phase and coordinates, respectively. The Gibbs function in a mesoscopic phase can also be expressed as $G(\sigma_o, \phi_s)$ in terms of renormalized variables.

1.3.2 Thermodynamic Changes and Nonlinear Dynamics

Physical properties of matter can change with the responsible dynamical system in equilibrium under isothermal and adiabatic conditions. Small deviations from equilibrium are always restorable in stable crystals, where the dynamical system is *dissipative* and *dispersive* under these conditions. Accordingly, the equation of motion during thermodynamic changes can be *nonlinear*, leading to the modulated lattice structure. While not particularly familiar in traditional physics, we need to deal with the nonlinearity in thermodynamics of crystalline states. *Soliton* theory [4] in nonlinear physics provides with collective motion in an accurate approximation better than the mean-field approach.

1.3.3 Fields within Thermodynamic Boundaries

For collective motion in crystals, it is convenient to use the *field* concept [5]. The superconducting transition in metals can be discussed with the field concept on condensation of electrons in the reciprocal lattice space. Characterized by *supercurrents* and *Meissner effects*, superconducting phase transitions are more complex than other ordering phenomena, although exhibiting similar specific heat anomalies arising from condensing charge carriers. Fröhlich's field-theoretical model of electron–phonon interaction provides a clear image of condensates to initiate a phase transition, on which the theory of Bardeen, Cooper, and Schrieffer can define order variables. The last three chapters in this book are for superconducting transitions in metals, which can be discussed by analogy of structural changes. Phase changes in crystals are generally associated with adiabatic potentials, where the *nonlinearity* arises from interactions with the lattice.

In this book, dynamical variables in crystals are discussed within thermal boundaries at constant T and p, which can be described as isothermal and adiabatic processes in crystals. The former remains the same as in traditional theories, whereas the latter process cannot be separated from the former, showing thermal relaxation to stabilize structure.

1.3.4 Adiabatic Potentials

Introduced early by Boltzmann and Ehrenfest, the concept of adiabatic invariance should be extended explicitly to crystalline states to deal with long-range correlations, which can be discussed by means of the *internal adiabatic potential* in the lattice. Recognized in practical applications of thermodynamics, it was however not seriously dealt with in most available references. We therefore emphasized its significance for driving condensates for structural order.

1.4 Sampling Modulated States

For critical fluctuations, the timescale t is a significant for sampling experiments of the spatial distribution of σ. For that purpose, the timescale t_o of sampling should be shorter than the timescale t of fluctuations. A Gibbs potential, if sampled at a given p–T condition, should be expressed by a space–time average over t_o, namely,

$$G_{p,T} = \frac{1}{t_o}\int_0^{t_o} dt\left(\frac{1}{V}\int_V g(\boldsymbol{r},t)d^3\boldsymbol{r}\right), \tag{1.7a}$$

where $g(\boldsymbol{r},t)$ is the density of Gibbs potential in the sampling volume V. Writing the time average as $\frac{1}{t_o}\int_0^{t_o} g(\boldsymbol{r},t)dt = \langle g(\boldsymbol{r},t)\rangle_t$, (1.7a) can be expressed by

$$G_{p,T} = \frac{1}{V}\int_V \langle g(\boldsymbol{r},t)\rangle_t d^3\boldsymbol{r}. \tag{1.7b}$$

The average density $\langle g(\boldsymbol{r},t)\rangle_t$ over t_o can be sampled from a specimen of volume V. It is noted that such a sampling results in a useful information only if $t>t_o$. Sampled mesoscopic $\langle\sigma(\boldsymbol{r},t)\rangle_t$ represents $\sigma(\phi_s)$, where $0\le\phi_s\le 2\pi$ and the distributed $\sigma(\phi_s)$ is explicit, provided that $t>t_o$.

Exercise 1

1. What is the mean-field average? If a physical event at each lattice point is random and independent, the average should always be zero, should it not? There must be some kind of correlations among them in order to have a nonzero average. Discuss this issue.
2. Review Ehrenfest's classification of phase transitions from a standard textbook on thermodynamics. In his definition of second-order transitions, vanishing first-order derivatives and nonzero second-order derivatives of the Gibbs function are necessary. On the other hand, as we mentioned in Sect. 1.2, nonzero correlations $\Delta\sigma_i\Delta\sigma_j$among internal variables are required, in addition to Ehrenfest's definition. Is this a conflict? This problem should be discussed to clear this issue before proceeding to the following chapters.
3. In traditional thermodynamics, the Gibbs potential was defined as a function of macroscopic variables. However, macroscopic heat and order parameter are associated with internal variables in a crystal, which are essentially related to thermal and adiabatic averages of fluctuating microscopic variables, respectively. In this context, the Gibbs function defined in this chapter is associated with all sites that may not be identical. Discuss this issue for thermodynamics of heterogeneous matters.

References

1. J.G. Kirkwood, I. Oppenheim, *Chemical Thermodynamics* (McGraw-Hill, New York, 1961)
2. M. Born, K. Huang, *Dynamical Theory of Crystal Lattices* (Oxford University Press, Oxford, 1968)
3. D. Chandler, *Introduction to Modern Statistical Mechanics* (Oxford University Press, New York, 1987)
4. G.L. Lamb, *Elements of Soliton Theory* (Wiley, New York, 1980)
5. H. Haken, *Quantenfeldtheorie des Festkörpers* (B. G. Teubner, Stuttgart, 1973)

Chapter 2
Phonons

Disregarding point symmetry, we can simplify the crystal structure by the space group [1, 2], representing the thermodynamic state in equilibrium with the surroundings at given values of p and T. In this approach, the restoring forces secure stability of the lattice, where the masses at lattice points are in harmonic motion. In this case, we realize that their *directional* correlations in the lattice are ignored so that a possible disarrangement in the lattice can cause structural instability.

In this chapter, we discuss a harmonic lattice to deal with basic excitations in equilibrium structure. Lattice vibrations in periodic structure are in propagation, specified by frequencies and wavevectors in virtually continuous spectra. Quantum mechanically, on the other hand, the corresponding *phonons* signify the dynamical state in crystals. In strained crystals, as modulated by correlated constituents, low-frequency excitations dominate over the distorted structure, which is however thermally unstable as discussed in this chapter.

2.1 Normal Modes in a Simple Crystal

A crystal of chemically identical constituent ions has a rigid periodic structure in equilibrium with the surroundings, which is characterized by *translational symmetry*. Referring to symmetry axes, physical properties can be attributed to the translational invariance, in consequence of energy and momentum conservations among constituents.

Constituents are assumed to be bound together by restoring forces in the lattice structure. Considering a cubic lattice of N^3 identical mass particles in a cubic crystal in sufficiently large size, we can solve the *classical* equation of motion with nearest-neighbor interactions. Although such a problem should be solved *quantum mechanically*, classical solutions provide also a useful approximation. It is noted that the lattice symmetry is unchanged with the nearest-neighbor interactions, assuring

M. Fujimoto, *Thermodynamics of Crystalline States*,
DOI 10.1007/978-1-4614-5085-6_2, © Springer Science+Business Media New York 2013

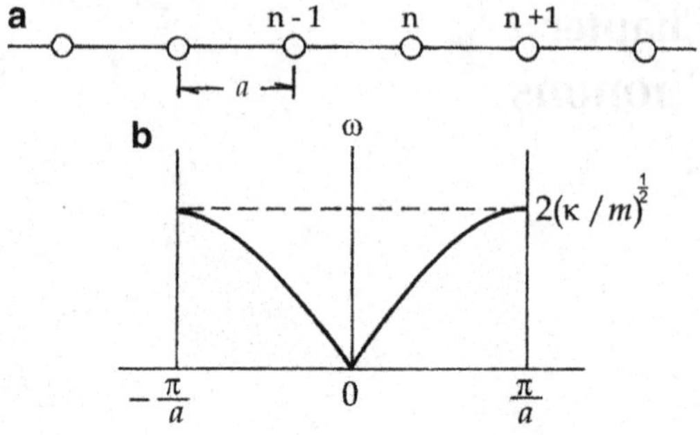

Fig. 2.1 (**a**) One-dimensional monatomic chain of the lattice constant a. (**b**) A dispersion curve ω *vs.* k of the chain lattice.

structural stability in this approach. In the harmonic approximation, we have *linear* differential equations, which can be separated into 3N independent equations; this one-dimensional equation describes *normal modes* of N constituents in collective motion along the symmetry axis x, y, or z [3]. Denoting the displacement by a vector $\boldsymbol{q}_n$ from a site n, we write equations of motion for the components $q_{x,n}$, $q_{y,n}$ and $q_{z,n}$ independently, that is,

$$\ddot{q}_{x,n} = \omega^2\left(q_{x,n+1} + q_{x,n-1} - 2q_{x,n}\right), \quad \ddot{q}_{y,n} = \omega^2\left(q_{y,n+1} + q_{y,n-1} - 2q_{y,n}\right) \quad \text{and}$$
$$\ddot{q}_{z,n} = \omega^2\left(q_{z,n+1} + q_{z,n-1} - 2q_{z,n}\right),$$

where $\omega^2 = \kappa/m$ and κ and m are the mass of a constituent particle and the force constant, respectively. As these equations are identical, we write the following equation for a representative component to name q_n for brevity:

$$\ddot{q}_n = \omega^2\left(q_{n+1} + q_{n-1} - 2q_n\right), \tag{2.1}$$

which assures lattice stability along any symmetry direction.

Defining the conjugate momentum by $p_n = m\dot{q}_n$, the *Hamiltonian* of a harmonic lattice can be expressed as

$$\mathsf{H} = \sum_{n=0}^{\mathrm{N}} \left\{\frac{p_n^2}{2m} + \frac{m\omega^2}{2}\left(q_{n+1} - q_n\right)^2 + \frac{m\omega^2}{2}\left(q_n - q_{n-1}\right)^2\right\}. \tag{2.2}$$

Each term in the summation represents one-dimensional infinite chain of identical masses m, as illustrated in Fig. 2.1a.

Normal coordinates and *conjugate momenta*, Q_k and P_k, are defined with the Fourier expansions

$$q_n = \frac{1}{\sqrt{\mathrm{N}}}\sum_{k=0}^{k_\mathrm{N}} Q_k \exp(ikna) \quad \text{and} \quad p_k = \sum_{k=0}^{k_\mathrm{N}} P_k \exp(ikna), \tag{2.3}$$

where a is the *lattice constant.* For each mode of q_n and p_n, the amplitudes Q_k and P_k are related as

$$Q_{-k} = Q_k{}^*,\ P_{-k} = P_k{}^* \quad \text{and} \quad \sum_{n=0}^{\mathrm{N}} \exp i(k-k')na = \mathrm{N}\delta_{kk'}, \tag{2.4}$$

where $\delta_{kk'}$ is *Kronecker's delta*, that is, $\delta_{kk'} = 1$ for $k = k'$, otherwise zero for $k = k'$.

Using normal coordinates Q_k and P_k, the Hamiltonian can be expressed as

$$\mathsf{H} = \frac{1}{2m}\sum_{k=0}^{2\pi/a}\left\{P_kP_k{}^* + Q_kQ_k{}^*m^2\omega^2\left(\sin^2\frac{ka}{2}\right)\right\}, \tag{2.5}$$

from which the equation of motion for Q_k is written as

$$\ddot{Q}_k = -m^2\omega^2 Q_k, \tag{2.6}$$

where

$$\omega_k = 2\omega\sin\frac{ka}{2} = 2\sqrt{\frac{\kappa}{m}}\sin\frac{ka}{2}. \tag{2.7}$$

As indicated by (2.7), the k-mode of coupled oscillators is *dispersive*, which are linearly independent from the other modes of $k' \neq k$. H is composed of N independent harmonic oscillators, each of which is determined by the normal coordinates Q_k and conjugate momenta P_k. Applying Born–von Kármán's boundary conditions to the periodic structure, k can take discrete values as given by $k = \frac{2\pi n}{\mathrm{N}a}$ and $n = 0, 1, 2, \ldots, \mathrm{N}$. Figure 2.1b shows the dispersion relation (2.7) determined by the characteristic frequency ω_k.

With initial values of $Q_k(0)$ and $\dot{Q}_k(0)$ specified at $t = 0$, the solution of (2.7) can be given by

$$Q_k(t) = Q_k(0)\cos\omega_k t + \frac{\dot{Q}_k(0)}{\omega_k}\sin\omega_k t.$$

Accordingly,

$$q_n(t) = \frac{1}{\sqrt{\mathrm{N}}}\sum_{k=0}^{k_{\mathrm{N}}}\sum_{n'=n,n\pm1}\left[q_{n'(0)}\cos\{ka(n-n') - \omega_k t\} + \frac{\dot{q}_{n'}(0)}{\omega_k}\sin\{ka(n-n') - \omega_k t\}\right], \tag{2.8}$$

where $a(n-n')$ represents distances between sites n and n' so that we write it as $x = a(n-n')$ in the following. The crystal is assumed as consisting of a large number of the cubic volume L^3 where $L = \mathrm{N}a$, if disregarding surfaces.

The periodic boundary conditions can then be set as $q_{n=0}(t) = q_{n=\mathrm{N}}(t)$ at an arbitrary time t. At a lattice point $x = na$ between $n = 0$ and N, (2.8) can be expressed as

$$q(x,t) = \sum_k [A_k \cos(\pm kx - \omega_k t) + B_k \sin(\pm kx - \omega_k t)],$$

where $A_k = \frac{q_k(0)}{\sqrt{\mathrm{N}}}$ and $B_k = \frac{\dot{q}_k(0)}{\omega_k\sqrt{\mathrm{N}}}$, and x is virtually continuous in the range $0 \leq x \leq L$, if L is taken as sufficiently long. Consisting of waves propagating in $\pm x$ directions, we can write $q(x,t)$ conveniently in complex exponential form, that is,

$$q(x,t) = \sum_k C_k \exp i(\pm kx - \omega_k t + \varphi_k), \tag{2.9}$$

where $C_k^2 = A_k^2 + B_k^2$ and $\tan\varphi_k = \frac{B_k}{A_k}$. For a three-dimensional crystal, these one-dimensional k-modes along the x-axis can be copied to other symmetry axes y and z; accordingly, there are 3N normal modes in total in a cubic crystal.

2.2 Quantized Normal Modes

The classical equation of motion of a harmonic crystal is separable to 3N independent normal propagation modes specified by $k_n = \frac{2\pi n}{a\mathrm{N}}$ along the symmetry axes. In quantum theory, the normal coordinate Q_k and conjugate momentum $P_k = -i\hbar\frac{\partial}{\partial Q_k}$ are *operators*, where $\hbar = \frac{h}{2\pi}$ and h is the Planck constant. For these normal and conjugate variables, there are commutation relations:

$$[Q_k, Q_{k'}] = 0, \ [P_k, P_{k'}] = 0 \quad \text{and} \quad [P_k, Q_{k'}] = i\hbar\delta_{kk'}, \tag{2.10}$$

and the Hamiltonian operator is

$$\mathsf{H}_k = \frac{1}{2m}\left(P_k P_k^\dagger + m^2\omega_k^2 Q_k Q_k^\dagger\right). \tag{2.11a}$$

Here, $P_k^\dagger$ and $Q_k^\dagger$ express *transposed* matrix operators of the complex conjugates $P_k{}^*$ and $Q_k{}^*$, respectively.

Denoting the eigenvalues of H_k by ε_k, we have the equation

$$\mathsf{H}_k \Psi_k = \varepsilon_k \Psi_k. \tag{2.11b}$$

For real eigenvalues ε_k, P_k and Q_k should be *Hermitian* operators, which are characterized by the relations $P_k^\dagger = P_{-k}$ and $Q_k^\dagger = Q_{-k}$, respectively. Defining operators

$$b_k = \frac{m\omega_k Q_k + iP_k^\dagger}{\sqrt{2m\varepsilon_k}} \quad \text{and} \quad b_k^\dagger = \frac{m\omega_k Q_k^\dagger - iP_k}{\sqrt{2m\varepsilon_k}}, \tag{2.12}$$

we can write the relation

$$\begin{aligned} b_k b_k^\dagger &= \frac{1}{2m\varepsilon_k}\left(m^2\omega_k^2 Q_k^\dagger Q_k + P_k^\dagger P_k\right) + \frac{i\omega_k}{2\varepsilon_k}\left(Q_k^\dagger P_k^\dagger - P_k Q_k\right) \\ &= \frac{\mathsf{H}_k}{\varepsilon_k} + \frac{i\omega_k}{2\varepsilon_k}(Q_{-k}P_{-k} - P_k Q_k). \end{aligned}$$

From this relation, we can be derive

$$\mathsf{H}_k = \hbar\omega_k\left(b_k^\dagger b_k + \frac{1}{2}\right), \quad \text{if} \quad \varepsilon_k = \frac{1}{2}\hbar\omega_k. \tag{2.13}$$

Therefore, H_k are commutable with the operator $b_k^\dagger b_k$, that is,

$$\left[\mathsf{H}_k,\ b_k^\dagger b_k\right] = 0,$$

and from (2.12)

$$\left[b_{k'},\ b_k^\dagger\right] = \delta_{k'k},\ [b_{k'},\ b_k] = 0 \quad \text{and} \quad \left[b_{k'}^\dagger,\ b_k^\dagger\right] = 0.$$

Accordingly, we obtain

$$\left[\mathsf{H}_k,\ b_k^\dagger\right] = \hbar\omega_k b_k^\dagger \quad \text{and} \quad [b_k,\ \mathsf{H}_k] = \hbar\omega_k b_k.$$

Combining with (2.11b), we can derive the relations

$$\mathsf{H}_k\left(b_k^\dagger \Psi_k\right) = (\varepsilon_k + \hbar\omega_k)\left(b_k^\dagger \Psi_k\right) \quad \text{and} \quad \mathsf{H}_k(b_k\Psi_k) = (\varepsilon_k - \hbar\omega_k)(b_k\Psi_k),$$

indicating that $b_k^\dagger \Psi_k$ and $b_k\Psi_k$ are eigenfunctions for the energies $\varepsilon_k + \hbar\omega_k$ and $\varepsilon_k - \hbar\omega_k$, respectively. In this context, $b_k^\dagger$ and b_k are referred to as *creation* and *annihilation* operators for the energy quantum $\hbar\omega_k$ to add and subtract in the energy ε_k; hence, we can write

$$b_k^\dagger b_k = 1. \tag{2.14}$$

Applying the creation operator $b_k^\dagger$ on the ground state function $\Psi_k n_k$-times, the eigenvalue of the wavefunction $\left(b_k^\dagger\right)^{n_k}\Psi_k$ can be given by $\left(n_k + \frac{1}{2}\right)\hbar\omega_k$, generating a state of $n_k$quanta plus $\frac{1}{2}\hbar\omega_k$. Considering a quantum $\hbar\omega_k$ like a particle, called a *phonon*, such an exited state with n_k identical phonons is multiply degenerate by *permutation* n_k! Hence, the normalized wavefunction of n_k phonons can be expressed by $\frac{1}{\sqrt{n_k!}}\left(b_k^\dagger\right)^{n_k}\Psi_k$. The total lattice energy in an excited state of n_1, n_2, ... phonons in the normal modes 1, 2,. .... can be expressed by

$$U(n_1,\ n_2,\ .....) = U_\mathrm{o} + \sum_k n_k\hbar\omega_k, \tag{2.15a}$$

where $U_\mathrm{o} = \sum_k \frac{\hbar\omega_k}{2}$ is the total zero-point energy. The corresponding wavefunction can be written as

$$\Psi(n_1, n_2, ...) = \frac{\left(b_1^\dagger\right)^{n_1}\left(b_2^\dagger\right)^{n_2}.....}{\sqrt{n_1!\ n_2!\ .....}}(\Psi_1\Psi_2.....), \tag{2.15b}$$

which describes a state of $n_1, n_2, ...$ phonons of energies $n_1\hbar\omega_{k_1}, n_2\hbar\omega_{k_2}, ...$ The total number N$=n_1 + n_2 + ...$ cannot be evaluated by the dynamical theory; however, we can determine the value in thermodynamics, as related to the level of thermal excitation at a given temperature.

2.3 Phonon Field and Momentum

In a one-dimensional chain of identical mass particles, the displacement mode q_k is independent from each other's modes, and hence representing normal modes in a three-dimensional crystal. However, this model is only approximate, in that these normal modes arise from the one-dimension harmonic chain model, where mutual interactions between different normal modes are prohibited. For propagation in arbitrary direction, the *vibrating field* offers more appropriate approach than the normal modes, where quantized phonons move in any direction like free particles in the field space.

Setting rectangular coordinates x, y, z along the symmetry axes of an orthorhombic crystal in classical theory, the lattice vibrations are described by a set of equations

$$\frac{{p_{x,n_1}}^2}{2m} + \frac{\kappa}{2}\left\{\left(q_{x,n_1} - q_{x,n_1+1}\right)^2 + \left(q_{x,n_1} - q_{x,n_1-1}\right)^2\right\} = \varepsilon_{x,n_1},$$

$$\frac{{p_{y,n_2}}^2}{2m} + \frac{\kappa}{2}\left\{\left(q_{y,n_2} - q_{y,n_2+1}\right)^2 + \left(q_{y,n_2} - q_{y,n_2-1}\right)^2\right\} = \varepsilon_{y,n_2}, \tag{2.16}$$

and

$$\frac{{p_{z,n_3}}^2}{2m} + \frac{\kappa}{2}\left\{\left(q_{z,n_3} - q_{z,n_3+1}\right)^2 + \left(q_{z,n_3} - q_{z,n_3-1}\right)^2\right\} = \varepsilon_{z,n_3},$$

where $\varepsilon_{x,n_1} + \varepsilon_{y,n_2} + \varepsilon_{z,n_3} = \varepsilon_{n_1 n_2 n_3}$ is the total propagation energy along the direction specified by the vector $q(n_1, n_2, n_3)$ and κ is the force constant.

The variables $q_{x,n_1}, q_{y,n_2}, q_{z,n_3}$ in (2.16) are components of a classical vector $\boldsymbol{q}(n_1, n_2, n_3)$, which can be interpreted quantum theoretically as *probability amplitudes* of components of the vector $\boldsymbol{q}$ in the vibration field. We can therefore write the wavefunction of the displacement field as $\Psi(n_1, n_2, n_3) = q_{x,n_1} q_{y,n_2} q_{z,n_3}$, for which these classical components are written as

$$q(x,t) = \sum_{k_x} C_{k,x} \exp i(\pm k_x x - \omega_{k_x} t + \varphi_{k_x}),$$

$$q(y,t) = \sum_{k_y} C_{k_y} \exp i\left(\pm k_y y - \omega_{k_y} t + \varphi_{k_y}\right),$$

$$q(z,t) = \sum_{k_z} C_{k_y} \exp i(\pm k_z z - \omega_{k_z} t + \varphi_{k_z}),$$

and hence, we have

$$\Psi(n_1, n_2, n_3) = \sum_{k} \mathcal{A}_k \exp i\left(\pm \boldsymbol{k}.\boldsymbol{r} - \frac{n_1 \varepsilon_{x,n_1} + n_2 \varepsilon_{y,n_2} + n_3 \varepsilon_{z,n_3}}{\hbar} t + \varphi_k \right).$$

Here, $\mathcal{A}_k = C_{k_x} C_{k_y} C_{k_z}$, $\varphi_k = \varphi_{k_x} + \varphi_{k_y} + \varphi_{k_z}$, and $\boldsymbol{k} = (k_x, k_y, k_z)$ are the amplitude, phase constant, and wavevector of $\Psi(n_1, n_2, n_3)$, respectively. Further writing

$$\frac{n_1 \varepsilon_{x,n_1} + n_2 \varepsilon_{y,n_2} + n_3 \varepsilon_{z,n_3}}{\hbar} = \omega_k(n_1, n_2, n_3) = \omega_k, \tag{2.17a}$$

the field propagating along the direction of a vector $\boldsymbol{k}$ can be expressed as

$$\Psi(\boldsymbol{k}, \omega_{\boldsymbol{k}}) = \mathcal{A}_k \exp i(\pm \boldsymbol{k}.\boldsymbol{r} - \omega_{\boldsymbol{k}} t + \varphi_{\boldsymbol{k}}), \tag{2.17b}$$

representing a phonon of energy $\hbar\omega_{\boldsymbol{k}}$ and momentum $\pm\hbar\boldsymbol{k}$. For a small $|\boldsymbol{k}|$, the propagation in a cubic lattice can be characterized by a constant speed v of propagation determined by $\omega_{\boldsymbol{k}} = v|\boldsymbol{k}|$, indicating no *dispersion* in this approximation.

The phonon propagation can be described by the vector $\boldsymbol{k}$, composing a *reciprocal lattice* space, as illustrated in two dimensions in Fig. 2.2 by

$$k_x = \frac{2\pi n_1}{L}, \; k_y = \frac{2\pi n_2}{L} \quad \text{and} \quad k_z = \frac{2\pi n_3}{L},$$

Fig. 2.2 Two-dimensional reciprocal lattice. A lattice point is indicated by (k_x, k_y). Two quarter-circles of radii k and $k + dk$ show surfaces of constant $\varepsilon_{\mathbf{k}}$ and $\varepsilon_{\mathbf{k}+d\mathbf{k}}$ for small $|\mathbf{k}|$ in $k_x k_y$ plane.

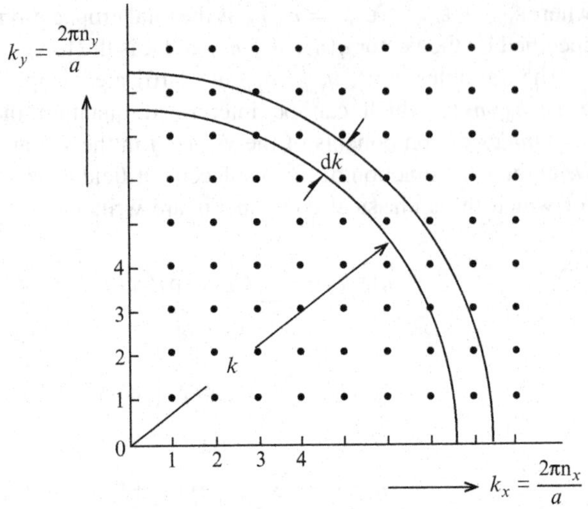

in a cubic crystal, where $a\mathrm{N} = L$. A set of integers (n_1, n_2, n_3) determines an energy and momentum of a phonon propagating in the direction of $\boldsymbol{k}$, where $\omega_{\boldsymbol{k}} = \omega_{\boldsymbol{k}}(n_1, n_2, n_3)$ and $|\boldsymbol{k}| = \frac{2\pi}{L}\sqrt{n_x^2 + n_y^2 + n_z^2}$. In the reciprocal space, all points on a spherical surface of radius $|\boldsymbol{k}|$correspond to the same energy $\hbar\omega_{\boldsymbol{k}}$, representing a sphere of a constant radius $|\boldsymbol{k}|$ and energy $\varepsilon_{\mathbf{k}}$. Quantum mechanically, we can write the phonon momentum as $\boldsymbol{p} = \hbar\boldsymbol{k}$ to supplement the energy $\hbar\omega_{\boldsymbol{k}}$, characterizing a *phonon particle*.

2.4 Thermal Equilibrium

In thermodynamics, a crystal must always be in *thermal contact* with the surroundings. At constant external pressure p, the quantized vibration field can be in equilibrium with the surroundings at a given temperature T. A large number of phonons are in collision-free motion, traveling in all directions through the lattice, colliding with surfaces to exchange their energy and momentum with the surroundings. Assuming the crystal volume as unchanged, the average of phonon energies can be calculated with the Boltzmann probability at T.

In equilibrium, the total energy of a crystal can be expressed as $U + U_s$, where U_s is the contribution from the heat reservoir and Urepresents the energy of a stable crystal. In this case, the total energy $U + U_s$ should be stationary with any thermodynamic variation around equilibrium. Using probabilities w and w_s for keeping the crystal in equilibrium with the surroundings, the product ww_sshould be calculated as maximizing $U + U_s$ to determine the most probable value. Setting this variation problem as

$$\delta(ww_s) = 0 \quad \text{and} \quad \delta(U + U_s) = 0$$

for arbitrary variations δw and δw_s, these variations can be calculated as

$$w_s \delta w + w \delta w_s = 0 \quad \text{and} \quad \delta U + \delta U_s = 0,$$

respectively. We can therefore write

$$\frac{\delta(\ln w)}{\delta U} = \frac{\delta(\ln w_s)}{\delta U_s},$$

which is a common quantity between U and U_s. Writing it as equal to $\beta = \frac{1}{k_B T}$, we can relate β to the conventional absolute temperature T. Therefore,

$$\frac{\delta(\ln w)}{\delta U} = \beta = \frac{1}{k_B T} \quad \text{or} \quad w = w_o \exp\left(-\frac{U}{k_B T}\right). \tag{2.18}$$

Here, w is called the *Boltzmann probability*; w_o is the integration constant that can be determined by assuming $U = 0$ at $T = 0$ K, where k_B is the Boltzmann constant. Quantum theoretically, however, $T = 0$ is fundamentally unreachable, as stated in the *third law of thermodynamics*. Accordingly, we write $U = 0 + U_o$ at 0 K, where $U_o = \frac{1}{2} N\hbar\omega_o$ is the zero-point energy.

Although we considered only vibrations so far, physical properties of a crystal are also contributed by other variables located at lattice points or at *interstitial* sites. Though primarily independent of lattice vibrations, these variables can interact with the lattice via phonon scatterings. If accessed by random collisions of phonons, energies ε_i of these variables are statistically available with Boltzmann's probabilities w_i, so we can write equations

$$U = \sum_i \varepsilon_i \quad \text{and} \quad w = \Pi_i w_i,$$

where

$$w_i = w_o \exp\left(-\frac{\varepsilon_i}{k_B T}\right) \quad \text{and} \quad \sum_i w_i = 1, \tag{2.19}$$

as these w_i are for *exclusive* events. In this case, the function $Z = \sum_i \exp\left(-\frac{\varepsilon_i}{k_B T}\right)$, called the *partition function*, is useful for statistical calculation. In such a system as called *microcanonical ensemble*, thermal properties can be calculated directly with Z.

The Boltzmann statistics is a valid assumption for a dynamical system under the *ergodic* hypothesis. Despite of the absence of rigorous proof, the Boltzmann statistics can usually be applied to phonon gas in a fixed volume. Thermodynamically, however, it is only valid for isothermal processes, because the volume is not always constant in adiabatic processes of internal origin. The anharmonic lattice cannot be *ergodic* in strict sense, whereas the harmonicity is essentially required for stable crystals at constant volume and pressure.

2.5 Specific Heat of a Monatomic Crystal

The specific heat at a constant volume $C_V = \left(\frac{\partial U}{\partial T}\right)_V$ is a quantity measurable with varying temperature under a constant external pressure p. The phonon theory is adequate for simple monatomic crystals, if characterized with no structural changes.

For such a crystal, the specific heat and internal energy are given by quantized phonon energies $\varepsilon_k = \left(n_k + \frac{1}{2}\right)\hbar\omega_k$, for which the wavevector $\boldsymbol{k}$ is distributed virtually in all directions in the reciprocal space. Assuming 3N phonons in total, the energies ε_k are *degenerate* with the density of k-states written as $g(k)$, which is a large number as estimated from a spherical volume of radius $|\boldsymbol{k}|$ in the reciprocal space. In this case, the partition function can be expressed as

$$Z_k = g(k)\exp\left(-\frac{\varepsilon_k}{\mathrm{k_B}T}\right) = g(k)\exp\left(-\frac{\hbar\omega_k}{2\mathrm{k_B}T}\right)\sum_{n_k=0}^{\infty}\exp\left(-\frac{n_k\hbar\omega_k}{\mathrm{k_B}T}\right),$$

where the infinite series on the right converges, if $\frac{\hbar\omega_k}{\mathrm{k_B}T}<1$. In fact, this condition is satisfied at any practical temperature T lower than the melting point so that Z_k is expressed as

$$Z_k = g(k)\frac{\exp\left(-\frac{\hbar\omega_k}{2\mathrm{k_B}T}\right)}{1-\exp\left(-\frac{\hbar\omega_k}{\mathrm{k_B}T}\right)}.$$

The total partition function is given by the product $Z = \Pi_k Z_k$, so that $\ln Z = \sum_k \ln Z_k$; the free energy can therefore be calculated as the sum of $\ln Z_k$, namely, $F = \mathrm{k_B}T\sum_k \ln Z_k$, where

$$\ln Z_k = -\frac{\hbar\omega_k}{2} + \mathrm{k_B}T\ln g(k) - \mathrm{k_B}T\ln\left\{1-\exp\left(-\frac{\hbar\omega_k}{\mathrm{k_B}T}\right)\right\}.$$

By definition, we have the relation $F = U - TS = U + T\left(\frac{\partial F}{\partial T}\right)_V$ from which we can derive the formula $U = -T^2\frac{\partial}{\partial T}\left(\frac{F}{T}\right)$. Using the above $\ln Z_k$, we can show that the internal energy is given by

$$U = U_o + \sum_k \frac{\hbar\omega_k}{\exp\frac{\hbar\omega_k}{\mathrm{k_B}T}-1} \quad \text{and} \quad U_o = \frac{1}{2}\sum_k \hbar\omega_k. \tag{2.20}$$

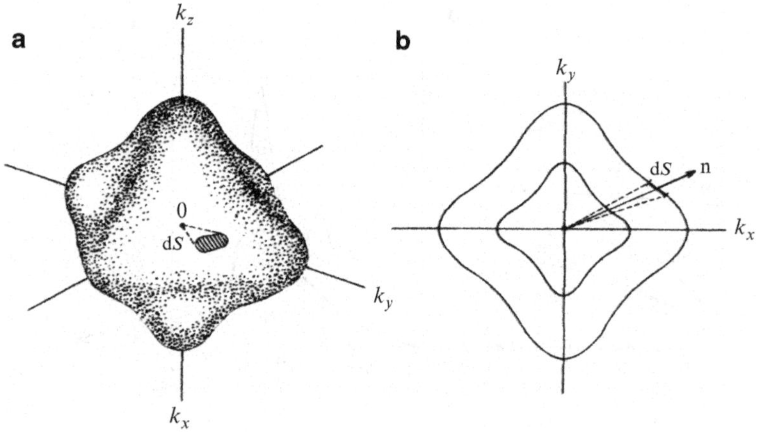

Fig. 2.3 (**a**) A typical constant-energy surface in three-dimensional reciprocal space, where dS is a differential area on the surface. (**b**) The two-dimensional view in the $k_x k_y$-plane.

The specific heat at constant volume can then be expressed as

$$C_V = \left(\frac{\partial U}{\partial T}\right)_V = \mathrm{k_B}\sum_k \frac{\left(\frac{\hbar\omega_k}{\mathrm{k_B}T}\right)^2 \exp\frac{\hbar\omega_k}{\mathrm{k_B}T}}{\left(\exp\frac{\hbar\omega_k}{\mathrm{k_B}T} - 1\right)^2}. \tag{2.21}$$

To calculate C_V with (2.21), we need to evaluate the summation with the number of phonon states on energy surface $\varepsilon_k = \hbar\left(\omega_k + \frac{1}{2}\right)$ in the reciprocal space. In anisotropic crystals, such a surface is not spherical, but a closed surface, as shown in Fig. 2.3a. In this case, the summation in (2.21) can be replaced by a volume integral over the closed surface, whose volume element is written as $\mathrm{d}^3\boldsymbol{k} = \mathrm{d}\boldsymbol{k}.\mathrm{d}\boldsymbol{S} = \mathrm{d}k_\perp|\mathrm{d}\boldsymbol{S}|$. Here, $\mathrm{d}k_\perp$ is the component of $\boldsymbol{k}$ perpendicular to the surface element $|\mathrm{d}\boldsymbol{S}| = \mathrm{d}S$, as illustrated two-dimensionally in Fig. 2.3b. We can write

$$\mathrm{d}\omega_k = \frac{\mathrm{d}\varepsilon_k}{\hbar} = \frac{1}{\hbar}|\mathrm{grad}_k\varepsilon(\boldsymbol{k})|\mathrm{d}k_\perp,$$

where $\frac{1}{\hbar}|\mathrm{grad}_k\varepsilon(\boldsymbol{k})| = v_\mathrm{g}$ represents the *group velocity* for propagation, and hence $\mathrm{d}k_\perp = \frac{\mathrm{d}\omega_k}{v_\mathrm{g}}$. Using these notations, we can reexpress (2.21) as

$$C_V = \mathrm{k_B}\int_{\omega_k} \frac{\left(\frac{\hbar\omega_k}{\mathrm{k_B}T}\right)^2 \exp\frac{\hbar\omega_k}{\mathrm{k_B}T}}{\left(\exp\frac{\hbar\omega_k}{\mathrm{k_B}T} - 1\right)^2} D(\omega_k)\mathrm{d}\omega_k, \tag{2.22a}$$

where

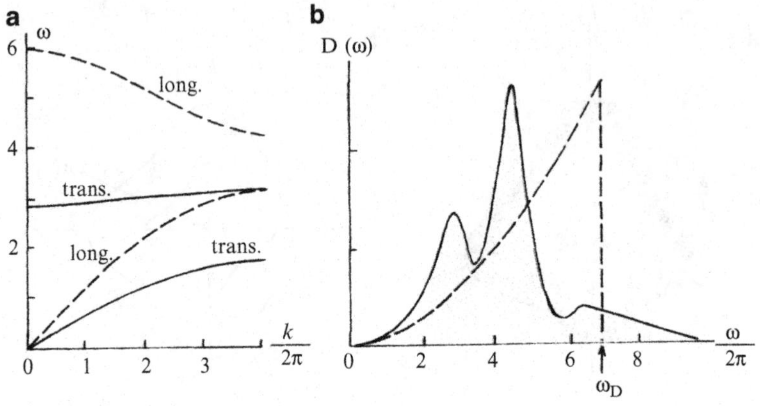

Fig. 2.4 (**a**) Examples of practical dispersion curves. Longitudinal and transverse dispersions are shown by solid and broken curves, respectively. (**b**) The solid curve shows an example of an observed density function, being compared with the broken curve of Debye model.

$$D(\omega_k) = \left(\frac{L}{2\pi}\right)^3 \oint_S \frac{dS}{v_g} \tag{2.22b}$$

is the density of phonon states on the surface S.

Tedious numerical calculations performed in early studies on representative crystals resulted in such curves as shown in Fig. 2.4a, for example, of a diamond crystal. However, in relation with dispersive longitudinal and transversal modes, the analysis was extremely difficult to obtain satisfactory comparison with experimental results. On the other hand, Einstein and Debye simplified the function $\mathcal{D}(\omega_k)$ independently, although somewhat oversimplified for practice crystals. Nevertheless, their models are proven to be adequate in many applications to obtain useful formula for $U$and C_V for simple crystals [4].

2.6 Approximate Models

2.6.1 Einstein's Model

At elevated temperatures T, we can assume that thermal properties of a crystal are dominated by n phonons of energy $\hbar\omega_o$. Einstein proposed that the dominant mode at a high temperature is of a single frequency ω_o, disregarding all other modes in the vibration spectrum. In this model, using the expression (2.22b) simplified as $\mathcal{D}(\omega_o) = 1$, we can express the specific heat (2.22a) and the internal energy as

$$C_V = 3\mathrm{Nk_B} \frac{\xi^2 \exp \xi}{(\exp \xi - 1)^2} \quad \text{and} \quad U = 3\mathrm{Nk_B}\left(\frac{1}{2}\xi + \frac{\xi}{\exp \xi - 1}\right), \tag{2.23}$$

respectively, where $\xi = \frac{\Theta_E}{T}$; the parameter $\Theta_E = \frac{\hbar\omega_0}{k_B}$ is known as the Einstein temperature. It is noted that in the limit $\xi \to 0$, we obtain $C_V \to 3\mathrm{Nk_B}$.

At high temperatures, U can be attributed to constituent masses, vibrating independently in *degrees of freedom* 2; hence, the corresponding thermal energy is $2 \times \frac{1}{2}k_B T$, and

$$U = 3\mathrm{Nk_B}T \quad \text{and} \quad C_V = 3\mathrm{Nk_B}. \tag{2.24}$$

This is known as the Dulong–Petit law, which is consistent with Einstein's model in the limit of $T \to \infty$.

2.6.2 Debye's Model

At lower temperatures, longitudinal vibrations at low frequencies are dominant modes, which are characterized approximately by a nondispersive relation $\omega = v_g k$. The speed v_g is assumed as constant on a nearly spherical surface for constant energy in the $\boldsymbol{k}$-space. Letting $v_g = v$, for brevity, (2.22b) can be expressed as

$$\mathcal{D}(\omega) = \left(\frac{L}{2\pi}\right)^3 \frac{4\pi\omega^2}{v^3}. \tag{2.25}$$

Debye assumed that with increasing frequency, the density $\mathcal{D}(\omega)$ should be terminated at a frequency $\omega = \omega_D$, called Debye *cutoff frequency*, as shown by the broken curve in Fig. 2.4b. In this case, the density function $\mathcal{D}(\omega) \propto \omega^2$ can be normalized as $\int_0^{\omega_D} \mathcal{D}(\omega)d\omega = 3\mathrm{N}$, so that (2.25) can be replaced by

$$\mathcal{D}(\omega) = \frac{9\mathrm{N}}{\omega_D^3}\omega^2. \tag{2.26}$$

Therefore, in the Debye model, we have

$$U = 3\mathrm{Nk_B}T \int_0^{\omega_D} \left(\frac{\hbar\omega}{2} + \frac{\hbar\omega}{\exp\frac{\hbar\omega}{k_B T} - 1}\right) \frac{3\omega^2 d\omega}{\omega_D^3}$$

and

$$C_V = 3Nk_B \int_0^{\omega_D} \frac{\exp\frac{\hbar\omega}{k_B T}}{\left(\exp\frac{\hbar\omega}{k_B T} - 1\right)^2} \left(\frac{\hbar\omega}{k_B T}\right)^2 \frac{3\omega^2 d\omega}{\omega_D^3}.$$

Defining Debye temperature $\frac{\hbar\omega_D}{k_B T} = \Theta_D$ and $\frac{\hbar\omega}{k_B T} = \xi$, similar to Einstein's model, these expressions can be simplified as

$$U = \frac{9}{8}Nk_B\Theta_D + 9Nk_B T\left(\frac{T}{\Theta_D}\right)^3 \int_0^{\frac{\Theta_D}{T}} \frac{\xi^3}{\exp\xi - 1} d\xi$$

and

$$C_V = 9Nk_B\left(\frac{T}{\Theta_D}\right)^3 \int_0^{\frac{\Theta_D}{T}} \frac{\xi^4 \exp\xi}{(\exp\xi - 1)^2} d\xi.$$

Introducing the function defined by

$$\eta\left(\frac{\Theta_D}{T}\right) = 3\left(\frac{T}{\Theta_D}\right)^3 \int_0^{\frac{\Theta_D}{T}} \frac{\xi^3 d\xi}{\exp\xi - 1}, \tag{2.27}$$

known as the Debye function, the expression $C_V = 3Nk_B\eta\left(\frac{\Theta_D}{T}\right)$ describes temperature-dependent C_V for $T<\Theta_D$. In the limit of $\frac{\Theta_D}{T} \to \infty$, however, these U and C_V are dominated by the integral

$$\int_0^{\infty} \frac{\xi^3 d\xi}{\exp\xi - 1} = \frac{\pi^4}{15},$$

and hence the formula

$$U = \frac{9}{8}Nk_B\Theta_D + 9Nk_B T\left(\frac{T}{\Theta_D}\right)^3 \frac{\pi^4}{15}$$

and

$$C_V = 9Nk_B\left(\frac{T}{\Theta_D}\right)^3 \frac{\pi^4}{15}. \tag{2.28}$$

Fig. 2.5 Observed specific heat $C_V/3R$ against T/Θ_D for representative metals. R is the molar gas constant. In the bottom-right corner, values of $_D$ for these metals are shown. The T^3-law and Dulong–Petit limits are indicated to compare with experimental results.

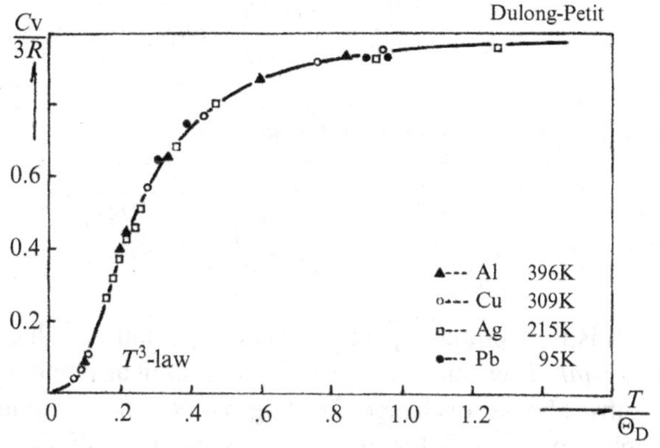

can be used at lower temperatures than Θ_D. In the Debye model, we have thus the approximate relation $C_V \propto T^3$ for $T < \Theta_D$, which is known as Debye T^3-law.

Figure 2.5 shows a comparison of observed values of C_V from representative monatomic crystals with the Debye and Dulong–Petit laws, valid at low and high temperatures, respectively, showing reasonable agreements.

2.7 Phonon Statistics Part 1

Quantizing the lattice vibration field, we consider a gas of phonons $(\hbar\omega_k, \hbar\boldsymbol{k})$. A large number of phonons exist in excited lattice states, behaving like classical particles. On the other hand, phonons are correlated at high densities, owing to their quantum nature of *unidentifiable particles*. Although dynamically unspecified, the *total* number of phonons is thermodynamically determined by the surface boundaries at T, where phonon energies are exchanged with heat from the surroundings. In equilibrium, the number of photons on each k-state can be either one of $n = 1, 2, \ldots, 3\text{N}$. Therefore, the Gibbs function can be expressed by $G(p, T, n)$, but the entropy fluctuates with varying n in the crystal. Such fluctuations can be described in terms of a thermodynamic probability $g(p, T, n)$, so that we consider that two phonon states, 1 and 2, can be characterized by probabilities $g(p, T, n_1)$ and $g(p, T, n_2)$ in an *exclusive event*, in contrast to the Boltzmann statistics for *independent* particles.

At constant p, the equilibrium between the crystal and reservoir can therefore be specified by minimizing the total probability $g(p, T, n) = g(p, T, n_1) + g(p, T, n_2)$, considering such binary correlations dominant under $n = n_1 + n_2 = \text{constant}$, leaving all other $n_i (i \neq 1, 2)$ as unchanged. Applying the variation principle for small arbitrary variations $\delta n_1 = -\delta n_2$, we can minimize $g(p, T, n)$ to obtain

$$(\delta g)_{p,T} = \left(\frac{\partial g_1}{\partial n_1}\right)_{p,T} \delta n_1 + \left(\frac{\partial g_2}{\partial n_2}\right)_{p,T} \delta n_2 = 0,$$

from which we derive the relation

$$\left(\frac{\partial g_1}{\partial n_1}\right)_{p,T} = \left(\frac{\partial g_2}{\partial n_2}\right)_{p,T}.$$

This is a common quantity between g_1 and g_2, which is known as the *chemical potential*. Therefore, we have equal chemical potentials $\mu_1 = \mu_2$ in equilibrium against phonon exchange. Writing the common potential as μ, a variation of the Gibbs potential G for an open system at equilibrium can be expressed for an arbitrary variation d n as

$$dG = dU - TdS + pdV - \mu dn, \tag{2.29}$$

where dn represents a macroscopic variation in the number of phonons n.

Consider a simple crystal, whose two thermodynamic states are specified by the internal energy and phonon number, (U_o, N_o) and $(U_o - \varepsilon, N_o - n)$, which are signified by probabilities g_o and g, as related to their entropies $S(U_o, N_o)$ and $S(U_o - \varepsilon,\ N_o - n)$, respectively. Writing the corresponding Boltzmann relations, we have

$$g_o = \exp\frac{S(U_o, N_o)}{k_B} \quad \text{and} \quad g = \exp\frac{S(U_o - \varepsilon, N_o - n)}{k_B}.$$

Hence,

$$\frac{g}{g_o} = \frac{\exp\{S(U_o - \varepsilon, N_o - n)/k_B\}}{\exp\{S(U_o, N_o)/k_B\}} = \exp\frac{\Delta S}{k_B},$$

where

$$\Delta S = S(U_o - \varepsilon, N_o - n) - S(U_o, N_o) = -\left(\frac{\partial S}{\partial U_o}\right)_{N_o} \varepsilon - \left(\frac{\partial S}{\partial N_o}\right)_{U_o} n.$$

Using (2.29), we obtain the relations

$$\left(\frac{\partial S}{\partial U_o}\right)_{N_o} = \frac{1}{T} \quad \text{and} \quad \left(\frac{\partial S}{\partial N_o}\right)_{U_o} = -\frac{\mu}{T}$$

so that

$$g = g_0 \exp\frac{\mu n - \varepsilon}{k_B T}. \tag{2.30}$$

For phonons, the energy ε is determined by any wavevector $\boldsymbol{k}$, where $|\mathbf{k}| = 1, 2, \ldots, 3N$, and N can take any integral number. The expression (2.30) is the *Gibbs factor*, whereas for classical particles, we use the Boltzmann factor instead. These factors are essential in statistics for *open* and *closed* systems, respectively. For phonons, it is convenient to use the notation $\lambda = \exp\frac{\mu}{k_B T}$, with which (2.30) can be written as $g = g_0 \lambda^n \exp\left(-\frac{\varepsilon}{k_B T}\right)$. The factor λ here implies a probability for the energy level ε to accommodate one phonon *adiabatically* [5], whereas the conventional Boltzmann factor $\exp\left(-\frac{\varepsilon}{k_B T}\right)$ is an *isothermal probability* of ε at T. Originally, the chemical potential μ was defined for an adiabatic equilibrium with an external chemical agent; however, for phonons λ is temperature dependent as defined by $\lambda = \exp\frac{\mu}{k_B T}$. Here, the chemical potential is determined as $\mu = -\left(\frac{\partial G}{\partial n}\right)_{p,T}$ from (2.29), which is clearly related with the internal energy due to phonon correlations in a crystal.

For phonon statistics, the energy levels are $\varepsilon_n = n\hbar\omega$, and the Gibbs factor is determined by $\varepsilon = \hbar\omega$ and n. The partition function can therefore be expressed as

$$Z_N = \sum_{n=0}^{N} \lambda^n \exp\left(-\frac{n\varepsilon}{k_B T}\right) = \sum_{n=0}^{N} \left\{\lambda \exp\left(-\frac{\varepsilon}{k_B T}\right)\right\}^n.$$

Considering $\lambda \exp\left(-\frac{\varepsilon}{k_B T}\right) < 1$, the sum of the infinite series evaluated for $N \to \infty$ is

$$Z = \frac{1}{1 - \lambda \exp\left(-\frac{\varepsilon}{k_B T}\right)}.$$

With this so-called grand partition function, the average number of phonons can be expressed as

$$\langle n \rangle = \lambda \frac{\partial \ln Z}{\partial \lambda} = \frac{1}{\frac{1}{\lambda}\exp\left(-\frac{\varepsilon}{k_B T}\right) - 1} = \frac{1}{\exp\frac{\varepsilon - \mu}{k_B T} - 1}. \tag{2.31}$$

This is known as the Bose–Einstein distribution. It is noted that the energy ε is basically dependent on temperature, whereas the chemical potential is small and temperature independent. Further, at elevated temperatures, we consider that for $\varepsilon > \mu$, (2.31) is approximated as $\langle n \rangle \approx \exp\frac{\mu - \varepsilon}{k_B T} \approx \exp\frac{-\varepsilon}{k_B T}$, which is the Boltzmann factor. However, there should be a critical temperature T_c for $\langle n \rangle = \infty$ to be determined by $\varepsilon(T_c) = \mu$, which may be considered for *phonon condensation*.

So far, phonon gas was specifically discussed, but the Bose–Einstein statistics (2.31) can be applied to all other identical particles characterized by even parity; particles obeying the Bose–Einstein statistics are generally characterized by *even parity* and called *Bosons*. Particles with *odd parity* will be discussed in Chap. 11 for electrons.

2.8 Compressibility of a Crystal

In the foregoing, we discussed a crystal under a constant volume condition. On the other hand, under constant temperature, the Helmholtz free energy can vary with a volume change ΔV, if the crystal is compressed by

$$\Delta F = \left(\frac{\partial F}{\partial V}\right)_T \Delta V,$$

where $p = -\left(\frac{\partial F}{\partial V}\right)_T$ is the *pressure* on the phonon gas in a crystal. At a given temperature, such a change ΔF must be offset by the external work $-p\Delta V$ by applying a pressure p, which is *adiabatic* to the crystal.

It is realized that volume-dependent energies need to be included in the free energy of a crystal in order to deal with the pressure from outside. Considering an additional energy $U_o = U_o(V)$, the free energy can be expressed by

$$\begin{aligned} F &= U_o + 9\mathrm{Nk_B}T\left(\frac{T}{\Theta_\mathrm{D}}\right)^3 \int_0^{\frac{\Theta_\mathrm{D}}{T}} \left[\frac{\xi}{2} + \ln\{1 - \exp(-\xi)\}\right]\xi^2 d\xi \\ &= U_o + 3\mathrm{Nk_B}T\eta\left(\frac{\Theta_\mathrm{D}}{T}\right), \end{aligned} \tag{2.32}$$

where $\eta\left(\frac{\Theta_\mathrm{D}}{T}\right)$ is the Debye function defined in (2.27), for which we have the relation

$$\left(\frac{\partial \eta}{\partial \ln \Theta_\mathrm{D}}\right)_T = -\left(\frac{\partial \eta}{\partial \ln T}\right)_V = -\frac{T}{\Theta_\mathrm{D}}\frac{\partial \eta}{\partial T}. \tag{2.33}$$

Writing $\zeta = \zeta(T, V) = T\eta\left(\frac{\Theta_\mathrm{D}}{T}\right)$ for convenience, we obtain

$$\left(\frac{\partial \zeta}{\partial V}\right)_T = -\frac{\gamma}{V}\left(\frac{\partial \zeta}{\partial \ln \Theta_\mathrm{D}}\right)_T = \frac{\gamma T}{V}\left(\frac{\partial \eta}{\partial \ln T}\right)_V,$$

where the factor

$$\gamma = -\frac{\mathrm{d}\ln\Theta}{\mathrm{d}\ln V}$$

is known as *Grüneisen's constant.* Using (2.30), the above relation can be reexpressed as

$$\left(\frac{\partial\zeta}{\partial V}\right)_T = \frac{\gamma}{V}\left\{T\left(\frac{\partial\zeta}{\partial T}\right)_V - \zeta\right\}.$$

From (2.29), we have $\mathrm{N}k_\mathrm{B}\zeta(T,V) = F - U_\mathrm{o}$; therefore, this can be written as

$$\left\{\frac{\partial(F-U_\mathrm{o})}{\partial V}\right\}_T = \frac{\gamma}{V}\left\{T\left(\frac{\partial(F-U_\mathrm{o})}{\partial T}\right)_V - F + U_\mathrm{o}\right\}. \tag{2.34}$$

Noting $U_\mathrm{o} = U_\mathrm{o}(V)$, the derivative in the first term of the right side is equal to $T\left(\frac{\partial F}{\partial T}\right)_V = -TS$; hence, the quantity in the curly brackets is $-U + U_\mathrm{o} = U_\mathrm{vib}$ that represents the energy of lattice vibration. From (2.31), we can derive the expression for pressure in a crystal, that is,

$$p = -\frac{\mathrm{d}U_\mathrm{o}}{\mathrm{d}V} + \frac{\gamma U_\mathrm{vib}}{V}, \tag{2.35}$$

which is known as Mie–Grüneisen's equation of state.

The compressibility is defined as

$$\kappa = -\frac{1}{V}\left(\frac{\partial V}{\partial p}\right)_T, \tag{2.36}$$

which can be obtained for a crystal by using (2.32). Writing (2.32) as $pV = -V\frac{\mathrm{d}U_\mathrm{o}}{\mathrm{d}V} + \gamma U_\mathrm{vib}$ and differentiating it, we can derive

$$p + V\left(\frac{\partial p}{\partial V}\right)_T = -\frac{\mathrm{d}U_\mathrm{o}}{\mathrm{d}V} - V\frac{\mathrm{d}^2 U_\mathrm{o}}{\mathrm{d}V^2} + \gamma\left(\frac{\partial U_\mathrm{vib}}{\partial V}\right)_T.$$

Since the atmospheric pressure is negligible compared with those in a crystal, we may omit p, and also from (2.31)

$$\left(\frac{\partial U_\mathrm{vib}}{\partial V}\right)_T = \frac{\gamma}{V}\left\{T\left(\frac{\partial U_\mathrm{vib}}{\partial T}\right)_V - U_\mathrm{vib}\right\}$$

in the above expression. Thus, the compressibility can be obtained from

Table 2.1 Debye temperatures Θ_D determined by thermal and elastic experiments[a]

	Fe	Al	Cu	Pb	Ag
Thermal	453	398	315	88	215
Elastic[b]	461	402	332	73	214

[a]Data: from Ref. [3]
[b]Calculated with elastic data at room temperature

$$\frac{1}{\kappa} = -V\left(\frac{\partial p}{\partial V}\right)_T = \frac{dU_0}{dV} + V\frac{d^2U_0}{dV^2} - \frac{\gamma^2}{V}(TC_V - U_0), \tag{2.37}$$

where $C_V = \left(\frac{\partial U_{vib}}{\partial T}\right)_V$ is the specific heat of lattice vibrations.

If $p = 0$, the volume of a crystal is constant, that is, $V = V_0$ and $\frac{dU_0}{dV} = 0$, besides $U_{vib} = \text{const.}$ of V. Therefore, we can write $\frac{1}{\kappa_0} = V_0\left(\frac{d^2U_0}{dV^2}\right)_{V=V_0}$, meaning a hypothetical compressibility κ_0 in equilibrium at $p = 0$. Then with (2.34) the *volume expansion* can be defined as

$$\beta = \frac{V - V_0}{V_0} = \frac{\kappa_0 \gamma U_{vib}}{V}. \tag{2.38}$$

Further, using (2.32)

$$\left(\frac{\partial p}{\partial T}\right)_V = \frac{\gamma}{V}\left(\frac{\partial U_{vib}}{\partial T}\right)_V = \frac{\gamma C_V}{V},$$

which can also be written as

$$\left(\frac{\partial p}{\partial T}\right)_V = \frac{\frac{1}{V}\left(\frac{\partial V}{\partial T}\right)_p}{-\frac{1}{V}\left(\frac{\partial V}{\partial p}\right)_T},$$

and hence we have the relation among γ, κ, and β, that is, $\gamma = -\frac{V}{C_V}\frac{\beta}{\kappa}$.

Such constants as Θ, κ, β, and γ are related with each other and are significant parameters to characterize the nature of crystals. Table 2.1 shows measured values of $\Theta_D$by thermal and elastic experiments on some representative monatomic crystals.

Exercise 2

1. It is important that the number of phonons in crystals can be left as arbitrary, which is thermodynamically significant for *Boson particles*. Sound wave propagation at low values of $\boldsymbol{k}$ and ω can be interpreted for transporting phonons, which is a typical example of low-level excitations, regardless of temperature. Discuss why undetermined number of particles is significant in Boson statistics. Can there be any other Boson systems where the number of particles if a fixed constant?

2. Einstein's model for the specific heat is consistent with assuming crystals as a uniform medium. Is it a valid assumption that elastic properties can be attributed to each unit cell? What about a case of nonuniform crystal? At sufficiently high temperatures, a crystal can be considered as uniform. Why? Discuss the validity of Einstein's model at high temperatures.
3. Compare the average number of phonons $\langle n \rangle$ calculated from (2.26) with that expressed by (2.31). Notice the difference between them depends on the chemical potential: either $\mu = 0$ or $\mu \neq 0$. Discuss the role of a chemical potential in making these two cases different.
4. The wavefunction of a phonon is expressed by (2.17b). Therefore in a system of many phonons, phonon wavefunctions should be substantially overlapped in the crystal space. This is the fundamental reason why phonons are unidentifiable particles; hence, the phonon system in crystals can be regarded as condensed liquid. For Boson particles ^{4}He, discuss if helium-4 gas can be condensed to a liquid phase at 4.2K.
5. Are the hydrostatic pressure p and compressibility discussed in Sect. 2.8 adequate for anisotropic crystals? Comment on these thermodynamic theories applied to anisotropic crystals.

References

1. M. Tinkham, *Group Theory and Quantum Mechanics* (McGraw-Hill, New York, 1964)
2. R.S. Knox, A. Gold, *Symmetry in the Solid State* (Benjamin, New York, 1964)
3. C. Kittel, *Quantum Theory of Solids*, (John Wiley, New York, 1963)
4. C. Kittel, *Introduction to Solid State Physics*, 6th edn. (Wiley, New York, 1986)
5. C. Kittel, H. Kroemer, *Thermal Physics* (Freeman, San Francisco, 1980)

Chapter 3
Order Variables and the Adiabatic Potential

Crystals constitute a representative group of condensed matter, composed of a large number of identical ions and molecules. Because of imprecise theories of cohesive forces available today, we considered their geometrical structure as granted [1]. Nevertheless, their correlation energy is essential for structural stability, for which we may consider permutable constituents quantum-mechanically. On the other hand, if symmetry in a chemically pure crystal pertains to the thermodynamic environment, the stability can be attributed to the invariance under symmetry operations like rotation, reflection, translation, etc.; the structural invariance determines the correlation energy in a stable crystal. However, the local symmetry can be disrupted below the critical temperature, if a displacement occurs in the constituent. Exhibiting local fluctuations, the strained lattice suffers instability, which is however relaxed thermally to a strain-free structure with decreasing temperature, as postulated by Born and Huang. Meanwhile, symmetry in stable crystals is well documented by X-ray crystallography, as classified by the group theory. In this chapter, leaving details of the symmetry group to references, we discuss the origin of *internal variables* at the critical temperature, which are known as *order variables* and responsible for structural transitions.

3.1 A One-dimensional Ionic Chain

The lattice excitation in a crystal is basically sinusoidal, characterized by the wavevector $\boldsymbol{k}$ and frequency ω. Following Born and von Kárman, we consider a one-dimensional infinite chain of positive and negative ions that are alternately arranged, as illustrated in Fig. 3.1. Disregarding Coulomb interactions, we assume harmonic interactions between neighboring ions as insignificant for the dynamics in a crystal. In this model, we consider only nearest neighbors, whose interaction can be described by potentials $\phi(r - r_o)$, where r_o is the static lattice constant. We can write the equation of motion for ionic masses m_+ and m_- as

M. Fujimoto, *Thermodynamics of Crystalline States*,
DOI 10.1007/978-1-4614-5085-6_3,

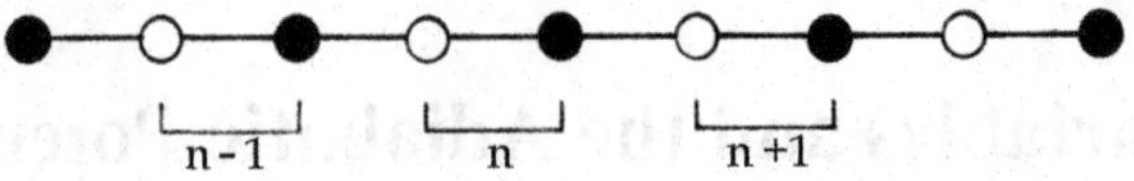

Fig. 3.1 Dipolar chain lattice in one dimension. Plus and minus ions are located in pair at each lattice sites $n-1, n-1$, n, n + 1.

$$m_+\ddot{u}_n^+ = -\phi'(r_0 + u_n^- - u_n^+) + \phi'(r_0 + u_n^+ - u_{n-1}^-) = \phi''(u_n^- + u_{n-1}^- - 2u_n^+)$$

and

$$m_-\ddot{u}_n^- = -\phi'(r_0 + u_{n+1}^- - u_n^+) + \phi'(r_0 + u_n^- - u_n^+) = \phi''(u_{n+1}^+ + u_n^+ - 2u_n^-),$$

where the suffix n indicates lattice sites for a pair of adjacent $\pm$ ions, as indicated in Fig. 3.1. Here, ϕ' and ϕ'' are derivatives with respect to the position x; the time derivatives are written as $\ddot{u}_n^\pm = \mathrm{d}^2u_n^\pm/\mathrm{d}t^2$. Such a structure is dynamically stable with harmonic interactions. Letting $u_n^\pm = u_0^\pm \exp i(kx_n - \omega t)$ and $|x_{n+1}^0 - x_n^0| = r_0$, we can derive time-independent relations between amplitudes $u_0^\pm$, namely

$$\{m_+\omega^2 - 2\phi''(r_0)\}u_0^+ + \phi''(r_0)\{1 + \exp(-ikr_0)\}u_0^- = 0$$

and (i)

$$\phi''(r_0)\{1 + \exp(ikr_0)\}u_0^+ + \{m_-\omega^2 - 2\phi''(r_0)\}u_0^- = 0.$$

Eliminating u_0^+ and u_0^- from these equations in (i), we obtain the *secular* equation

$$\begin{vmatrix} m_+\omega^2 - 2\phi''(r_0) & \phi''(r_0)\{1 + \exp(-ikr_0)\} \\ \phi''(r_0)\{1 + \exp(ikr_0)\} & m_-\omega^2 - 2\phi''(r_0) \end{vmatrix} = 0.$$

Solving it for ω^2, we obtain

$$\omega^2 = \frac{\phi''(r_0)}{m_+m_-}\left\{(m_+ + m_-) \pm \sqrt{(m_+ + m_-)^2 - 4m_+m_-\sin^2\frac{kr_0}{2}}\right\}. \tag{ii}$$

Using (ii) in (i), the amplitude ratio is given by

$$\frac{u_0^+}{u_0^-} = \frac{-m_-\{1 + \exp(-\mathrm{i}kr_0)\}}{(m_+ - m_-) \pm \sqrt{(m_+ + m_-)^2 - 4m_+m_-\sin^2\frac{kr_0}{2}}}. \tag{iii}$$

For the plus sign in (ii) and (iii), the solution is characterized by $\omega^2 \neq 0$ at $k = 0$, whereas for the minus sign, we have $\omega^2 = 0$ at $k = 0$, respectively. In the former case,

$$\omega = \sqrt{\frac{2(m_+ + m_-)\phi''(r_0)}{m_+ m_-}} \neq 0 \quad \text{at} \quad k = 0, \tag{iv}$$

whereas in the latter we have

$$\omega \approx k\sqrt{\frac{\phi''(r_0)}{2(m_+ + m_-)}} \quad \text{for} \quad k \approx 0. \tag{v}$$

These two cases are generally referred to as *optic* and *acoustic* modes, respectively; linearly independent in the harmonic approximation.

The optic mode signified by the *reduced mass* $\mu = m_+ m_- / (m_+ + m_-)$ and nonzero frequency at $k = 0$, and from (iii) we derive the characteristic relation

$$\frac{u_0^+}{u_0^-} = -\frac{m_-}{m_+} \quad \text{or} \quad m_+ u_0^+ + m_- u_0^- = 0. \tag{vi}$$

This relation indicates that the center of mass of $\pm$ ions is unchanged in the optic mode. Define the center-of-mass coordinate by $u_0^{cm} = \frac{m_+ u_0^+ + m_- u_0^-}{m_+ + m_-}$, the optic mode can be expressed in propagation as $u_n^{cm} = u_0^{cm} \exp i(kx - \omega t)$; the amplitude u_0^{cm} can be determined, if the initial conditions are given. In contrast, for an acoustic mode, the relation (v) indicates that the total mass $m_+ + m_-$ determines the frequency.

In the above model, we realize that the acoustic mode represents the vibrating lattice as a whole, whereas the optic mode describes pairs of $m_\pm$ ions that can be regarded as order variables. The ionic chain is signified by different masses, m_+ and m_-, for which Fig. 3.2 sketches the dispersion relations (iii) with respective to various values of m_+/m_-, for which the wavelength λ is utilized as related by the relation $k = 2\pi/\lambda$, instead of the wavevector k; in this figure, the lattice position is specified as related to λ. We note that there are frequency gaps between optic and acoustic modes for a given mass ratio m_+/m_-, except for the specific case $m_+/m_- = 1$. We therefore realized that the two modes are not quite distinguishable, if the mass ratio is close to 1. Otherwise, a well-defined gap exists, for which dielectric and elastic properties of a crystal can be studied separately.

3.2 Displacive Variables

We consider that lattice strains occur with rearranged order variables, which can however be removed thermally to attain unstressed structure, as proposed by *Born-Huang's principle* in their general theory of stable crystals. The optic mode in Sect. 3.1 is primarily independent of lattice deformation, as it refers to the center-of-mass coordinate of the ionic pair. In contrast, the acoustic mode is responsible

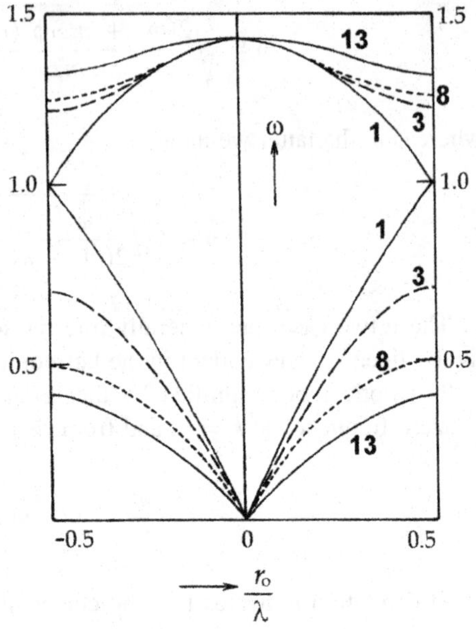

Fig. 3.2 Dispersion curves in one-dimensional dipolar lattice. Inserted *thick numbers* indicate mass ratios m_-/m_+. Frequency gaps between acoustic and optic modes, except for $m_-/m_+ = 1$.

for the strains, displacing centers of masses. These two modes are independent in harmonic approximation, however become significant, if interactions beyond nearest neighbors are considered. Nonetheless, aside from long-distant interactions, it is necessary to identify order variables in a practical system.

In optic modes, ionic displacements accompany with electric dipole moments as expressed by $p_n = e\left(u_n^+ - u_n^-\right)$, namely

$$p_n = p(x_n, t_n) = e\left(u_o^+ - u_o^-\right) \exp i(kx_n - \omega t_n). \tag{3.1}$$

Experimentally, the time average $\langle p(x_n, t_n)\rangle_t$ over the timescale t_o of observation is detectable by applying an electric field externally, although this average can vanish, if $\left\langle u_o^+ - u_o^-\right\rangle_t = 0$.

The acoustic mode is related with the *combined mass* $m_+ + m_-$, which can be subjected to elastic experiments. In nonpolar crystals, $u_n^+ - u_n^-$ can be considered for order variables to modulate the lattice elastically.

In dielectric crystals, an external electric field can act on ionic charges, nevertheless hardly displaceable from lattice points. There are however mobile charges or masses in complex crystals, as discussed in the following examples. Displacing a mobile charge, there should always be an oppositely charged *hole* left behind, generating an electric dipole. In this case, the optic mode is dynamically determined by the reduced mass of displaced charge and hole.

In a stable crystal, the symmetry is unchanged with varying temperature; on the other hand, a complex point structure in a constituent can be deformed, as shown in the following specific examples. In a dielectric crystal, a structural change may generate a local electric dipole moment, which can act as a *polar order variable*. On the other hand, in a magnetic crystal, such a dipolar variable is a magnetic dipole moment, originating from a displaced spin in a magnetic ion or complex. In general, however, order variables may not always be visualized crystallographically. In X-ray diffraction experiments, observed *diffuse diffraction patterns* in some crystals can be interpreted as arising from fluctuating atoms at lattice sites.

If order variables σ_n are correlated with $\sigma_{n'}$ at another site n', the density of order variables is modified in their thermodynamic phase. We assume that the correlation potential is minimized for a modulated structure at a given temperature; although we simply describe the crystal as *heterogeneous* macroscopically. At any rate, modulated densities signify *mesoscopic* crystalline phases.

It is significant quantum-mechanically that these *microscopic* σ_n do not appear *simultaneously at all lattice sites* at a critical temperature, so that their correlation energy is obscured by uncertainties $\Delta\sigma_n$, exhibiting critical anomalies as observed with varying temperature. Leaving the fluctuations to later discussions, in the section we discuss a few examples of order variables in representative crystalline systems.

Example 1 Perovskites

In crystals of perovskites [1], the constituent of chemical formula $A(BO_3)$ is composed of a negative octahedral ion $(BO_6)^{2-}$ that are surrounded by eight positive ions $(BO_6)^{2-}$ at the corners of a cubic cell, as illustrated in Fig. 3.3. The *bipyramidal* negative group has an additional degree of freedom, in which the central ion B^{4+}can be displaced between two positions marked 1 and 1′ in Fig. 3.3a, or the group can be rotated in direction 1 or 1′ in Fig. 3.3b. Related by *inversion* with respect to the center, such a linear or angular displacement is responsible for a binary transition between inversion-related macroscopic domains, as observed in $BaTiO_3$ and $SrTiO_3$ crystals, respectively. In the former, the center shifts along the $+z$ or $-z$ direction, generating an electric dipole moment, whereas in the latter the structure is strained along the z-axis, while keeping the center unchanged. It is significant that the z-direction is parallel to one of the symmetry axes, keeping the lattice in minimal strains. We can define an order variable $\sigma_z(t)$ as related to displacements in both cases. Such order variables identified in perovskite family are detectable, if the inversion time t is longer than the timescale t_o of observation, i.e., $t>t_o$. We know that $BaTiO_3$ is a polar ferroelectric crystal, whereas $SrTiO_3$ is nonpolar below their transition temperatures. These transitions analyzed with the model illustrated in Fig. 3.3a, b are typical examples of displacive variables.

In the above, we assumed the z-direction to be the inversion axis. However, the x-, y-, and z-directions are all equivalent in cubic crystals so that the crystal below T_c should be divided into three tetragonal domains strained along the corresponding symmetry axes.

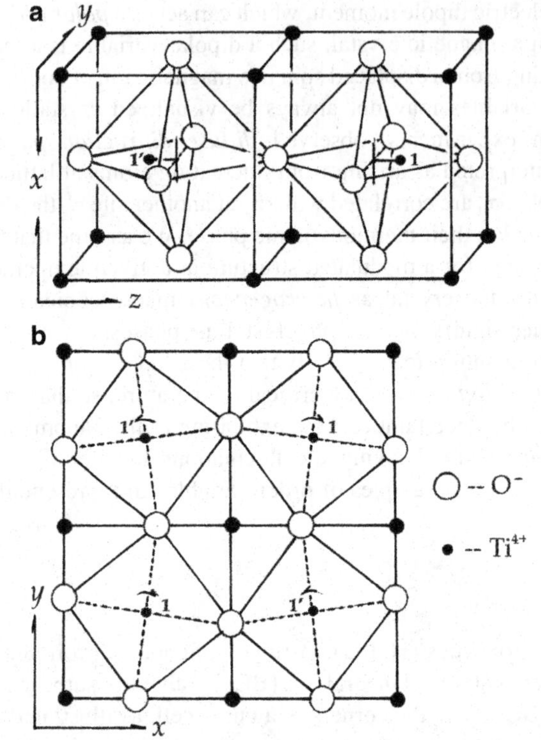

Fig. 3.3 Structure of a perovskite crystal. (**a**) $BaTiO_3$ type; (**b**) $SrTiO_3$ type. An Ti^{4+} ion in the bipyramidal TiO_6^{2-} complex is in motion between 1 and 1′ related by inversion.

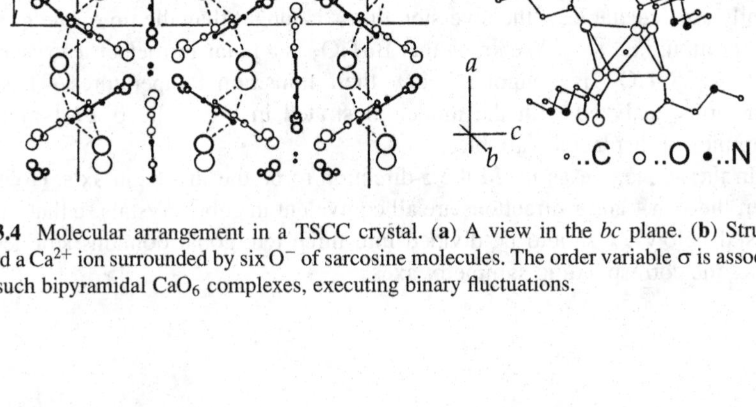

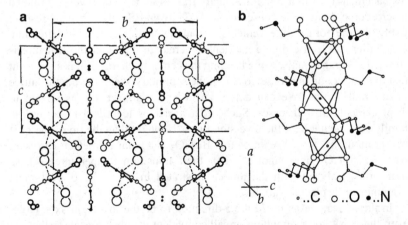

Fig. 3.4 Molecular arrangement in a TSCC crystal. (**a**) A view in the *bc* plane. (**b**) Structure around a Ca^{2+} ion surrounded by six O^- of sarcosine molecules. The order variable σ is associated with such bipyramidal CaO_6 complexes, executing binary fluctuations.

Example 2 Tris-Sarcosine Calcium Chloride (TSCC)

Sarcosine is an amino acid, $H_2C - NH_2 - CH_2 - CO_2H$, crystallizing with $CaCl_2$ molecules in *quasi-orthorhombic* structure, when prepared from aqueous solutions at room temperature. Crystals of TSCC are *twin* in quasi-hexagonal form, consisting of three triangular domains with the monoclinic axes in common, as will be shown by the photograph in Fig. 8.7 in Chap. 8, constituting a spontaneously strained structure. Composed of slightly monoclinic along the *a*-axis, a single-domain sample can be easily cut from a twin crystal for studying ferroelectric order; a *ferroelectric phase transition* occurs at 120 K under atmospheric pressure. Figure 3.4a shows a result of X-ray crystallographic studies by Kakudo and his collaborators [2], where a *quasi-hexagonal* molecular arrangement is evident in the *bc* plane. Figure 3.4b illustrated $Ca^{2+}O_6^-$ complex, in which the central Ca^{2+} ion is surrounded in bipyramidal form by six O^- ions that belong to sarcosine molecules. In paramagnetic resonance experiments with Mn^{2+} impurity ions substituted for Ca^{2+}, the order variable σ_b was identified as such a $Ca^{2+}(\text{sarcosine})_6^-$ complex that is deformed along the *b*-axis.

Example 3 Potassium Dihydrogen Phosphate (KDP)

Figure 3.5 shows the molecular arrangement in orthorhombic crystals of potassium dihydrogen phosphate, $K_2H_2PO_4$ [3]. In this structure, *tetrahedral* PO_4^{3-} ions are linked to four neighboring PO_4^{3-} via four hydrogen bonds, as shown in the figure. Along each bond, a proton oscillates between two oxygen ions in tetrahedral PO_4^{3-} ions, forming electric dipoles in the proton configuration in these hydrogen bonds. In this model, the order variable should be related with such a complex structure of proton configuration with PO_4^{3-} ions.

3.3 Born-Oppenheimer's Approximation

Born and Huang discussed a classical displacement of an electron from the constituent ion in a crystal with Born-Oppenheimer's approximation [4]. Although such a displacement is hypothetical, the result of calculations clearly indicates the presence of its origin, as it should exist physically. Owing to Newton's action–reaction principle, electronic and ionic masses are displaced by their mutual force, as signified by their reduced mass. In this case, the lattice should provide such an interaction as originated from the correlations among constituents.

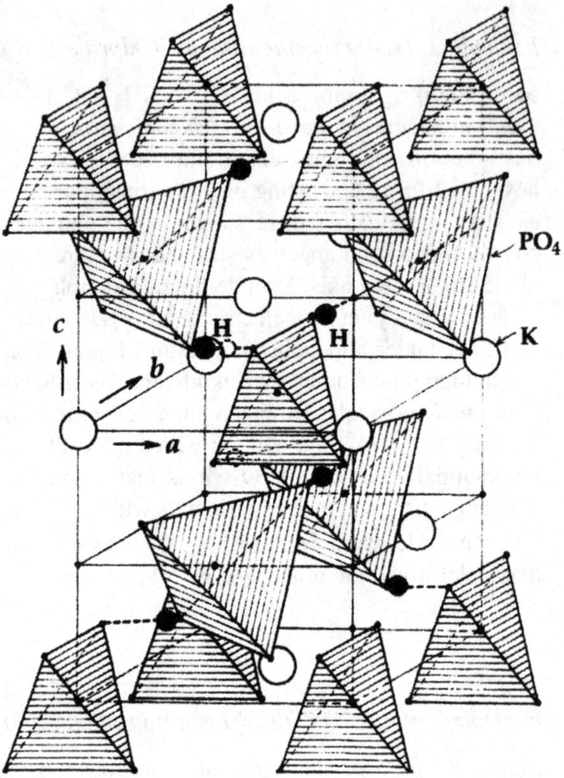

Fig. 3.5 Structure of a KDP crystal. The order variable can be associated with a pyramidal PO_4 ion with four surrounding protons: two protons at closer and two others at distant positions to PO_4.

We consider here an ion of mass m and charge e, displacing hypothetically from the complex ion of mass M and charge $-e$ in a crystal, as inspired by Example 1 in the previous section. Such a classical displacement arising from correlations in a crystal can be considered for the dipolar order variable. A displacing mass m from the constituent mass $M - m \approx M$ constitutes a reduced mass that is nearly identical to m, if $m \ll M$. Such a hole in the point mass $M - m$ may be a little oversimplified; however, Born and Huang defined a parameter $\kappa = (m/M)^{1/4}$ for their convenience in perturbation calculation, emphasizing the significance of the mass ratio m/M.

To simplify the problem, we assume one-dimensional displacements, x_n and X_n, at a representative site n of the displacing part and the rest of the constituent ion, respectively, on the x-axis. For the displacing ion, we have the Hamiltonian

$$\mathsf{H}_o = -\sum_n \frac{\hbar^2}{2m}\frac{\partial^2}{\partial x_n^2} + U,$$

where $U = \sum_n U_n$ is the potential energy related to the mutual *correlations* in a crystal. Then the perturbed Hamiltonian H by the ion M can be expressed as

$$\mathsf{H} = \mathsf{H}_o - \sum_n \frac{\hbar^2}{2M} \frac{\partial^2}{\partial X_n^2}.$$

Introducing the parameter κ defined by $\kappa^4 = m/M$, H can be written as

$$\mathsf{H} = \mathsf{H}_o + \kappa^4 \mathsf{H}_1, \quad \text{where} \quad \mathsf{H}_1 = -\sum_n \frac{\hbar^2}{2m} \frac{\partial^2}{\partial \mathrm{X}_n^2}.$$

Omitting the index n for brevity, the unperturbed and perturbed wave functions can be written as $\phi_i(x, X)$ and $\psi_i(x, X)$, respectively, at a representative lattice site. Namely

$$\mathsf{H}_o \phi_i(x, X) = \varepsilon_i \phi_i(x.X) \tag{3.2a}$$

and

$$\mathsf{H}\psi_i(x, X) = E_i \psi_i(x, X), \tag{3.2b}$$

where ε_i and E_i are unperturbed and perturbed eigenvalues.

Corresponding to x, the coordinate X can be written as $X - X_o = \kappa u$, where κu represents an effective shift of M from the original position X_o. For a fixed value of x, we first expand unperturbed functions of (3.2a) in terms of κu, which are expressed as

$$\varepsilon_i(x, X_o + \kappa u) = \varepsilon_i^{(0)} + \kappa \varepsilon_i^{(1)} + \kappa^2 \varepsilon_i^{(2)} + \cdots,$$

$$\phi_i(x, X_o + \kappa u) = \phi_i^{(0)} + \kappa \phi_i^{(1)} + \kappa^2 \phi_i^{(2)} + \ldots, \tag{3.3}$$

and

$$\mathsf{H}_o\left(x, \frac{\partial}{\partial x}, X_o + \kappa u\right) = \mathsf{H}_o^{(0)} + \kappa \mathsf{H}_i^{(1)} + \kappa^2 \mathsf{H}_i^{(2)} + \cdots,$$

where $\phi_i^{(0)}, \phi_i^{(1)}, \phi_i^{(2)}, \ldots$ are mutually orthogonal. In these expansions (3.2a), if coefficients of terms $\kappa^0, \kappa^1, \kappa^2, \ldots$ can be set equal to zero, we obtain relations

$$\left(\mathsf{H}_o^{(0)} - \varepsilon_i^{(0)}\right) \phi_i^{(0)} = 0, \tag{3.3a}$$

$$\left(\mathsf{H}_o^{(0)} - \varepsilon_i^{(0)}\right) \phi_i^{(1)} = -\left(\mathsf{H}_o^{(1)} - \varepsilon_i^{(1)}\right) \phi_i^{(0)}, \tag{3.3b}$$

$$\left(\mathsf{H}_o^{(0)} - \varepsilon_i^{(0)}\right) \phi_i^{(2)} = -\left(\mathsf{H}_o^{(1)} - \varepsilon_i^{(1)}\right) \phi_i^{(1)} - \left(\mathsf{H}_o^{(2)} - \varepsilon_i^{(2)}\right) \phi_i^{(0)}, \tag{3.3c}$$

$$\ldots\ldots\ldots\ldots\ldots ,$$

which are to be solved for $\phi_i^{(0)}$, $\phi_i^{(1)}$, $\phi_i^{(2)}$, ...

For convenience, we replace the term $\kappa^4\mathsf{H}_1$ in the perturbed H as

$$\kappa^4\mathsf{H}_1 = \kappa^2\mathsf{H}_1^{(2)} \quad \text{where} \quad \mathsf{H}_1^{(2)} = -\sum_u \frac{\hbar^2}{2m}\frac{\partial^2}{\partial u^2},$$

We then have H expressed as

$$\mathsf{H} = \mathsf{H}_\mathrm{o}^{(0)} + \kappa\mathsf{H}_\mathrm{o}^{(1)} + \kappa^2\left(\mathsf{H}_\mathrm{o}^{(2)} + \mathsf{H}_1^{(2)}\right) + \kappa^3\mathsf{H}_\mathrm{o}^{(3)} + \cdots$$

For perturbed states, we write expansions of eigenvalues and eigenfunctions, i.e.,

$$E_i = E_i^{(0)} + \kappa E_i^{(1)} + \kappa^2 E_i^{(2)} + \cdots$$

and (3.4)

$$\psi_i = \psi_i^{(0)} + \kappa\psi_i^{(1)} + \kappa^2\psi_i^{(2)} + \cdots$$

Noting $E_i^{(0)} = \varepsilon_i^{(0)}$, we obtain the following relations for perturbed functions:

$$\left(\mathsf{H}_\mathrm{o}^{(0)} - \varepsilon_i^{(0)}\right)\psi_i^{(0)} = 0, \tag{3.4a}$$

$$\left(\mathsf{H}_\mathrm{o}^{(0)} - \varepsilon_i^{(0)}\right)\psi_i^{(1)} = -\left(\mathsf{H}_\mathrm{o}^{(1)} - E_i^{(1)}\right)\psi_i^{(0)}, \tag{3.4b}$$

$$\left(\mathsf{H}_\mathrm{o}^{(0)} - \varepsilon_i^{(0)}\right)\psi_i^{(2)} = -\left(\mathsf{H}_\mathrm{o}^{(1)} - E_i^{(1)}\right)\psi_i^{(1)} - \left(\mathsf{H}_\mathrm{o}^{(2)} + \mathsf{H}_1^{(2)} - E_i^{(2)}\right)\psi_i^{(0)}, \tag{3.4c}$$

...............,

First, comparing (3.3a) and (3.4a), we see that $\psi_i^{(0)}(x) = \phi_i^{(0)}(x, X_\mathrm{o})$ is the solution at $u = 0$. Therefore, for $u \neq 0$ we can write that

$$\psi_i^{(0)}(x,u) = \chi^{(0)}(u)\phi_i^{(0)}(x), \tag{3.5}$$

where $\chi^{(0)}(u)$ is an arbitrary function of u. Using (3.5), the inhomogeneous equation (3.4b) can be solved, if

$$\int \phi_i^{(0)}\left(\mathsf{H}_\mathrm{o}^{(1)} - E_i^{(1)}\right)\psi_i^{(0)}\mathrm{d}x = \chi^{(0)}(u)\int \phi_i^{(0)}\left(\mathsf{H}_\mathrm{o}^{(1)} - E_i^{(1)}\right)\phi_i^{(0)}\mathrm{d}x = 0.$$

In fact, from (3.3b), we have

$$\int \phi_i^{(0)}\left(\mathsf{H}_\mathrm{o}^{(0)} - \varepsilon_i^{(0)}\right)\phi_i^{(1)}\mathrm{d}x = -\int \phi_i^{(0)}\left(\mathsf{H}_\mathrm{o}^{(1)} - \varepsilon_i^{(1)}\right)\phi_i^{(0)}\mathrm{d}x = 0,$$

because $\phi_i^{(0)}$ and $\phi_i^{(1)}$ are orthogonal. Further, comparing the above results, we obtain $E_i^{(1)} = \varepsilon_i^{(1)}$. It is noted that $\varepsilon_i^{(1)}$ is the coefficient of the term of κ in the expansion of $\varepsilon_i(X_0 + \kappa u)$, that is,
$\kappa\varepsilon_i^{(1)} = (\partial\varepsilon_i/\partial X_i)_{X_i=X_0}\kappa u_i = 0$ at $u_i = 0$. Hence, we have the relations

$$E_i^{(1)} = \varepsilon_i^{(1)} = 0 \quad \text{and} \quad \mathsf{H}_o^{(1)} = 0,$$

in addition to the solution $\phi_i^{(1)}$ of (3.4b). Therefore, up to this accuracy of κ, the wavefunction is written as

$$\psi_i^{(1)} = \chi^{(0)}(u)\phi_i^{(1)}(x) + \chi^{(1)}(u)\phi_i^{(0)}(x), \tag{3.6}$$

where $\chi^{(0)}(u)$ and $\chi^{(1)}(u)$ are arbitrary functions of u.

Using (3.5), (3.6), and $E_i^{(1)} = 0$ in (3.4c), we obtain

$$\left(\mathsf{H}_o^{(0)} - \varepsilon_i^{(0)}\right)\phi_i^{(2)} = -\mathsf{H}_o^{(1)}\left(\chi^{(0)}\phi_i^{(1)} + \chi^{(1)}\phi_i^{(0)}\right) - \left(\mathsf{H}_o^{(2)} + \mathsf{H}_1^{(2)} - E_i^{(2)}\right)\chi^{(0)}\phi_i^{(0)}.$$

Subtracting $\chi^{(1)} \times$ (3.3a) and $\chi^{(0)} \times$ (3.4a) from this expression, we derive

$$\begin{aligned}&\left(\mathsf{H}_o^{(0)} - \varepsilon_i^{(0)}\right)\left(\psi_i^{(2)} - \chi^{(0)}\phi_i^{(2)} - \chi^{(1)}\phi_i^{(1)}\right)\\ &= -\left(\mathsf{H}_i^{(2)} + \varepsilon_i^{(2)} - E_i^{(2)}\right)\chi^{(0)}\phi_i^{(0)},\end{aligned} \tag{3.7}$$

which is soluble, provided that $\int \phi_i^{(0)}\left(\mathsf{H}_i^{(2)} + \varepsilon_i^{(2)} - E_i^{(2)}\right)\chi^{(0)}\phi_i^{(0)}\mathrm{d}x = 0$. Noting that the factor $\left(\mathsf{H}_i^{(2)} + \varepsilon_i^{(2)} - E_i^{(2)}\right)\chi^{(0)}$ in the integrand is independent of x, the solubility condition is satisfied when the function $\chi^{(0)}(u)$ can be obtained from the equation

$$\left(\mathsf{H}_i^{(2)} + \varepsilon_i^{(2)} - E_i^{(2)}\right)\chi^{(0)}(u) = 0. \tag{3.8}$$

This implies that the perturbed motion of M is harmonic and independent of x. In this first-order approximation, the lattice structure remains unmodified, and the electronic function is written as $\psi_i(x,u) = \psi_i^{(0)} + \kappa\psi_i^{(1)}$, which is the *harmonic approximation.*

In the second order of κ^2, we have

$$\psi_i(x,u) = \psi_i^{(0)} + \kappa\psi_i^{(1)} + \kappa^2\psi_i^{(2)}, \tag{3.9}$$

for which $\psi_i^{(2)}$ can be calculated as follows. If the condition (3.8) is met, (3.7) is simplified as

$$\left(\mathsf{H}_{\mathrm{o}}^{(2)}-\varepsilon_i^{(2)}\right)\left(\psi_i^{(2)}-\chi^{(0)}\psi_i^{(2)}-\chi^{(1)}\psi_i^{(1)}\right)=0.$$

This should be consistent with (3.3a) and (3.4a), which is expressed as $\left(\mathsf{H}_{\mathrm{o}}^{(2)}-\varepsilon_i^{(2)}\right)\left(\chi^{(2)}\phi_i^{(0)}\right)=0$, where $\chi^{(2)}(u)$ is another arbitrary function of u. Accordingly, $\chi^{(2)}\phi_i^{(0)}=\psi_i^{(2)}-\chi^{(0)}\phi_i^{(2)}-\chi^{(1)}\phi_i^{(1)}$, and we can write

$$\psi_i^{(2)}(x,u)=\chi^{(0)}(u)\phi_i^{(2)}(x,u)+\chi^{(1)}(u)\phi_i^{(1)}(x,u)+\chi^{(2)}(u)\phi_i^{(0)}(x). \tag{3.10}$$

Using (3.5), (3.6), and (3.10) in (3.9),

$$\psi_i(x,u)=\chi^{(0)}(u)\left\{\phi_i^{(0)}(x)+\kappa\phi_i^{(1)}(x,u)+\kappa^2\phi_i^{(2)}(x,u)\right\}+\kappa\chi^{(1)}(u)$$
$$\times\left\{\phi_i^{(1)}(x)+\kappa\phi_i^{(1)}(x,u)\right\}+\kappa^2\chi^{(2)}(u)\phi_i^{(0)}(x).$$

By adding such higher-order terms as $\kappa^3\chi^{(2)}(u)\phi_i^{(1)}(x,u)$, $\kappa^3\chi^{(1)}(u)\phi_i^{(2)}(x,u)$, and $\kappa^4\chi^{(2)}(u)\phi_i^{(2)}(x,u)$, this expression can be written as

$$\psi_i(x,u)\approx\left\{\chi^{(0)}(u)+\kappa\chi_i^{(1)}(u)+\kappa^2\chi_i^{(2)}(x,u)\right\}\phi_i(x,u)$$

where

$$\phi_i(x,u)=\phi_i^{(0)}(x)+\kappa\phi_i^{(1)}(x,u)+\kappa^2\phi_i^{(2)}(x,u),$$

indicating that the ionic motion represented by the bracketed factor function $\{\ldots\}$, as if the moving ion is independent of the steady ion. In this *adiabatic* approximation, the state $\phi_i(x,u)$ modified by u should be in propagation in the lattice, as will be discussed in Sect. 3.4. Born and Huang proceeded further to higher-order calculations; however, we shall not discuss it beyond the adiabatic approximation.

In the adiabatic approximation, the perturbed Hamiltonian can generally be written as

$$\mathsf{H}=\mathsf{H}_{\mathrm{o}}+\kappa^2\mathsf{H}_1^{(2)}+\Delta U,$$

where

$$\Delta U=\kappa^2E_i^{(2)}(u)+\kappa^3E_i^{(3)}(u)+\kappa^4\left\{E_i^{(4)}(u)+C\right\}. \tag{3.11}$$

Here, C is a constant term arising from $\partial^2\varepsilon_i^{(2)}/\partial u^2$, as verified by Born-Huang's book. We shall call ΔU an *adiabatic potential*, originating from the interaction with the lattice. In thermodynamics, such processes can be understood as adiabatic, for which an internal potential ΔU is responsible. In real crystals, the variable u should be characterized as a classical vector for a structural change, for which ΔU should express a potential for the crystal field in three-dimensional space. Considering that κu represents a vector displacement, such a crystal field potential can be expressed, for example, in orthorhombic crystals, as $\Delta U = Au_x^2 + Bu_y^2 + Cu_z^2$ in harmonic approximation. In adiabatic approximation, in contrast, ΔU is anharmonic, acting for driving order variables to displace adiabatically.

3.4 Equilibrium Crystals and the Bloch Theorem

In Sect. 3.3, we discussed an order variable as arising from a partial displacement from the constituent ion in a crystal. Quantum mechanically, it is significant to realize that such local displacements at all lattice sites do not occur *simultaneously*. On the other hand, we postulated classical displacements at all sites, disrupting the lattice symmetry as a whole. Nevertheless, space-time coordinates of such displacements are distributed at the transition threshold, thereby initiating a phase transition at random phase, exhibiting transition *anomalies*. We write classical displacement as $u_n = u_{no} \exp i\phi(x_n, t_n)$ for symmetry change, where the phases ϕ (x_n, t_n) are randomly distributed with finite amplitudes u_{no}. Further we discussed in Sect. 3.5 that such distributed $\boldsymbol{u}_i$ are synchronized to establish strain-free symmetry in a crystal with decreasing temperature. Lattice symmetry characterizes a stable structure, if there are no strains, for which the Bloch theorem and its corollaries are applicable. In this section, the Bloch theory is reviewed as a prerequisite for the modulated lattice.

3.4.1 Reciprocal Lattice and Renormalized Coordinates

In a stable crystal, we assume that such site events as expressed by displacements $\boldsymbol{u}(\boldsymbol{r}_n, t)$ are identical at almost all sites except on surfaces and at defect positions. In this approximation, $\boldsymbol{u}(\boldsymbol{r}_n, t)$ are considered as invariant with lattice translations $\boldsymbol{r}_n \rightarrow \boldsymbol{r}_m$ at any time t, if these imperfections are disregarded. In fact, an infinite periodic structure is assumed for crystals in sufficiently large size, allowing to disregard surfaces.

Owing to short timescale of X-ray collisions, we can specify lattice points geometrically by vectors $\boldsymbol{r}_i$, referring in principle to the timescale t_o of observation. The vector

$$\boldsymbol{R} = n_1\boldsymbol{a}_1 + n_2\boldsymbol{a}_2 + n_3\boldsymbol{a}_3 \tag{3.12}$$

describes a translation of lattice points in an orthorhombic crystal, where $\boldsymbol{a}_1$, $\boldsymbol{a}_2$ and $\boldsymbol{a}_3$ are basic unit translations along the symmetry axes 1, 2, and 3, respectively, and

n_1, n_2 and n_3 are integers attached to the lattice points. Using (3.12), the spatial invariance of $\boldsymbol{u}_i$ can be expressed as

$$\boldsymbol{u}(\boldsymbol{r}_n) = \boldsymbol{u}(\boldsymbol{r}_n + \boldsymbol{R}) \tag{3.13}$$

at arbitrary time t with respect to renormalized coordinates $\boldsymbol{r}_n$. Here, the coordinates X in the previous section are replaced by $\boldsymbol{R}$. Physically, a Fourier series

$$\boldsymbol{u}(\boldsymbol{r}_n, t) = \sum_{\boldsymbol{G}} \boldsymbol{u}_{\mathrm{G}} \exp i(\pm \boldsymbol{G} \cdot \boldsymbol{r}_n - \omega t) \tag{3.14}$$

expresses a lattice *excitation,* if the vector $\boldsymbol{G}$ is determined as related to the periodic boundary condition. The amplitude $\boldsymbol{u}_{\mathrm{G}}$ is the Fourier transform $\boldsymbol{u}_{\mathrm{G}} = \boldsymbol{u}(\boldsymbol{r}_n, t) \exp i(\mp \boldsymbol{G} \cdot \boldsymbol{r}_n - \omega t)$. The expression (3.14) is invariant for inversion $\boldsymbol{G} \to -\boldsymbol{G}$, as the symmetry group includes *inversion symmetry*.

Using (3.12), we can derive the relation $\exp i(\boldsymbol{G} \cdot \boldsymbol{R}) = 1$, namely $\boldsymbol{G}$ is related with $\boldsymbol{R}$ as

$$\boldsymbol{G} \cdot \boldsymbol{R} = 2\pi \times (0 \, \text{or integers}), \tag{3.15}$$

indicating that

$$\boldsymbol{G} \parallel \boldsymbol{R} \quad \text{and} \quad |\boldsymbol{G}||\boldsymbol{R}| = 2\pi \times \text{integer}.$$

Corresponding to the unit translations $\boldsymbol{R} = \boldsymbol{a}_1, \boldsymbol{a}_2$ and $\boldsymbol{a}_3$, we define the vectors

$$\boldsymbol{a}_1{}^* = \frac{2\pi}{\Omega}(\boldsymbol{a}_2 \times \boldsymbol{a}_3), \; \boldsymbol{a}_2{}^* = \frac{2\pi}{\Omega}(\boldsymbol{a}_3 \times \boldsymbol{a}_1), \; \boldsymbol{a}_3{}^* = \frac{2\pi}{\Omega}(\boldsymbol{a}_1 \times \boldsymbol{a}_2), \tag{3.16}$$

where $\Omega = (\boldsymbol{a}_1, \, \boldsymbol{a}_2, \, \boldsymbol{a}_3)$ is the volume of the primitive cell. The space spanned by vectors $\boldsymbol{a}_1{}^*$, $\boldsymbol{a}_2{}^*$ and $\boldsymbol{a}_3{}^*$ multiplied by integers constitutes the *reciprocal lattice*. Consider a rectangular volume $V = L_1L_2L_3$ of sufficiently large edges L_1, L_2 and L_3 in a crystal, we take specifically $|\boldsymbol{R}| = L_1, L_2$ and L_3, and find that the relation $\boldsymbol{G} \cdot \boldsymbol{R} = 2\pi \times$ integer is satisfied by $G_1 = (2\pi/L_1)n_1$, $G_2 = (2\pi/L_2)n_2$, and $G_3 = (2\pi/L_3)n_3$, respectively, where n_1, n_2, n_3 are integers. We can verify that these vectors, $\boldsymbol{a}_1{}^*$, $\boldsymbol{a}_2{}^*$ and $\boldsymbol{a}_3{}^*$, are orthonormal in the space of $\boldsymbol{a}_1$, $\boldsymbol{a}_2$ and $\boldsymbol{a}_3$, namely, $\boldsymbol{a}_1{}^* \cdot \boldsymbol{a}_1 = 2\pi$, $\boldsymbol{a}_1{}^* \cdot \boldsymbol{a}_2 = 0$, etc. Thus, the propagation of $\boldsymbol{u}_i(\boldsymbol{r}_i, t)$ in the crystal space can be specified by the vector

$$\boldsymbol{G} = h\boldsymbol{a}_1{}^* + k\boldsymbol{a}_2{}^* + l\boldsymbol{a}_3{}^* \tag{3.17}$$

in the direction (h, k, l), which indicate a lattice point in the reciprocal lattice. We note there is the relation

$$hn_1 + kn_2 + ln_3 = 2\pi \times \text{integer}. \tag{3.18}$$

Realizing that the quantity $\boldsymbol{G}^2$ is invariant for rotation of the crystal space, the vector $\boldsymbol{G}$ is associated with the kinetic energy in propagation, hence referred to as the *wavevector*.

The adiabatic potential $\Delta U(u_n)$ is responsible for the displacement u_n, whose phases $\phi_n = \boldsymbol{G} \cdot \boldsymbol{r}_n - \omega t_n$ are in common in thermodynamic states. Using Born-von Kárman boundary conditions, the phase ϕ_n can be an arbitrary angle in the whole angular range $0 \leq \phi_n \leq 2\pi$. In this context, the site position $\boldsymbol{r}_i$ can be replaced by a continuous arbitrary position $\boldsymbol{r}$, thereby using a continuous phase angle $\phi(\boldsymbol{r}, t) = \boldsymbol{G} \cdot \boldsymbol{r} - \omega t$, where $0 \leq \phi < 2\pi$. We shall call $\phi(\boldsymbol{r}, t)$ as a *renormalized phase*. In some cases, the angular range $(0, 2\pi)$ is converted conveniently to the equivalent wavelength $\lambda = 2\pi/|\boldsymbol{G}|$ in the range $0 \leq \lambda < L$. The renormalized variable $\boldsymbol{r}$ is a representative lattice point in a sufficiently large crystal, whose surfaces are ignored. For the renormalized phase, the time t is insignificant in that the lattice translation is a simple Galilean transformation in a steady crystalline phase. Experimentally however, we can observe renormalized phases in standing order waves formed by reflections from boundaries.

3.4.2 The Bloch Theorem

We write the Hamiltonian equation for a nearly free electron in a crystal as

$$\mathrm{H}\psi(\boldsymbol{r}) = \left\{\frac{p^2}{2m} + U(\boldsymbol{r})\right\}\psi(\boldsymbol{r}) = E\psi(\boldsymbol{r}), \tag{3.19a}$$

where $U(\boldsymbol{r})$ is an adiabatic potential at a renormalized $\boldsymbol{r}$ that satisfies the periodic condition

$$U(\boldsymbol{r}) = U(\boldsymbol{r} + \boldsymbol{R}). \tag{3.19b}$$

We assume a similar periodicity for the wavefunction

$$\psi(\boldsymbol{r}) = \psi(\boldsymbol{r} + \boldsymbol{R}), \tag{3.20}$$

indicating that $U(\boldsymbol{r})$ and $\psi(\boldsymbol{r})$ are in phase. In the above, the position $\boldsymbol{r}$ may be considered a representative lattice point conveniently as a renormalized position in an idealized crystal.

Using periodicity along the direction of a basic vector $\boldsymbol{a}$, we can write $\boldsymbol{R} = n\boldsymbol{a}$, where $n = 1, 2, \ldots, \mathrm{N}$; N is the number of sites on the chain of the length $L = \mathrm{N}a$. The functions $\psi(\boldsymbol{r})$ at all sites are identical, so that from (3.20) we can write

$$\psi(\boldsymbol{r}+n\boldsymbol{a}) = c^n\psi(\boldsymbol{r}),$$

where $c^{\mathrm{N}} = 1$ if $n = \mathrm{N}$, so that $c = \exp i(2\pi\mathrm{N})$. Considering $L = na$ and so $2\pi n/L = |\boldsymbol{G}|$ in this case,

$$\psi(\boldsymbol{r}+n\boldsymbol{a}) = \{\exp i\boldsymbol{G}\cdot(n\boldsymbol{a})\}\psi(\boldsymbol{r}).$$

Writing the Fourier expansion as $\psi(\boldsymbol{r}) = \sum_G \varphi_G \exp i\boldsymbol{G}.\boldsymbol{r}$ with the transform φ_G, we can express the translational symmetry of the wavefunction by

$$\psi(\boldsymbol{r}+\boldsymbol{R}) = \sum_G \varphi_G \exp i\boldsymbol{G}\cdot(\boldsymbol{r}+\boldsymbol{R}), \tag{3.21}$$

which is generally called the *Bloch theorem* [5].

In the above, we only considered the periodic potential for (3.18), but in the presence of kinetic energy, the wavefuntion $\psi(\boldsymbol{r}+\boldsymbol{R})$ is expressed in Fourier expansion as

$$\psi(\boldsymbol{r}+\boldsymbol{R}) = \sum_{k+G} \varphi_{k+G} \exp i(\boldsymbol{k}+\boldsymbol{G})\cdot(\boldsymbol{r}+\boldsymbol{R})$$

which is the same as

$$\psi(\boldsymbol{r}+\boldsymbol{R}) = \sum_{k+G} \varphi_{k+G} \exp i(\boldsymbol{k}+\boldsymbol{G})\cdot\boldsymbol{r} = \psi(\boldsymbol{r}) = \psi(\boldsymbol{r}).$$

On the other hand, $\psi(\boldsymbol{r}) = \sum_k \varphi_k \exp i\boldsymbol{k}\cdot\boldsymbol{r}$; hence we can derive the expression

$$\varphi_k = \varphi_{k+G}. \tag{3.22}$$

Quantum-mechanically, the momentum p is an operator $\boldsymbol{p} = -i\hbar\nabla$; hence (3.18) can be written for $\psi(\boldsymbol{r})$ as

$$\left\{\frac{(\boldsymbol{p}+\hbar\boldsymbol{k})^2}{2m} + U(\boldsymbol{r})\right\}\varphi_k(\boldsymbol{r}) = \varepsilon_k\varphi_k(\boldsymbol{r}),$$

which can be expressed as

$$\left\{-\frac{\hbar^2}{2m}(\nabla^2 + 2i\boldsymbol{k}\cdot\nabla) + U(\boldsymbol{r})\right\}\varphi_k(\boldsymbol{r}) = \left(\varepsilon_k - \frac{\hbar^2\boldsymbol{k}^2}{2m}\right)\varphi_k(\boldsymbol{r}). \tag{3.23}$$

Accordingly, $\varepsilon_k - \hbar^2\boldsymbol{k}^2/2m = \text{const.}$ can be plotted to express a constant-energy surface in the reciprocal space. Corresponding to (3.22), we have the relation for the energy eigenvalues,

$$\varepsilon_k = \varepsilon_{k+G}. \tag{3.24}$$

Based on (3.23) and (3.24), polyhedral surfaces can be assigned for $|\boldsymbol{G}|$ in the reciprocal space to deal with the essential momentum space without repetition, which is called *Brillouin zones.*

3.4.3 Lattice Symmetry and the Brillouin Zone

In the foregoing, we discussed an ionic excitation with the Bloch theorem, referring to a representative lattice point. Such one-dimensional excitation in propagation through the lattice should be a standing wave, in order to be detected experimentally. In fact, it is in standing wave in thermodynamic states, as composed by $\pm\boldsymbol{k}$ waves that are subjected to Born-von Kárman's periodic boundary conditions. Therefore, we can use an arbitrary position $\boldsymbol{r}$ in-between adjacent boundaries for the phase in the range of $0 \le \phi(\boldsymbol{r}) < 2\pi$, which we call the Brillouin zone. With the Bloch theory, the wavefunction $\varphi_k(\boldsymbol{r})$can be written as

$$\varphi_k(\boldsymbol{r}+\boldsymbol{R}) = (\exp i\boldsymbol{k}\cdot\boldsymbol{R})\varphi_k(\boldsymbol{r}), \tag{3.25}$$

where an operator $T = \exp i\boldsymbol{k}\cdot\boldsymbol{R}$ can be regarded as a operator for translation by $\boldsymbol{R}$. If $\boldsymbol{k} = \boldsymbol{G}$, a group of such translations constitutes the space group in the Brillouin zone [5, 6].

In addition to the space group, the unit cell needs to be specified by a *point group* composed of geometrical rotation, reflection, etc. for comprehensive characterization of the structure. Elements of the point group constitute an orthogonal set for invariant structure. With a point operation S, we have the relation

$$S\varphi_k(\boldsymbol{r}) = \varphi_k(S^{-1}\boldsymbol{r}), \tag{3.26}$$

indicating that S characterizes inversion $\boldsymbol{r} \to -\boldsymbol{r}$.

For a crystal in thermal and adiabatic equilibrium, the Gibbs potential is a function of p, T, and $\varphi(\boldsymbol{r})$, which must be invariant under symmetry operations T, S, and $\boldsymbol{k} = \boldsymbol{G}$. The spatial inversion $\boldsymbol{r} \to -\boldsymbol{r}$ can be substituted for the phase inversion $\phi \to -\phi$, disregarding *time inversion*. This is consistent with the thermodynamic laws.

Crystal symmetry characterizes the geometrical structure of a crystal, whereas the reciprocal lattice represents symmetry of kinetic excitation of the lattice. Physically, such excitations are considered to arise from lattice vibrations signified by k and ε_k. Equations (3.24) and (3.25) are for excitations in the reciprocal lattice, where the vectors $\boldsymbol{k}$ and $\boldsymbol{k}+\boldsymbol{G}$ specify dynamically identical states. Instead of repeating the identical relations, we can reduce the lattice to the Brillouin zone, for

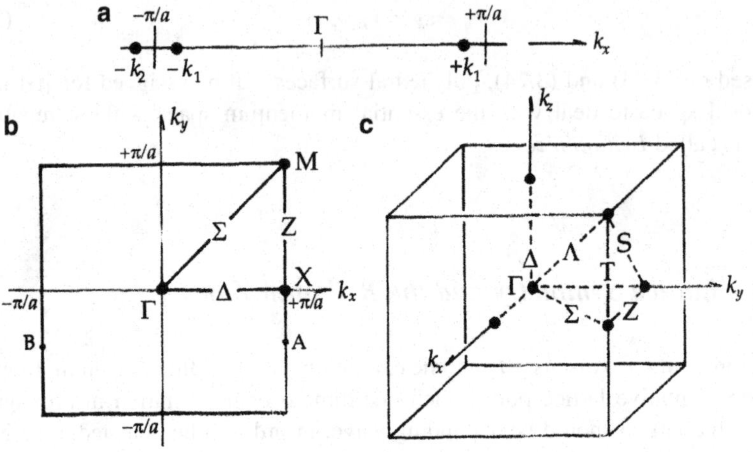

Fig. 3.6 (**a**) Brillouin zone in one dimension, $-\pi/a<k_x<\pi/a$; the zone center is Γ. (**b**) Two-dimensional Brillouin zone in the k_xk_y plane. (**c**) Brillouin zone in a cubic lattice.

example, the first zone at $\boldsymbol{G}=0$ surrounded by $G_1=|\boldsymbol{a}_1{}^*|$, $G_2=|\boldsymbol{a}_2{}^*|$, and $G_3=|\boldsymbol{a}_3{}^*|$.

Figure 3.6a shows the Brillouin zone in one dimension, where the center is at $G=0$, and the boundaries between the first and second are determined by $k=\pm G/2=\pm\pi/a$, which are signified by reflections. In terms of wavefunctions $\psi_k(\boldsymbol{r})$, the boundaries are connected as

$$\psi_k\left(\frac{\pi}{a}\right)=\psi_k\left(-\frac{\pi}{a}\right) \quad \text{for} \quad k=\pm\frac{\pi}{a},$$

and the standing waves can be written as

$$\psi_{k=\frac{\pi}{a}}(x)=\sin\frac{\pi x}{a}\varphi_{k=\frac{\pi}{a}}(x) \quad \text{or} \quad \psi_{k=\frac{\pi}{a}}(x)=\cos\frac{\pi x}{a}\varphi_{k=\frac{\pi}{a}}(x) \quad \text{at} \quad k=\frac{\pi}{a}.$$

Also significant at the boundary is that the energy eigenvalue is continuous and

$$\nabla_k\varepsilon_k=0 \quad \text{at} \quad k=\pm\frac{\pi}{a}. \tag{3.27}$$

Figure 3.6b illustrates the Brillouin zone of a square lattice. Symmetry operations of the point group are expressed as 4*mm*: a fourfold axis of two sets of mirror planes m_x, m_y and two diagonal planes m_d, $m_{d'}$. Significant points Γ, M, X and lines Δ, Z, Σ are indicated in the zone. The point Γ is at the origin $\boldsymbol{k}=0$, transferring into itself under all operations in the group. The point M transforms into itself or into opposite corner under the same operations; these corners are related with each other by the same reciprocal vector, and hence the four corners are equivalent.

The point X is invariant under the operations $2_z, m_x, m_y$, where a reflection m_x and a rotation 2_zin succession carried a point $\left(\frac{\pi}{a}, 0\right)$ into $\left(-\frac{\pi}{a}, 0\right)$.

The lines Δ, Σ, Z and are invariant under mirror reflections m_x, m_d, m_y, respectively. It is noted that by m_x, an arbitrary point $\left(\frac{\pi}{a}, -k_y\right)$ on the line Z can be carried to the point $\left(-\frac{\pi}{a}, -k_y\right)$ that differs by the vector $G = \left(\frac{2\pi}{a}, 0, 0\right)$, implying that (3.27) can be applied to all points on the zone boundaries.

For a simple cubic lattice, similar points and lines as in the square lattice are shown in Fig. 3.6c, where the basic feature of the Brillouin zone can be confirmed by using point group operations $\frac{4}{m}, \bar{3}, \frac{2}{m}$. There are four specific points R, M, X, Γ and six specific lines $\Delta, S, T, \Sigma, Z, \Lambda$, as shown in the figure, representing invariant points and lines under the group operations similar to those in the square lattice. However, we shall not discuss crystal symmetry any further, leaving the detail to standard reference books on the group theory [7].

3.5 Phonon Scatterings

3.5.1 *Phonon Scatterings in Crystals*

All the order variables at lattice sites together compose an excited *field* in a crystal, which are signified by a wavefunction $\psi(\boldsymbol{r}_i, t)$ with Born-von Kárman' periodic boundaries. Instead of discrete space-time $\boldsymbol{r}_i$, t, we can use *renormalized* space-time $\boldsymbol{r}$, t within periodic boundaries, we can thereby express the excited state by an arbitrary phase $\phi = \boldsymbol{k} \cdot \boldsymbol{r} - \omega t$ in the range $0 \leq \phi < 2\pi$, where the corresponding density is given by $|\psi(\boldsymbol{r}, t)|^2 = \boldsymbol{\sigma}(\phi) = \boldsymbol{\sigma}_o \exp i\phi$

The wave of $\boldsymbol{\sigma}(\phi)$ is driven by the correlation potential $U\{\sigma(\phi')\}$ most efficiently if $\phi = \phi'$, in which case the crystal is regarded as in equilibrium with the surroundings. Therefore, we can write the field equation

$$\frac{\partial^2 \boldsymbol{\sigma}_k}{\partial {\phi_k}^2} = -\nabla_k U(\boldsymbol{\sigma}_k) \tag{3.28}$$

at a given temperature, where ∇_k is the gradient operator in the reciprocal lattice. Equation (3.28) is an inhomogeneous linear differential equation, describing a propagating $\boldsymbol{\sigma}_k(\phi_k)$ in space-time, where the potential $U(\boldsymbol{\sigma}_k)$ modulates the amplitude $\boldsymbol{\sigma}_{ko}$.

In thermodynamics, the time-dependent equation (3.28) provides solutions determined within thermal surface boundaries, which are to be observable in thermal experiments. We realize that *thermal* measurements can be performed with equipment of infinite timescale $t_o = \infty$; while in *sampling* experiments the timescale t_o is finite. Embedded in a crystal, the timescale for phonon scatterings from $U(\boldsymbol{u}_k)$ is important for visualizing modulated structure in time-dependent experiments.

In the power expansion of $U(\boldsymbol{u_k})$, first-order terms proportional to $\boldsymbol{u_k}$ are predominant time-dependent perturbations, for which off-diagonal matrix elements $\langle \boldsymbol{K},\Omega;N|u_k|\boldsymbol{K}',\Omega';N'\rangle$ play an essential role between two multiphonon states $\langle \boldsymbol{K},\Omega;N|$ and $|\boldsymbol{K}',\Omega';N'\rangle$. Here, N and N' are the densities of phonons in these states. Considering a scattering of a single phonon of $(\boldsymbol{K},\Omega)$ to another one of $(\boldsymbol{K}',\Omega')$, we have the laws of momentum-energy conservation

$$\boldsymbol{K}' = \boldsymbol{K} \pm \boldsymbol{k} \quad \text{and} \quad \Omega' = \Omega \pm \omega \tag{3.29}$$

for inelastic scatterings, where the $\pm$ signs are interpreted as random emission and absorption of phonons by the lattice.

In sampling experiments, the observed response signified by integrated off-diagonal elements over the volume V_s has a common factor

$$\int_V \exp i(\boldsymbol{K}\prime - \boldsymbol{K} \mp \boldsymbol{k})\cdot \boldsymbol{r}\,\mathrm{d}^3\boldsymbol{r} \times \frac{1}{2t_0}\int_{-t_0}^{+t_0} \exp i(-\Omega'+\Omega\pm\omega)i\,\mathrm{d}t,$$

where the exponential functions are simply equal to 1, if $\boldsymbol{k}$ and ω satisfy the conservation laws (3.29), otherwise expressing distributed $(\boldsymbol{k},\omega)$ over the sample volume. One-dimensional condensates give a clear-cut example for distributed $k \pm \Delta k$, $\Omega \pm \Delta\Omega$ to be integrated over the time range $-t_0 \le t \le +t_0$, resulting in the anomaly

$$\langle \cos(\Delta k \cdot x + \phi_s)\rangle \times \frac{\sin(\Delta\Omega \cdot t_0)}{(\Delta\Omega \cdot t_0)}, \tag{3.30}$$

indicating the spatial modulation $\langle \cos(\Delta k \cdot x + \phi_s)\rangle$ averaged over V_s. This is detectable, if $\Delta\Omega \cdot t_0 \sim 1$, otherwise averaged out. In Chap. 10, we discuss such sampling examples in practical crystals.

We have so far discussed the nature of order variables and corresponding lattice variables to study for thermodynamic processes in crystals. However, prior to proceeding further, in Chap. 4 existing statistical theories of order–disorder phenomena are reviewed, from which since many useful ideas have emerged.

3.5.2 The Born-Huang Relaxation and Soft Modes

Thermodynamically, the field of order variables represents an energy transfer process to the lattice. Using the equipartition theorem, the phonon densities N and N' for scatterings can be assumed as proportional to temperatures T and T', respectively. Therefore, assuming $T>T'$, we can define the emission and absorption probabilities of a phonon as proportional to T' and T, respectively. Writing the transition probability as proportional to

$$\frac{|\langle \boldsymbol{K}, \Omega; 1|u_k|\boldsymbol{K}', \Omega'; 1\rangle|^2}{\hbar(\Omega - \Omega')} = \frac{1}{\tau},$$

we have

$$\frac{\mathrm{d}(N - N')}{\mathrm{d}T} = -\frac{N - N'}{\tau}, \tag{3.31}$$

indicating how the order energy is transferred to the surroundings; τ is the thermal relaxation time.

In modified crystals, dynamical equations are often dissipative, which should therefore be attributed to inelastic scatterings by phonons, while in idealized crystals with symmetrical lattices, we obtain solutions for free propagation. Born and Huang interpreted that the structural strains should be minimized to attain thermal equilibrium. Therefore, we can postulate that the uncertain phase $\Delta\varphi_n$ in critical anomalies should be temperature dependent in thermodynamic environment, in order to be consistent with Born-Huang's principle. However, renormalized $\Delta\varphi$ can dynamically be oscillatory, which has been detected as a *soft mode* characterized by a temperature-dependent frequency. In this book, we discuss the problem of soft modes in Chap. 9.

The function $\boldsymbol{\sigma}_k(\phi)$ is defined for distributed order variables $\boldsymbol{\sigma}_i$ at lattice sites. Using the Born-von Kárman boundaries, the renormalized $\sigma_k(\phi)$ describes a propagating wave with a phase ϕ and amplitude $\boldsymbol{\sigma}_{\mathrm{k0}}$. It is significant that such displacements in crystals have space-time uncertainties at lattice sites, $\Delta\phi_i$. The lattice structure are stressed randomly by $\boldsymbol{u}_k$, and the strain energy can be dissipated by thermal relaxation to the surroundings to achieve stability. As postulated by Born and Huang, such a process for a stable structure occurs spontaneously for a stable equilibrium state, as expressed by $\Delta\phi_i \rightarrow 0$. For such a phasing process, the Gibbs potential including the strain energy should therefore be minimized with respect to the amplitude $|\boldsymbol{\sigma}_k|$ and phase ϕ. In practice, such a phasing process can be observed as a relaxation

$$\frac{\partial\, \boldsymbol{\sigma}_k}{\partial t} = -\frac{1}{\tau}\, \boldsymbol{\sigma}_k \tag{3.32}$$

where τ is a relaxation time constant.

Exercise 3

1. In Chap. 2, we considered lattice vibrations in terms of normal modes, whereas the order variables $\boldsymbol{\sigma}_k$ represent vector displacements defined without any reference to symmetry axes. In the classical model, the vector $\boldsymbol{\sigma}_k$ should consist of longitudinal and transversal components with respect to the direction for

propagation, from which $\boldsymbol{\sigma}_k$ makes an angle ϕ expressed by the phase. In real crystals, these $\boldsymbol{\sigma}_k(\phi)$ and $\boldsymbol{\sigma}_k(\phi')$ are in parallel propagation, interacting like proportional to $\sigma_\perp(\phi)\sigma_\perp(\phi')$.

Minimizing such interactions by assuming $\phi = \phi'$, such a phasing process was postulated in order to be consistent with the Born-Huang principle. Discuss this postulate to see if physically acceptable, as this principle is required for lattice stability.

2. Can the Born-von Kárman boundary conditions be applied to crystal surfaces? If yes, explain how you can use the conditions on surfaces.
3. Assuming an asymmetric boundary condition for lattice waves $\boldsymbol{u}_k(\phi)$, discuss the phonon scatterings from a crystal surface with nonvanishing matrix elements $\langle \boldsymbol{K}|\boldsymbol{u}_k|\boldsymbol{K}'\rangle$. How you can explain heat exchange between a crystal and its surroundings?
4. What is "the renormalized variable"? It is a variable defined with respect to the propagation vector $\boldsymbol{k}$, while $\boldsymbol{k}$ is related to the kinetic energy of lattice excitation. Is it somewhat ambiguous that a crystal is divided into repeated regions of length $2\pi/\lambda$, where $\lambda = 2p/|\boldsymbol{k}|$? In such a unit $0 \le |\boldsymbol{r}| < \lambda$, $|\boldsymbol{r}|$ is taken as an arbitrary position, and the corresponding phase $0 \le \phi < 2\pi$. Why such $\boldsymbol{r}$ can be considered as a representative position? Discuss this definition for idealized crystals.
5. Discuss why the adiabatic energy in a crystal can be transferred to the surroundings. Explain what kind of mechanism is necessary for exchanging thermal energies with the surroundings.
6. Discuss about the physical origin of the crystal field that represents the geometrical lattice structure.

References

1. H.D. Megaw, *Crystal Structures: A Working Approach* (Saunders, Philadelphia, 1973)
2. T. Asida, S. Bando, M. Kakudo, Acta Cryst. **B28**, 1131 (1972)
3. J.M. Ziman, *Models of Disorder* (Cambridge Univ. Press, London, 1958)
4. M. Born, K. Huang, *Dynamical Theory of Crystal Lattices* (Oxford Univ. Press, 1968)
5. C. Kittel, *Introduction to Solid State Physics*, 6th edn. (Wiley, New York, 1986)
6. C. Kittel, *Quantum Theory of Solids* (Wiley, New York, 1963)
7. M. Tinkham, *Group Theory of Quantum Mechanics* (McGraw-Hill, New York, 1964)

Chapter 4
Mean-Field Theories of Binary Ordering

Order–disorder phenomena in alloys and magnetic crystals can be discussed by statistical mechanics in mean-field accuracy, for which the lattice structure is assumed as insignificant. Statistical theories deal with essential features of ordering, although some details are left unexplained in the mean-field approximation. It is a serious drawback of mean-field theories that we cannot deal with the symmetry change that characterizes structural phase transitions in a crystal. Nevertheless, existing theories are reviewed in this chapter, realizing that probabilities are an important concept for order–disorder phenomena. In any case, the mean-field theory is basically inappropriate for long-range interactions, unless crystals are characterized by uniform densities.

4.1 Probabilities in Binary Alloys

Physical properties of binary alloys, such as β-brass CuZn, are well documented, exhibiting order–disorder transitions at specific temperatures T_c. In their statistical theory of binary alloys of AB atoms, Bragg and Williams [1–3] considered *probabilities* for pairs of *like-atoms*, A–A and B–B, and for *unlike-pairs* A–B or B–A to locate at nearest-neighbor lattice sites, thereby introduced the concept of *short-* and *long-range order*. Judging from observed temperature $T_c = 450\,°C$ at the threshold, the lattice of β-brass may not be so rigid that a crystal is dominated by constituents in diffusive motion. Nonetheless, such probabilities are verified in later discussions as determined by the Boltzmann statistics. At any rate, we define probabilities $p_n(A)$ or $p_n(B)$ for a lattice site n to be occupied either by an atom A or by an atom B, respectively. For such exclusive events, we have the relation

$$p_n(A) + p_n(B) = 1, \tag{4.1a}$$

M. Fujimoto, *Thermodynamics of Crystalline States*,
DOI 10.1007/978-1-4614-5085-6_4,

thereby defining the order variable as

$$\sigma_n = p_n(\mathrm{A}) - p_n(\mathrm{B}). \tag{4.1b}$$

For these probabilities, we have $0 \le p_n(\mathrm{A}),\ p_n(\mathrm{B}) \le 1$ so that the order variable σ_n is in the range $-1 \le \sigma_n \le +1$. The completely *disordered state* characterized by $p_n(\mathrm{A}) = p_n(\mathrm{B}) = \frac{1}{2}$ can therefore be expressed by $\sigma_n = 0$. In contrast, a state of *complete order*, where all sites are occupied by either A or B, can be signified by $\sigma_n = +1$ or -1. Experimentally, such an ordered crystal consists of two ordered A-*domain* and B-*domain* in equal volumes; otherwise the whole crystal is an *alloy* that is composed of sublattices of A and B atoms. We cannot distinguish these two cases, unless either one of the states is in lower energy than the other, as determined experimentally. States in partial order should be described by distributed variables σ_n among lattice sites n. Assuming random distribution of σ_n, we can consider the average $\langle\sigma_n\rangle = \frac{1}{N}\sum \sigma_n$ over all lattice sites, expressing the macroscopic degree of order. In this case, we define the *order parameter* as $\eta = \langle\sigma_n\rangle$, which is called the *mean-field approximation*. Although adequate in some cases, such a mean-field average should be evaluated for validity experimentally.

In a random system, it is a usual practice to use the *correlation function* defined by

$$\Gamma = \langle(\sigma_m - \eta)(\sigma_n - \eta)\rangle = \langle\sigma_m\sigma_n\rangle - \eta^2, \tag{4.2}$$

using the relations $\langle\sigma_m\rangle = \langle\sigma_n\rangle = \eta$ for a uniform system; the function Γ indicates the presence of correlations expressed by $\langle\sigma_m\sigma_n\rangle$ for $m \ne n$. If there are correlation energy ε_{mn} between σ_n at the site n and σ_m at the nearest-neighbor sites m, we have a *cluster* of these σ_m formed within short distances, whose correlation energy E_n can be expressed as

$$E_n = \sum_m \left\{\varepsilon_{mn}^{\mathrm{AA}} p_m(\mathrm{A})p_n(\mathrm{A}) + \varepsilon_{mn}^{\mathrm{BB}} p_m(\mathrm{B})p_n(\mathrm{B}) + \varepsilon_{mn}^{\mathrm{AB}} p_m(\mathrm{A})p_n(\mathrm{B}) + \varepsilon_{mn}^{\mathrm{BA}} p_m(\mathrm{B})p_n(\mathrm{A})\right\}.$$

Using (4.1a) and (4.1b), we have

$$p_n(\mathrm{A}) = \frac{1}{2}(1 + \sigma_n) \quad \text{and} \quad p_n(\mathrm{B}) = \frac{1}{2}(1 - \sigma_n),$$

and similar expressions for $p_m(\mathrm{A})$ and $p_m(\mathrm{B})$ at a site m. Substituting the probabilities in E_n by order variables σ_n and σ_m, E_n can be written as

$$E_n = \sum_m E_{mn}$$

where

$$E_{mn} = \frac{1}{2}\left(2\varepsilon_{mn}^{\mathrm{AB}} + \varepsilon_{mn}^{\mathrm{AA}} + \varepsilon_{mn}^{\mathrm{BB}}\right) + \frac{1}{4}\left(\varepsilon_{mn}^{\mathrm{AA}} - \varepsilon_{mn}^{\mathrm{BB}}\right)(\sigma_m + \sigma_n)$$
$$+ \frac{1}{4}\left(2\varepsilon_{mn}^{\mathrm{AB}} - \varepsilon_{mn}^{\mathrm{AA}} - \varepsilon_{mn}^{\mathrm{BB}}\right)\sigma_m\sigma_n.$$

Writing $-K_{mn} = \frac{1}{4}\left(\varepsilon_{mn}^{AA} - \varepsilon_{mn}^{BB}\right)$ and $-J_{mn} = \frac{1}{4}\left(2\varepsilon_{mn}^{AB} - \varepsilon_{mn}^{AA} - \varepsilon_{mn}^{BB}\right)$, E_{mn} can be expressed as

$$E_{mn} = \text{const.} - K_{mn}(\sigma_n + \sigma_n) - J_{mn}\sigma_m\sigma_n.$$

Since $\varepsilon_{mn}^{AA} \approx \varepsilon_{mn}^{BB}$ for like atoms, we can assume $K_{mn} \approx 0$ for a binary system. Ignoring the insignificant constant term, E_{mn} is dominated by the last term; hence we can write

$$E_{mn} = -J_{mn}\sigma_m\sigma_n. \tag{4.3}$$

Although the factor J_{mn} is unknown by the first principle, the product $\sigma_m\sigma_n$ in (4.3) represents essential correlations between correlated order variables. We may consider that the mean-field average $\langle E_{mn}\rangle$ is proportional to the mean-field average $\langle\sigma_m\sigma_n\rangle$ for $m \neq n$. Equation (4.3) is the general form of correlation energy in short range, where the factor J_{mn} indicates the strength of correlation.

4.2 The Bragg-Williams Theory

Bragg and Williams assumed that the number of unlike pairs A–B is essential for the degree of disorder, which can be evaluated by using mean-field averages of probabilities

$$\langle p_n(\mathrm{A})\rangle = \frac{1}{2}(1+\eta) \quad \text{and} \quad \langle p_n(\mathrm{B})\rangle = \frac{1}{2}(1-\eta).$$

In a crystal of N lattice sites in total, we consider that each lattice point is surrounded by z nearest sites. In this case, the number of unlike pairs can be expressed by

$$\mathrm{N_{AB}} = 2\mathrm{N}z\langle p_n(\mathrm{A})\rangle\langle p_m(\mathrm{B})\rangle = \frac{1}{2}\mathrm{N}z(1-\eta^2).$$

This indicates that $\mathrm{N_{AB}} = \frac{1}{2}\mathrm{N}z$ for the disordered state at $\eta = 0$, whereas for the completely ordered states, $\mathrm{N_{AB}} = 0$, corresponding to $\eta = \pm 1$. Therefore, at a given ordering stage, $\mathrm{N_{AB}}$ is a function of η, and the short-range energy $E(\eta)$ is referred to $E(0) = 0$ at the disordered state. On the other hand, in the completely ordered case, the order energy is given by $E(\pm 1) = -\frac{1}{2}\mathrm{N}Jz$. Accordingly, in a partially ordered state, the short-range energy for $0<|\eta|<1$ is given by

$$E(\eta) = -\frac{1}{2}\mathrm{N}Jz\eta^2, \tag{4.4}$$

where J represents the average value $\langle J_{mn}\rangle$ in the cluster. Nevertheless, (4.4) may also be considered for the *long-range interaction* energy, if $\langle J_{mn}\rangle$ is an average over the whole crystal.

Thermodynamic states specified by the ordering energy $E(\eta)$ can be calculated statistically with a probability for N_{AB} unlike-pairs to be found at a given temperature T, which is determined by the number of combinations $g(\eta)$ for A and B atoms to be placed independently among N sites. Namely,

$$g(\eta) = \begin{pmatrix} N \\ N\langle p_n(A)\rangle \end{pmatrix}\begin{pmatrix} N \\ N\langle p_n(B)\rangle \end{pmatrix} = N^2\begin{pmatrix} 1 \\ \frac{1+\eta}{2} \end{pmatrix}\begin{pmatrix} 1 \\ \frac{1-\eta}{2} \end{pmatrix}.$$

Using (2.19) for $g(\eta)$ and $E(\eta)$, the macroscopic properties can be determined by the probability

$$\mathcal{W} = w_L w(\eta) = w_L g(\eta)\exp\left\{-\frac{E(\eta)}{k_B T}\right\},$$

where the lattice contribution can be included in the factor w_L, which however is *trivial* under a constant volume condition. The thermodynamic equilibrium can then be calculated by minimizing Helmholtz' free energy $F = -k_B T \ln \mathcal{W}$. Solving the equation $\left(\frac{\partial F}{\partial \eta}\right)_V = 0$, we have

$$\frac{\partial}{\partial \eta}\left\{\ln w_L + \ln g(\eta) + \frac{NzJ\eta^2}{2k_B T}\right\} = 0$$

For a large N, the second term can be evaluated by using Stirling's formula, i.e.,

$$\frac{\partial \ln g(\eta)}{\partial \eta} = -\frac{N}{2}\ln\frac{1+\eta}{1-\eta} \quad \text{and} \quad \frac{zJ\eta}{k_B T} = \ln\frac{1+\eta}{1-\eta},$$

from which we obtain

$$\eta = \tanh\frac{zJ\eta}{2k_B T}. \tag{4.5}$$

Equation (4.5) can be solved graphically for η, as shown in Fig. 4.1, where the straight line $\eta = \frac{2k_B T}{zJ} y$ and the curve $\eta = \tanh y$ cross at a point P to determine the value of the order parameter η at a given temperature T. It is noted that the origin $\eta = 0$ is the only crossing point if $\eta = y$, where the straight line is tangential to the curve, determining the transition temperature $T_c = \frac{zJ}{2k_B}$. Also noted is that there is one crossing point at all temperatures below T_c, while no solution is found at temperatures above T_c.

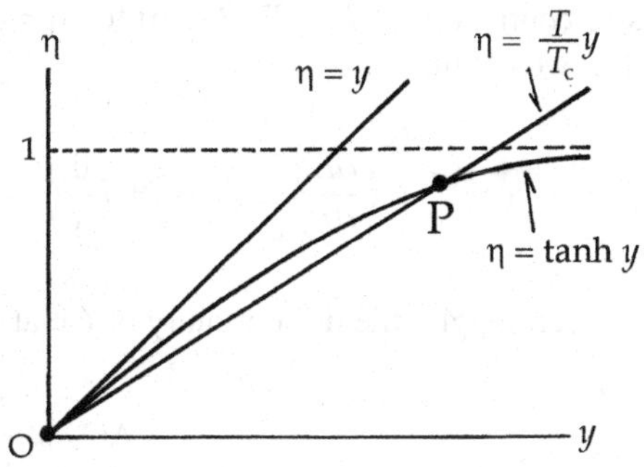

Fig. 4.1 Graphic solution of (4.5). Crossing points P between the straight line $\eta = y$ and the curve $\eta = \tanh y$. At $T = T_c$, the crossing point $\eta = 0$ is at O. For $T < T_c$, there is one crossing point P. No solution for $T > T_c$.

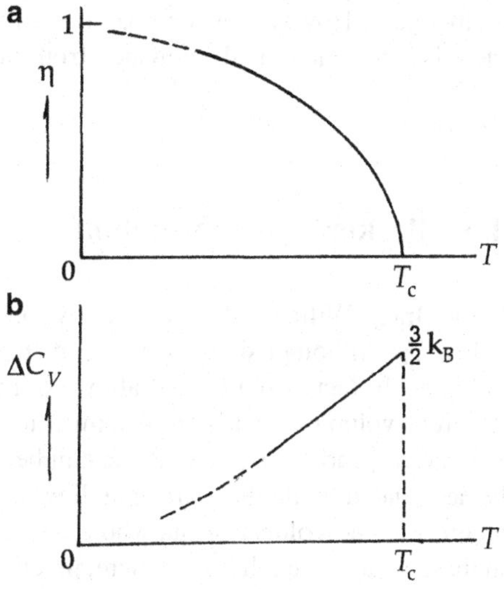

Fig. 4.2 Mean-field approximation. (**a**) Order parameter $\eta = \eta(T)$. (**b**) Specific heat discontinuity at T_c.

Writing $y = \frac{T_c}{T}\eta$ for convenience, the function $\tanh y$ can be expanded into a series, which may be truncated at low power if the temperature is close to T_c, i.e.,

$$y^2 = 3\left(\frac{T}{T_c}\right)^2\left(\frac{T_c}{T} - 1\right),$$

and hence

$$\eta^2 = \frac{3(T_c - T)}{T_c}.$$

Therefore, η is characterized by parabolic temperature dependence, i.e., $\eta \propto \sqrt{T_c - T}$ for $T \leq T_c$, as illustrated in Fig. 4.2a. In this analysis, the specific heat C_V

is discontinuous at $T = T_c$; $E = 0$ for $\eta = 0$ for $T > T_c$, whereas by differentiating (4.4) we obtain

$$\left(\frac{\partial E}{\partial T}\right)_V = -\frac{1}{2}\mathrm{N}zJ\frac{\mathrm{d}\eta^2}{\mathrm{d}T} = \frac{3}{2}\mathrm{N}k_\mathrm{B} \quad \text{for} \quad T < T_c.$$

Accordingly, the discontinuity of C_V at $T = T_c$ is given by

$$\Delta C_V = \frac{3}{2}\mathrm{N}k_\mathrm{B} \tag{4.6}$$

Figure 4.2b shows the specific curve in the vicinity of T_c sketched against temperature. However, as compared with the experimental curve of Cu-Zn alloy, there is a considerable discrepancy from the mean-field theory.

4.3 Becker's Interpretation

In the Bragg-Williams theory, an alloy can be segregated into two component parts below T_c. Although signified by $\pm\eta$, these parts may be fragmented into small volumes. In fact, a segregated alloy exhibits fragmented domains of two kinds in different volumes on lowering temperature. Domains are characterized by their volumes V_A and V_B in practice; the numbers $\mathrm{N_A}$ and $\mathrm{N_B}$ in the theory must therefore be associated with the corresponding volumes by the relation $V_A + V_B = V$, where V is the volume of the whole alloy. Following Becker's book [4], domains in these volumes are discussed here, practically interpreting by means of segregated or mixed volumes.

Writing $\mathrm{N_A} = \gamma\mathrm{N}$, we have $\mathrm{N_B} = (1-\gamma)\mathrm{N}$, where γ is a fractional parameter in the range $0 \le \gamma \le 1$, which is utilized as the order parameter for a binary alloy. It is noted that $\gamma = \frac{V_A}{V}$ and $1 - \gamma = \frac{V_B}{V}$ in a rigid structure, where one can assume $\mathrm{N_A} \propto V_A$ and $\mathrm{N_B} \propto V_B$. Then, the ordering energy (4.4) can be written as

$$E(\gamma) = -\frac{1}{2}\mathrm{N}Jz\gamma(1-\gamma).$$

In the whole system, this energy state should be degenerated by the statistical weight determined by the number of combinations to choose $\mathrm{N_A}$ and $\mathrm{N_B}$ atoms from the total N atoms, that is

$$w(\gamma) = \binom{\mathrm{N}}{\mathrm{N}\gamma}\binom{\mathrm{N}}{\mathrm{N}(1-\gamma)}$$

Therefore, the Helmholtz free energy can be written as

$$F = \mathrm{Nk_B}Tf(\gamma),$$

where

$$f(\gamma) = \gamma \ln \gamma + (1-\gamma)\ln(1-\gamma) + \frac{Jz}{\mathrm{k_B}T}\gamma(1-\gamma).$$

In equilibrium, two domains can be distinguished by different sets of parameters (N_1, γ_1) and (N_2, γ_2), with which the free energy of an alloy can be expressed as

$$F = \mathrm{k_B}T\{\mathrm{N}_1 f(\gamma_1) + \mathrm{N}_2 f(\gamma_2)\},$$

where

$$\mathrm{N}_1 + \mathrm{N}_2 = \mathrm{N} \quad \text{and} \quad \mathrm{N}_1\gamma_1 + \mathrm{N}_2\gamma_2 = \mathrm{N}\gamma.$$

For thermal equilibrium between two domains, F should be minimized against any variations $\delta\mathrm{N}_1$ and $\delta\mathrm{N}_2$ using Lagrange's multipliers λ and μ. Namely, from $\delta F = 0$ we obtain

$$f(\gamma_1) + \lambda + \mu\gamma_1 = 0 \quad \text{and} \quad f(\gamma_2) + \lambda + \mu\gamma_2 = 0$$

and write

$$f'(\gamma_1) + \mu = 0 \quad \text{and} \quad f'(\gamma_2) + \mu = 0$$

to assure the minimum against variations $\delta\gamma_1$ and $\delta\gamma_2$. Eliminating λ and μ from these relations, we can derive

$$f'(\gamma_1) = f'(\gamma_2) \quad \text{and} \quad f(\gamma_1) - f(\gamma_2) = f'(\gamma)(\gamma_1 - \gamma_2).$$

Figure 4.3a shows Becker's numerical analysis for the function $f(\gamma)$, where considering $s = \frac{zJ}{\mathrm{k_B}T}$ as an adjustable parameter, curves are drawn for $s = 0$, 2, 3, and 6. For $s = 2$, i.e., $T = \frac{zJ}{2\mathrm{k_B}} = T_\mathrm{c}$, the curve shows a single minimum at $\gamma_1 = \gamma_2 = \frac{\gamma}{2}$. On the other hand, for $s = 3$, two different minima at $\gamma_1 \neq \gamma_2$ are found with $f'(\gamma_1) = f'(\gamma_2)$ symmetrically as shown: one close to $\gamma = 0$ and the other close to $\gamma = 1$. For a larger value of s, e.g., $s = 6$, these are very close to 0 and 1, respectively. Figure 4.3b is the plot for T vs. γ or η, where the left and the right branches below T_c are for the domains A and B; the volume ratios $\frac{V_\mathrm{A}}{V}$ and $\frac{V_\mathrm{B}}{V}$ are related as $V_\mathrm{A} + V_\mathrm{B} = V$.

Although domain volumes exhibit an *exclusive event* for ordering process, Becker's treatment is not significantly different from Bragg-Williams' theory, as in both theories the density is assumed to remain unchanged throughout the ordering process. We realized that Becker considered the lattice structure by assuming constant atomic volumes to simplify the theory at a constant volume condition.

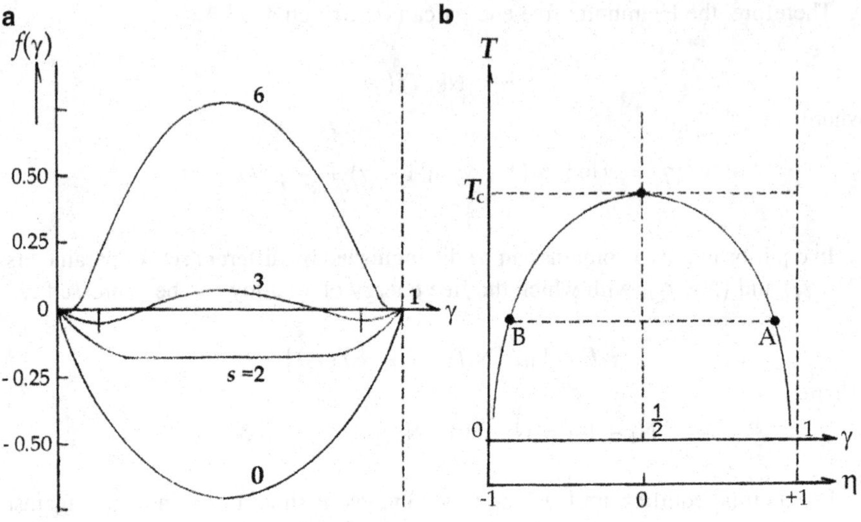

Fig. 4.3 Becker's graphic solutions. (**a**) $f(\gamma)$ vs. γ. (**b**) T vs. γ and η.

4.4 Ferromagnetic Order

Heisenberg (1929) postulated that electron exchange between magnetic ions is responsible for spin ordering in a ferromagnetic crystal, which is expressed by a Hamiltonian

$$\mathsf{H}_{n,n+1} = -2J_{n,n+1}\boldsymbol{s}_n \cdot \boldsymbol{s}_{n+1}, \tag{4.7}$$

where $J_{n,n+1}$ is the *exchange integral* between valence electrons of neighboring ions at n and $n+1$ sites. $\mathsf{H}_{n,n+1}$ expressed a correlation between neighboring electron spins $\boldsymbol{s}_n$ and $\boldsymbol{s}_{n+1}$. Instead of spin $\boldsymbol{s}_n$, order vectors, defined as $\boldsymbol{\sigma}_n = \gamma \boldsymbol{s}_n$, where γ is the gyromagnetic ratio, can be used for (4.7) to express the correlation energy between sites n and $n+1$. Rewriting $n+1$ and $2J_{nm}\gamma^{-2}$ as m and J_{nm}, respectively, we have

$$\mathsf{H}_{nm} = -J_{nm}\boldsymbol{\sigma}_n \cdot \boldsymbol{\sigma}_m \tag{4.8}$$

for a magnetic crystal, where the correlation $\boldsymbol{\sigma}_n \cdot \boldsymbol{\sigma}_m$ is in the form of a scalar product of vectors.

Heisenberg's model (4.8) suggests the spin exchange as the basic mechanism. Leaving the detail to Chap. 11, here we consider that (4.8) represents the basic magnetic correlation. In an uniform magnetic field $\boldsymbol{B}$ applied externally, the magnetic energy of $\boldsymbol{\sigma}_n$ can be written as

$$\mathsf{H}_n = -\boldsymbol{\sigma}_n \cdot \boldsymbol{B} + \sum_m J_{nm}\boldsymbol{\sigma}_n \cdot \boldsymbol{\sigma}_m, \tag{4.9}$$

where the summation can include all $\boldsymbol{\sigma}_m$ interacting with $\boldsymbol{\sigma}_n$. We can therefore consider that $\boldsymbol{\sigma}_n$ is acted by an effective magnetic field $\boldsymbol{B} + \boldsymbol{B}_n$, where $\boldsymbol{B}_n = \left\langle \sum_m J_{nm}\boldsymbol{\sigma}_m \right\rangle$ represents the mean-field arising from all other $\boldsymbol{\sigma}_m \cdot (m \neq n)$. Weiss was the first to consider such an internal field as $\boldsymbol{B}_n$, which he proposed to be proportional to $\langle \boldsymbol{\sigma}_n \rangle$, namely, $\boldsymbol{B}_n = \lambda \boldsymbol{\sigma}_n \rangle$, called the *molecular field*, to express all spin correlations in the long range in the mean-field approximation. In this book, we shall refer to it generally as the *Weiss field*. In a magnetic crystal, $\langle \boldsymbol{\sigma}_n \rangle$ per unit volume is defined as the magnetization $\boldsymbol{M}$ and the effective magnetic field $\boldsymbol{B} + \lambda \boldsymbol{M}$, acting on $\langle \boldsymbol{\sigma}_n \rangle = \boldsymbol{M}$. Hence, we write

$$\boldsymbol{M} = \chi_o(\boldsymbol{B} + \lambda \boldsymbol{M}), \tag{4.10}$$

where χ_o is a paramagnetic susceptibility, for which we have Curie's law $\chi_o = \frac{C}{T}$ at a temperature T, where C is the Curie's constant. From (4.10), we obtain the ferromagnetic susceptibility

$$\chi = \frac{M}{B} = \frac{C}{T - C\lambda} = \frac{C}{T - T_c}. \tag{4.11}$$

This is known as the Curie-Weiss law, where $T_c = C\lambda$ can be defined as the transition temperature, which is related to the long-range parameter λ, according to Weiss' postulate. Equation (4.11) is applicable to temperatures T close to T_c, when $T<T_c$. In the mean-field approximation, the ferromagnetic phase transition is characterized by the singularity of χ at $T = T_c$.

The Weiss field has a significant effect in the ordering process in general. The energy of binary ordering $E = -\frac{1}{2}\mathrm{N}zJ\eta^2$ in (4.4) can be written as $E = -\eta X_{\mathrm{int}}$, where $X_{\mathrm{int}} = \frac{1}{2}zJ\eta$. By writing $\frac{1}{2}zJ = \lambda$, the expression (4.4) analogous to the magnetization energy $-\boldsymbol{M} \cdot \boldsymbol{B}_n = -\lambda \boldsymbol{M}^2$.

Rewriting (4.5) as

$$\eta = \tanh\frac{X_{\mathrm{int}}}{\mathrm{k_B}T} = \frac{\exp\left(-\frac{(-1)X_{\mathrm{int}}}{\mathrm{k_B}T}\right) - \exp\left(-\frac{(+1)X_{\mathrm{int}}}{\mathrm{k_B}T}\right)}{\exp\left(-\frac{(-1)X_{\mathrm{int}}}{\mathrm{k_B}T}\right) + \exp\left(-\frac{(+1)X_{\mathrm{int}}}{\mathrm{k_B}T}\right)},$$

we can interpret it as

$$\eta = \langle \boldsymbol{\sigma}_n \rangle = \langle p_n(\mathrm{A}) \rangle - \langle p_n(\mathrm{B}) \rangle, \tag{4.12}$$

where

$$\langle p_n(\mathrm{A}) \rangle = \frac{1}{Z} \exp\left(-\frac{(-1)X_{\mathrm{int}}}{\mathrm{k_B}T} \right), \quad \langle p_n(\mathrm{B}) \rangle = \frac{1}{Z} \exp\left(-\frac{(+1)X_{\mathrm{int}}}{\mathrm{k_B}T} \right),$$

and

$$Z = \exp\left(-\frac{(-1)X_{\mathrm{int}}}{\mathrm{k_B}T} \right) + \exp\left(-\frac{(+1)X_{\mathrm{int}}}{\mathrm{k_B}T} \right).$$

Equation (4.12) indicates that an order parameter is given by the difference of Boltzmann's probabilities ± 1 for two inversion states that are separated by the Weiss field X_{int}. Accordingly, Bragg-Williams' probabilities for binary states are the Boltzmann probabilities at a given temperature, where we can consider a Weiss field of long-range order. Discussing on the statistical distribution of X_{int} at a given temperature, the Bragg-Williams' theory is clearly identical to Weiss' theory of ferromagnetism.

4.5 Ferromagnetic Transition in Applied Magnetic Field

Binary ferromagnetic order can be modified by an externally applied field B. Using Weiss' field, we consider that magnetic spins s_{n} are in the effective magnetic field $B + B_{\mathrm{n}}$, and the magnetic ordering energy can be expressed by

$$E_{\pm} = \lambda M^2 \pm MB = -M(B_{\mathrm{int}} \mp B).$$

Denoting the numbers of $+$ spins and $-$ spins in unit volume of the crystal by N_+ and N_-, the order parameter can be expressed by

$$\eta = \frac{\mathrm{N}_+ - \mathrm{N}_-}{\mathrm{N}_+ + \mathrm{N}_-}$$

where $\mathrm{N}_+ + \mathrm{N}_- = \mathrm{N}$ is the total number of spins. Following Becker's argument in Sect. 4.3, for the order parameter, we can use domain volumes V_+ and V_-, instead of $\mathrm{N}_+$and N_-, if the crystal volume is unchanged.

Writing that $\mathrm{N}_{\pm} = \frac{\mathrm{N}}{2}(1 \pm \eta)$, we can express

$$M = \mathrm{N}\gamma\eta \quad \text{and} \quad E(\pm\eta) = -\frac{1}{2}\mathrm{N}Jz\eta^2 \mp \mathrm{N}\gamma\eta B.$$

And the thermal properties can be determined by the free energy $F = -k_B T \ln Z$, where the function $Z = Z(\eta)$ is given by

$$Z = Z(+\eta)Z(-\eta) = \binom{N}{N_+}\binom{N}{N_-}\exp\left(-\frac{E(+\eta)+E(-\eta)}{k_B T}\right).$$

The most probable values of N_+ and N_- can be determined in such a way that $\ln Z(+\eta)$ and $\ln Z(-\eta)$ are maximized, respectively. Using the relation $dN_+ = -dN_-$ from a constant N, we therefore have

$$\frac{d}{dN_+}\ln\binom{N}{N_+} = \frac{d}{dN_+}(-N_+\ln N_+ + N_-\ln N_-) = \frac{1}{k_B T}\frac{dE(+\eta)}{dN_+}$$

and

$$\frac{d}{dN_+}\ln\binom{N}{N_-} = -\frac{d}{dN_-}\ln\binom{N}{N_-} = -\frac{1}{k_B T}\frac{dE(-\eta)}{dN_-}.$$

From these, we obtain

$$\ln\frac{N_-}{N_+} = \frac{2}{Nk_B T}\frac{dE(\pm\eta)}{d\eta},$$

and hence

$$\ln\frac{1-\eta}{1+\eta} = -\frac{2}{k_B T}\left(\frac{1}{2}Jz\eta \pm \gamma B\right),$$

which can be solved for η, obtaining

$$\eta = \tanh\left(\frac{Jz\eta}{2k_B T} \pm \frac{\gamma B}{k_B T}\right). \tag{4.13}$$

This is identical in form to (4.5) if $B = 0$; therefore, (4.13) can also be solved graphically. Namely, writing $y = \frac{Jz}{2k_B T}\eta + \frac{\gamma B}{k_B T}$, (4.13) is expressed as $\eta = \tanh y$; the value of η can be determined in the η–y plane from the crossing point between the straight line and the tanh curve, as illustrated in Fig. 4.4. Although we can define $T_c = \frac{Jz}{2k_B}$ for $B = 0$, the crossing point does not correspond to $\eta = 0$ if $B \neq 0$, in which case T_c does not signify a transition. However, the intercept $\frac{\gamma B}{k_B T}$ on the η-axis is small and insignificant in a practical magnetic field, where the transition temperature T_c can be determined as permitted within the accuracy of $\frac{\gamma B}{k_B T} \approx 0$.

Probabilities p_A and p_B introduced in the Bragg-Williams theory are those determined by the Boltzmann statistics, where binary energies should thermally be

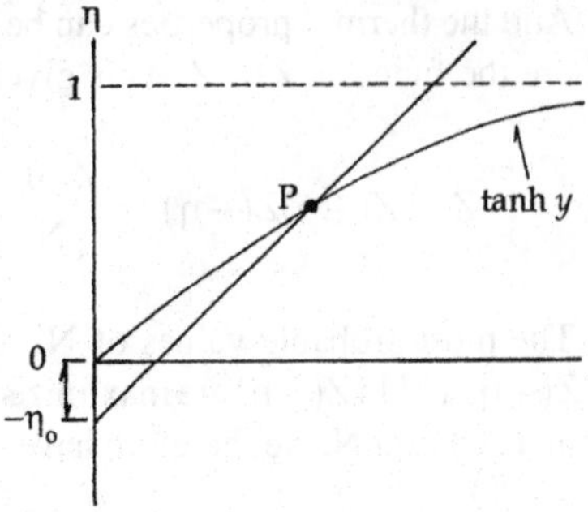

Fig. 4.4 Graphic solution of (4.10). $\eta = -\eta_o$ at T_c and one crossing point for $T<T_c$.

accessible via random phonons in a crystal. Although implicit in the theory, the consequence of statistical theory is based on the presence of randomness. On the other hand, the mean-field approach is not necessarily required for the statistical theory, if a better one is available instead.

Exercise 4

1. In this chapter, order variables σ_n and their short-range correlations (4.5) are defined statistically. In contrast, Heisenberg's spin–spin interactions were derived quantum-mechanically, which is interpreted as an adiabatic potential in thermodynamics. Despite different approaches, we have a consistent result expressed by (4.5). However, using statistical results, like (4.5), with a mean-field average and long-range correlations may not be an acceptable idea, because of lacking randomness in the latter case. Discuss physically on this issue.
2. Can the mean-field average be consistent with the Boltzmann statistics? If Weiss' molecular field is interpreted as the mean-field average, $\langle \boldsymbol{\sigma}_n \rangle$ should be a mean-field average of $\boldsymbol{\sigma}_n$. In this sense, it appears to be illogical to assume that $\langle \boldsymbol{\sigma}_n \rangle = \eta$. How do you resolve this matter?

References

1. J.M. Ziman, *Models of Disorder* (Cambridge Univ. Press, London, 1979)
2. L.D. Landau, E.M. Lifshitz, *Statistical Physics, English Trans* (Pergamon Press, London, 1958)
3. G.H. Wannier, *Statistical Physics* (Wiley, New York, 1966)
4. R. Becker, *Theory of Heat*, 2nd edn. revised by G. Liebfried, (Springer, New York, 1967)

Chapter 5
Pseudospin Clusters

Order variables for partial displacements of constituents at lattice sites are *clustered* in short range in crystals at the critical temperature. Disrupting local symmetry, a cluster propagates through the lattice, with increasing amplitude due to increasing correlations on lowering temperature. According to the Born-Huang principle, the corresponding condensate propagates in a direction specified by a wavevector for minimal structural strains. In this chapter, we define a *pseudospin* vector for inversion symmetry and calculate a short-range correlation energy for a cluster composed of the nearest- and next-nearest-neighbors, to determine the direction of propagation.

5.1 Pseudospins for Binary Displacements

In a crystal characterized by the *space group*, lattice sites are all identical, where the *point group* configures constituent molecules or complexes in the structure. Masses of constituents are primarily in harmonic vibration at site positions in stable structure. On the other hand, restricted by the point group, the mobile part of the constituent is characterized by the internal degree of freedom, independent from lattice vibrations. In perovskite crystals, such parts of constituents, called order variables, are in slow motion in the vicinity of T_c, as evidenced by *diffused* X-ray diffraction patterns in Fig. 5.1 [1], showing a smudged pattern from a perovskite of $NiNbO_3$ at 700°C, where diffraction spots are diffused by slow motion of order variables.

Among displacive types of phase transitions in perovskite crystals, $SrTiO_3$ and $BaTiO_3$ exhibit typical structural changes as illustrated in Fig. 3.3a, b. Here, we consider that the Ti^{4+} ion in Fig. 3.3a can fluctuate at a fast rate between two positions 1 and 1′ in the octahedral SiO_6^{2-} group along the z-direction with an equal probability, which is invariant for fast inversion above T_c. In Fig. 3.3b shown is

M. Fujimoto, *Thermodynamics of Crystalline States*,
DOI 10.1007/978-1-4614-5085-6_5,

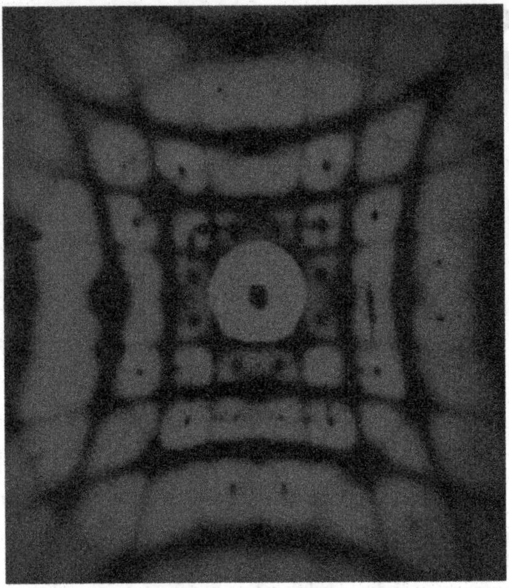

Fig. 5.1 A diffuse X-ray diffraction photograph from $NiNbO_3$ at 700°C. From R. Comes, R. Courrat, F. Desnoyer, M. Lambert, and A. M. Quittet, Ferroelectrics **12**, 3 (1976).

similar inversion in fast rotation of TiO_6^{2-} by small positive and negative angles around the z-axis. In this case, a vector variable $\boldsymbol{\sigma}_n$ parallel to the z-axis represents such inversion symmetry at a site n. In these perovskite structures, a variable $\boldsymbol{\sigma}_n$, inverting its direction between 1 and 1′ as a function of time, can be the order variable that is primarily independent from $\boldsymbol{\sigma}_m$ at a different site $m \neq n$. These variables can be correlated, if sites m and n are in close distance, if observed in longer time than the timescale of inversion. Such a correlated motion is determined by an *adiabatic potential U* originated from the lattice. Following Landau [2], we can write the dynamic equation of motion

$$i\hbar \frac{\partial \boldsymbol{\sigma}_n}{\partial t} = [\mathsf{H}_n, \boldsymbol{\sigma}_n], \tag{5.1}$$

where the Hamiltonian H_n represents kinetic energies of $\boldsymbol{\sigma}_n$ in the adiabatic potential U_n of correlations near the critical temperature.

To observe the motion of $\boldsymbol{\sigma}_n$ expressed by classical displacements at the transition threshold, it is necessary to consider the timescale of observation t_o. We therefore replace the time derivative in (5.1) by its time average of $\boldsymbol{\sigma}_n$ over t_o, i.e.,

$$\left\langle \frac{\partial \boldsymbol{\sigma}_n}{\partial t} \right\rangle_t = \frac{1}{t_o} \int_0^{t_o} \left(\frac{\partial \boldsymbol{\sigma}_n}{\partial t} \right) dt.$$

This however vanishes, if $t_o \to \infty$, which is the normal assumption for equilibrium states to be calculated in statistical mechanics. In this case, (5.1) can be written as

$$\left[\langle \mathsf{H}_n \rangle_t, \langle \boldsymbol{\sigma}_n \rangle_t\right] = 0, \tag{5.2}$$

indicating that the observable $\langle \boldsymbol{\sigma}_n \rangle_t$ is *commutable* with $\langle \mathsf{H}_n \rangle_t$ in equilibrium states.

If t_o is finite, however, the relaxation formula

$$\left\langle \frac{\partial \boldsymbol{\sigma}_n}{\partial t} \right\rangle_t = -\frac{1}{\tau} \langle \boldsymbol{\sigma}_n \rangle_t \tag{5.3}$$

can be applied to (5.1), where τ is a thermal *relaxation time* as defined in (3.31); therefore, we have the relation

$$-\frac{i\hbar \langle \boldsymbol{\sigma}_n \rangle_t}{\tau} = \left[\langle \mathsf{H}_n \rangle_t, \langle \boldsymbol{\sigma}_n \rangle_t\right].$$

Assuming such a thermal relaxation at the transition threshold, Landau replaced the commutator on the right by the product of quantum-mechanical uncertainties, $-\langle \Delta\varepsilon_n \rangle_t \langle \Delta\boldsymbol{\sigma}_n \rangle_t$. Here the uncertainties are originated from the fact that these do not simultaneously occur at all sites n. Indicated by a finite relaxation time τ, we assume that uncertainties should be relaxed to the lattice by transferring fluctuation energy. Therefore, we can write the uncertainty relation

$$\Delta\varepsilon \frac{\Delta \langle \boldsymbol{\sigma}_n \rangle_t}{\langle \boldsymbol{\sigma}_n \rangle_t} \approx \frac{\hbar}{\tau}. \tag{5.4}$$

On the other hand, (5.3) holds for $\langle \Delta\varepsilon_n \rangle_t = 0$ corresponding to $\langle \mathsf{H}_n \rangle_t = 0$, which represents an equilibrium crystal. On the other hand, Landau showed that nonzero fluctuations should characterize order variables $\langle \boldsymbol{\sigma}_n \rangle_t$ at the threshold of binary transitions.

If $\Delta\langle \boldsymbol{\sigma}_n \rangle_t > \langle \boldsymbol{\sigma}_n \rangle_t$, where the uncertainty $\Delta\langle \boldsymbol{\sigma}_n \rangle_t$ is overwhelming, and the variable $\langle \boldsymbol{\sigma}_n \rangle_t$ exhibits a quantum-mechanical character. On the other hand, if $\Delta\langle \boldsymbol{\sigma}_n \rangle_t < \langle \boldsymbol{\sigma}_n \rangle_t$, considering negligible fluctuations $\Delta\langle \boldsymbol{\sigma}_n \rangle_t \approx 0$ at T_c, $\langle \boldsymbol{\sigma}_n \rangle_t$ can be regarded as a classical vector. Writing the thermal relation $\Delta\varepsilon \approx k_B T_c$, (5.4) provides a criterion for $\langle \boldsymbol{\sigma}_n \rangle_t$ to be justified with respect to the critical temperature T_c, i.e.,

$$\tau T_c \gg \frac{\hbar}{k_B} \sim 10^{-11}\,\mathrm{s\,K}. \tag{5.5}$$

In perovskites, phase transitions occur in the range $100\,\mathrm{K} < T_c < 200\,\mathrm{K}$ and the relaxation time is estimated as 0.5×10^{-13} s. In this case, the value of τT_c is roughly 500 times larger than the Landau's limit 10^{-11} in (5.5), confirming that $\langle \boldsymbol{\sigma}_n \rangle_t$ is a classical displacement at and below T_c; hence the transition is called *displacive*. On the other hand, in crystals of hydrogen-bonding KDP, we have $T_c \approx 1{,}000\,\mathrm{K}$ and $\tau \approx 10^{-13}\,\mathrm{s}$, so that $\tau T_c \approx 10^{-11}$, indicating that $\langle \boldsymbol{\sigma}_n \rangle_t$ is barely of quantum character. In KDP, the order variable is therefore rated as quantum mechanical, although not precisely identified as in perovskites.

Characterized by inversion, the average $\langle \boldsymbol{\sigma}_n \rangle_t$ exhibits a behavior similar to a conventional spin of $\pm\frac{1}{2}$, so it is called a *pseudospin* by analogy. For $\Delta\varepsilon \neq 0$, the inversion can be analyzed in terms of slow tunneling motion through the potential in H, so that the phase transition is discontinuous in principle, owing to fluctuations of the order of $\Delta\varepsilon$. Considering $\Delta\varepsilon \approx \Delta U$, the critical uncertainties indicate emerging of a pseudopotential U.

5.2 A Tunneling Model

The pseudospin inversion is essentially a quantum-mechanical tunneling through a potential $\mathrm{U}(z_n)$ of the order of $\Delta\varepsilon$, where z_n is the position of the pseudospin on the direction of fluctuation. In this section, we discuss such a model of inversion in one dimension, following Blinc and Zeks [3].

Omitting the suffix n for brevity, we assume that the unperturbed state for $U(z) = 0$ can be determined by the wave equation

$$\mathsf{H}_o \varphi_o = \varepsilon_o \varphi_o,$$

where the eigenvalue is positive, i.e., $\varepsilon_o > 0$, representing the fluctuating kinetic energy. Figure 5.1 shows such a particle in the effective potential of so-called a *double well*, where the central hump represents the perturbing potential $U(z)$. The ground energy ε_o is therefore doubly degenerate, because of the symmetry. The wavefunctions in the right and left wells are denoted by $\varphi_o(+z)$ and $\varphi_o(-z)$, respectively. Owing to the perturbing potential $U(z)$, the degenerate level ε_o is split into two energies, corresponding to symmetric and antisymmetric combinations of these wavefunctions. Marked by suffixes $+$ and $-$, we have

$$\psi_+ = \frac{\varphi_o(+z) + \varphi_o(-z)}{\sqrt{2}} \quad \text{and} \quad \psi_- = \frac{\varphi_o(+z) - \varphi_o(-z)}{\sqrt{2}},$$

which are normalized as

$$\psi_+^* \psi_+ + \psi_-^* \psi_- = \varphi_o^*(+z)\varphi_o(+z) + \varphi_o^*(-z)\varphi_o(-z) = 1.$$

Here, $\varphi_o^*(+z)\varphi_o(+z)$ and $\varphi_o^*(-z)\varphi_o(-z)$ represent the probabilities $p_n(+z)$ and $p_n(-z)$ for the particle to be at $+z$ and $-z$, respectively, so that $p_n(+z) + p_n(-z) = 1$.

The perturbed Hamiltonian can then be expressed by

$$\begin{aligned}\mathsf{H} &= \varepsilon_+ \psi_+{}^* \psi_+ + \varepsilon_- \psi_-{}^* \psi_- \\ &= \varepsilon_o\left(\psi_+{}^*\psi_+ + \psi_-{}^*\psi_-\right) - \frac{U(0)}{2}\left(\psi_+{}^*\psi_+ - \psi_-{}^*\psi_-\right),\end{aligned}$$

hence $\varepsilon_\pm = \varepsilon_o \pm \frac{U(0)}{2}$ and $\varepsilon_+ - \varepsilon_- = U(0)$. Therefore, we can define

$$\sigma_z = \psi_+{}^*\psi_+ - \psi_-{}^*\psi_- = p_n(+z) - p_n(-z), \tag{5.6a}$$

as the z-component of a vector $\boldsymbol{\sigma}_n$. For the transverse components, we consider

$$\sigma_x = \varphi_o{}^*(+z)\varphi_o(-z) + \varphi_o{}^*(-z)\varphi_o(+z) \tag{5.6b}$$

and

$$\sigma_y = \varphi_o{}^*(+z)\varphi_o(-z) - \varphi_o{}^*(-z)\varphi_o(+z), \tag{5.6c}$$

for these components we can hold the commutation relations

$$[\sigma_x,\ \sigma_y] = i\sigma_z, \qquad [\sigma_y,\ \sigma_z] = i\sigma_x \quad \text{and} \quad [\sigma_z,\ \sigma_x] = i\sigma_y. \tag{5.6d}$$

These relations are required for $(\sigma_x, \sigma_y, \sigma_z)$ to constitute a quantum-mechanical vector $\boldsymbol{\sigma}_n$.

5.3 Pseudospin Correlations

As discussed in Sect. 4.1, correlations between pseudospins in a crystal are signified by the distance between right and left wells, or left and right, of adjacent pseudospins.

Considering two pseudospins at adjacent sites n and $n + 1$, the correlation energy can be written as

$$\mathsf{H}_{n,n+1} = \sum_{\alpha\beta,\gamma\delta} \left(\psi_{n,\alpha}{}^*\psi_{n,\beta} V_{\alpha\beta,\gamma\delta} \psi_{n+1,\gamma}{}^*\psi_{n+1,\delta}\right), \tag{5.7}$$

where $V_{\alpha\beta;\gamma\delta}$ are interaction tensor elements between the density matrices $\psi_{n,\alpha}{}^*\psi_{n,\beta}$ and $\psi_{n+1,\gamma}{}^*\psi_{n+1,\delta}$. Indexes α, β and γ, δ represent either $+$ or $-$ at each of the sites n and $n + 1$ in the chain. By symmetry, the interaction elements $V_{\alpha\beta;\gamma\delta}$ can be signified by

$$V_{++--} = V_{--++},\ V_{+-+-} = V_{-+-+} = V_{-++-},\ \text{etc.}$$

The density matrix elements can be written explicitly as

$$\psi_{n+}^{*}\psi_{n-} = \frac{1}{2}\{\varphi_o^*(+z)\varphi_o(+z) + \varphi_o^*(-z)\varphi_o(-z) + \varphi_o^*(+z)\varphi_o(-z) + \varphi_o^*(-z)\varphi_o(+z)\} = \frac{1}{2}(1+\sigma_{nz}),$$

$$\psi_{n+}^{*}\psi_{n-} = \frac{1}{2}\{\varphi_o^*(+z)\varphi_o(+z) - \varphi_o{}^*(-z)\varphi_o(-z) - \varphi_o{}^*(+z)\varphi_o(-z) + \varphi_o{}^*(-z)\varphi_o(+z)\} = \frac{1}{2}(\sigma_{nx}-\sigma_{ny}),$$

etc., and similarly for $\psi_{n+1,+}{}^*\psi_{n+1,+}$, $\psi_{n+1,+}{}^*\psi_{n+1,-}$, etc.

Therefore, $\mathsf{H}_{n,n+1}$ can be written in terms of components of pseudospins $\boldsymbol{\sigma}_n$ and $\boldsymbol{\sigma}_{n+1}$ as

$$\mathsf{H}_{n,n+1} = -J_{n,n+1}\sigma_{n,z}\sigma_{n+1,z} - K_{n,n+1}\sigma_{n,x}\sigma_{n+1,x}, \tag{5.8}$$

where $J_{n,n+1} = 2V_{+-+-}$ and $K_{n,n+1} = 2V_{+-+-} - V_{++++} - V_{----}$.

The correlation energy $\mathsf{H}_{n,n+1}$ can then be included in the Hamiltonian $\mathsf{H} = \mathsf{H}_n + \mathsf{H}_{n+1}$, and

$$\mathsf{H} = \mathsf{H}_n + \mathsf{H}_{n+1} + \mathsf{H}_{n.n+1} = \varepsilon_o - \frac{U}{2}\left(\sigma_{n.z} + \sigma_{n+1,z}\right) - J_{n,n+1}\sigma_{n,z}\sigma_{n+1,z} - K_{n,n+1}\sigma_{n.x}\sigma_{n+1,x}. \tag{5.9}$$

The second term on the right is the average energy of pseudospins, as if the potential U were applied externally; the interaction energy between σ_n and σ_{n+1} is determined by parameters $J_{n,n+1}$ and $K_{n,n+1}$ of the third and fourth terms, respectively. Equation (5.9) is analogous to the classical correlations discussed in 4.1.

Nevertheless, for small ε_o and U at the threshold of a structural change, $\boldsymbol{\sigma}_n$ and $\boldsymbol{\sigma}_{n+1}$ are classical vectors, as justified for perovskites in Sect. 5.1. In this case, we can assume that $V_{++++} = V_{----}$ by symmetry and obtain $J_{n,n+1} = K_{n,n+1}$. Therefore, the interaction terms in (5.9) can be given essentially in the scalar product form as

$$\mathsf{H}_{n,n+1} = -J_{n,n+1}\boldsymbol{\sigma}_n \cdot \boldsymbol{\sigma}_{n+1}, \tag{5.10}$$

representing the correlation energy between classical vectors $\boldsymbol{\sigma}_n$ and $\boldsymbol{\sigma}_{n+1}$.

In a binary crystal for $T>T_c$, we consider that pseudospins are in harmonic motion at a lattice point n with $U=0$. Hence, we can assume $\langle p_n(+)\rangle_t = \langle p_n(-)\rangle_t$, which is the same as the equal probability for $\sigma_{n,z} = \pm\frac{1}{2}$ in a random motion. On the other hand, if electrically dipolar pseudospins are in an applied field $\boldsymbol{E}$, we have to consider the potential $U = -e\langle\boldsymbol{\sigma}_n\rangle_t \cdot \boldsymbol{E}$. In this case, the thermal averages of these probabilities are unequal and proportional to the Boltzmann factor $\exp\left(-\frac{\pm U}{k_B T}\right)$, and the order parameter is given by

$$\eta = \langle\sigma_{n,z}\rangle_t = \langle p_n(+z)\rangle_t - \langle p_n(-z)\rangle_t. \tag{5.11}$$

Signified by a change in lattice symmetry, the structural change can be attributed to *displacive* vectors in finite magnitude, as exemplified by typical structural changes in perovskites. We consider that a center potential $U(z_n)$ emerges at the critical temperature T_c from the deformed structure, as discussed in Chap. 3.

5.4 Pseudospin Clusters in Crystals

5.4.1 Classical Pseudospins in Crystals

Order variables $\boldsymbol{\sigma}_n$ at the critical temperature should behave like classical vectors, which are driven by an adiabatic potential U_n of the lattice. Consequently, the lattice points displace by $\mathbf{u}_n$ from original sites by the force $-\nabla_n U_n$, counteracting against the force $\nabla_n U_n$ on $\boldsymbol{\sigma}_n$ with respect to the center of mass, according to Newton's action-reaction principle. Thus, a classical pair of displacements $(\boldsymbol{\sigma}_n, \boldsymbol{u}_n)$ should always be in phase during a structural change. We shall call such a pair of displacements a *condensate* by analogy of condensing gas.

These $\boldsymbol{\sigma}_n$ and $\boldsymbol{u}_n$ are in propagation in a direction in the crystal space, hence expressed by Fourier series. The variable $\boldsymbol{\sigma}_n$ can then be expressed by

$$\boldsymbol{\sigma}_n = \sum_{\boldsymbol{k}} \boldsymbol{\sigma}_{\boldsymbol{k}} \exp i(\boldsymbol{k} \cdot \boldsymbol{r}_n - \omega t_n), \tag{5.12}$$

representing a wave packet defined by the summation over possible values of $\boldsymbol{k}$, because of critical uncertainties. By virtue of space reversal symmetry, (5.12) should be invariant for inversion $\boldsymbol{r}_n \to -\boldsymbol{r}_n$; hence, the reversal $\boldsymbol{\sigma}_{\boldsymbol{k}} \to -\boldsymbol{\sigma}_n$ can be replaced by $\boldsymbol{k} \to -\boldsymbol{k}$. On the other hand, time-reversal symmetry is abandoned in all thermal applications. For the summation in (5.12), we can therefore consider only $\pm \boldsymbol{k}$ and $t_n > 0$. In a periodic crystal between boundaries set at $\boldsymbol{r}_{n=\mathrm{o}}$ and $\boldsymbol{r}_{n=L}$, these $\pm \boldsymbol{k}$ waves should therefore be standing waves of phases $\phi_\pm = \pm \boldsymbol{k} \cdot \boldsymbol{r}_n - \omega t_n$. We can therefore deal with pseudospin waves that are *pinned* at $\phi = 0$ and 2π, and stationary in equilibrium crystals.

Setting uncertainties aside, we have the relation

$$\boldsymbol{\sigma}_{\boldsymbol{k}} = -\boldsymbol{\sigma}_{-\boldsymbol{k}},$$

and express (5.12) as composed of $\pm \boldsymbol{k}$ waves

$$\boldsymbol{\sigma}_n = \boldsymbol{\sigma}_{+k} \exp i(+\boldsymbol{k} \cdot \boldsymbol{r}_n - \omega t_n) + \boldsymbol{\sigma}_{-k} \exp i(-\boldsymbol{k} \cdot \boldsymbol{r}_n + \omega t_n) \tag{5.13a}$$

where these amplitudes can be written by Fourier transforms

$$\boldsymbol{\sigma}_{\pm k} = \sum_n \boldsymbol{\sigma}_n \exp i(\mp \boldsymbol{k} \cdot \boldsymbol{r}_n \pm \boldsymbol{\omega} t_n). \tag{5.13b}$$

It is noted that for the phase variable $\phi_n = \boldsymbol{k} \cdot \boldsymbol{r}_n - \omega t_n$ at arbitrary site, the space-time $(\boldsymbol{r}_n, t_n)$ is a representative one in the range $0 \leq \phi_n \leq 2\pi$, for which the phase reversal $\phi_n \rightarrow -\phi_n$ can apply. Therefore, it is convenient to use renormalized space-time $(\boldsymbol{r}, t)$ and phase ϕ in the range $0 \leq \phi \leq 2\pi$, as already discussed in Chap. 1.

Physically, it is significant that the lattice is strained by such displacements $\boldsymbol{\sigma}_n$ and $\boldsymbol{u}_n$. For strained crystals, Born and Huang [4] proposed that the total strain energy should be minimized for stability of the equilibrium crystal. We may assume that the combined local displacement as $a\boldsymbol{\sigma}_n + b\boldsymbol{u}_n$ with constants a and b is responsible for strain energy in a crystal as a whole, which can be written as

$$\sum_{n,n'} (a\boldsymbol{\sigma}_n + b\boldsymbol{u}_n)^*(a\boldsymbol{\sigma}_{n'} + b\boldsymbol{u}_{n'})$$
$$= \sum_{n,n'} a^*a\, \boldsymbol{\sigma}_n{}^* \cdot \boldsymbol{\sigma}_{n'} + \sum_{n,n'} b^*b\, \boldsymbol{u}_n{}^* \cdot \boldsymbol{u}_{n'} + \sum_{n,n'} (a^*b\, \boldsymbol{\sigma}_n.\boldsymbol{u}_{n'} + b_n{}^* a_{n'} \boldsymbol{u}_n{}^* \boldsymbol{\sigma}_{n'}). \tag{5.14}$$

Minimizing this for a strain-free crystal, these variables at these sites n and n' should be *in phase* in all terms in (5.14). In such a cluster, the comprising $\boldsymbol{\sigma}_i$s are all ordered in parallel, so that the magnitude $\sum_{i\ \mathrm{cluster}} |\boldsymbol{\sigma}_i|$ represents the cluster. Noting that all cluster members have the same pseudospin $\boldsymbol{\sigma}_o \exp i\phi$, a cluster complex bears a larger pseudospin $z\boldsymbol{\sigma} \exp i\phi$, where z is the number of cluster members and ϕ in common.

Noted that those variables characterized by phases $\phi_n = \boldsymbol{k} \cdot \boldsymbol{r}_n - \omega t_n$ cannot be simultaneous at n in the critical region, for which we consider for Born-Huang's principle to drive a thermal process to a *coherent* phase. Accordingly, we may write the relaxation equation

$$\frac{\mathrm{d}}{\mathrm{d}t}(\boldsymbol{\sigma}_n - \boldsymbol{\sigma}_o) = -\frac{\boldsymbol{\sigma}_n - \boldsymbol{\sigma}_o}{\tau}, \tag{5.15}$$

where τ is a time constant for random phases of $\boldsymbol{\sigma}_n$ to become in phase, reaching the equilibrium value $\boldsymbol{\sigma}_o$ in strain-free structure. Synonymous with Born-Huang's original proposal, (5.15) is postulated for modulating condensates.

5.4.2 *Clusters of Pseudospins*

It is a postulate that pseudospins at nearest-neighbor sites become clustered in phase at minimum strains in a crystal. Supported by Born-Huang' principle, we may consider it as a law of nature, similar to *seeds* for condensation. We can consider that ordering in crystalline states proceeds from such a cluster seed, if the temperature is lowered, for which the correlation energy is responsible. The lattice strains

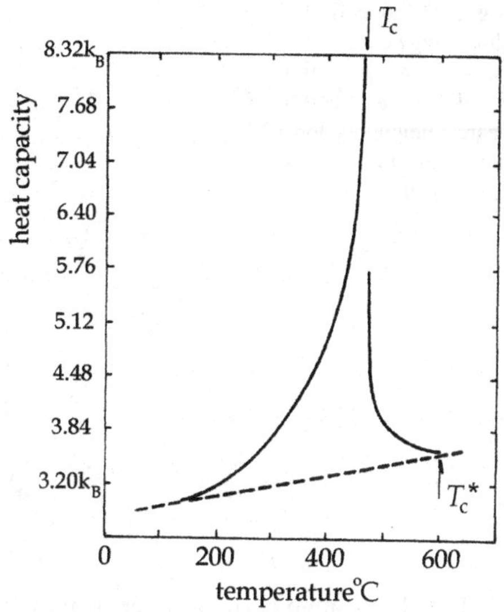

Fig. 5.2 An example of a transition anomaly from β-brass, which are characterized by a sharp rise at T_c, a narrow starting range from the threshold T_c*, and a gradual tail after the transition. From F. C. Nix and W. Shockley, Rev. Mod. Phys. **10**, 1 (1938).

may become appreciable when forming clusters in crystals, as evidenced by a sharp rise in the specific heat C_p–T curve at T_c, as shown in Fig. 5.2 [5]. The gradual slope in the noncritical region below T_c can be attributed to another phasing process, which will be discussed in the later chapter.

In this subsection, the pseudospin of a cluster is written specially as *italic* $\boldsymbol{\sigma}_n$, instead of individual $\boldsymbol{\sigma}_n$. The correlation energy of a cluster can then be expressed as

$$\mathsf{H}_n = -\sum_m J_{nm}\boldsymbol{\sigma}_n \cdot \boldsymbol{\sigma}_m, \tag{5.16}$$

where the index m refers to interacting sites of clustered pseudospins, each centered at $\boldsymbol{\sigma}_n$. Here, we use the same notation for J_{nm}, for simplicity, with unspecified distance between $\boldsymbol{\sigma}_n$ and $\boldsymbol{\sigma}_m$.

We assume that pseudospins in a cluster become all in phase at temperatures below T_c. For clustered pseudospins, we write $\boldsymbol{\sigma}_m = \boldsymbol{\sigma}_0 \boldsymbol{e}_m$ and $\boldsymbol{\sigma}_n = \sigma_0 \boldsymbol{e}_n$, where σ_0 is a finite amplitude, and $\boldsymbol{e}_m$ and $\boldsymbol{e}_n$ are unit vectors. Figure 5.3 sketches clustered pseudospins in a perovskite crystal. With respect to the center of TiO_6^{2-} ion, the short-range correlations with the nearest and next-nearest neighbors are indicated by J and J', respectively, in the bc plane. Leaving σ_0 as a finite constant, these unit vectors can be expressed as

$$\boldsymbol{e}_m = \boldsymbol{e}_{+q} \exp i(\boldsymbol{q} \cdot \boldsymbol{r}_m - \omega t_m) + \boldsymbol{e}_{-q} \exp i(-\boldsymbol{q} \cdot \boldsymbol{r}_m + \omega t_m)$$

and

$$\boldsymbol{e}_n = \boldsymbol{e}_{+q} \exp i(\boldsymbol{q} \cdot \boldsymbol{r}_n - \omega t_n) + \boldsymbol{e}_{-q} \exp i(-\boldsymbol{q} \cdot \boldsymbol{r}_n + \omega t_n)$$

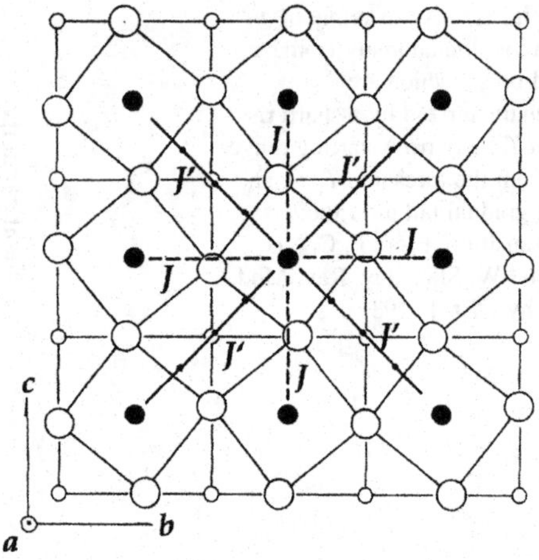

Fig. 5.3 Model of the short-range cluster in perovskites. Interactions J and J' are assigned for 6 nearest neighbors and 12 next-nearest neighbors, respectively.

The short-range correlation energy (5.16) can then be expressed as

$$\mathsf{H}_n = -2\sigma_0^2 \sum_m \boldsymbol{e}_{+\boldsymbol{q}} \cdot \boldsymbol{e}_{-\boldsymbol{q}} \exp i\{\boldsymbol{q}\cdot(\boldsymbol{r}_m - \boldsymbol{r}_n) - \omega(t_m - t_n)\}.$$

In practice, the time average $\langle \mathsf{H}_n \rangle_t = \frac{1}{2t_o}\int_{-t_o}^{+t_o} \mathsf{H}_n \mathrm{d}(t_m - t_n)$ over the timescale of observation $t_o \times 2$ is a measurable quantity, so that the correlation energy can be detected as

$$\langle \mathsf{H}_n \rangle_t = -\sigma_0^2 \Gamma_t \boldsymbol{e}_{+\boldsymbol{q}} \cdot \boldsymbol{e}_{-\boldsymbol{q}} \sum_m J_{mn} \exp i\boldsymbol{q}\cdot(\boldsymbol{r}_m - \boldsymbol{r}_n), \qquad (5.17)$$

where

$$\Gamma_t = \frac{1}{2t_o}\int_{-t_o}^{+t_o} \exp\{-i\omega(t_m - t_n)\}\mathrm{d}(t_m - t_n) = \frac{\sin \omega t_o}{\omega t_o}$$

is the time correlation function, whose value is close to 1, if $\omega t_o < 1$. Equation (5.17) can therefore be expressed for convenience as

$$\langle \mathsf{H}_n \rangle_t = -\sigma_o^2 \Gamma_t \boldsymbol{e}_{+\boldsymbol{q}} \cdot \boldsymbol{e}_{-\boldsymbol{q}} J_n(\boldsymbol{q}) \qquad (5.18\text{a})$$

where

$$J_n(\boldsymbol{q}) = \sum_m J_{mn} \exp i\boldsymbol{q}\cdot(\boldsymbol{r}_m - \boldsymbol{r}_n). \qquad (5.18\text{b})$$

This $\langle \mathsf{H}_n \rangle_t$ in (5.18a) can be minimized, if a specific wavevector $\boldsymbol{q} = \boldsymbol{k}$ can be found to satisfy the equation,

$$\nabla_q J_n(\boldsymbol{q}) = 0, \tag{5.18c}$$

where ∇_q is a gradient operator in the reciprocal space. Depending on the values of J_{mn}, such a wavevector $\boldsymbol{k}$ can specify the direction for the cluster to propagate, as will be shown in the next section for representative cases.

Although expressed in a complex form for mathematical convenience, σ_n must be real, so that the Fourier transform of the classical vectors σ_n should have a relation $\sigma_{+q}{}^* = \sigma_{-q}$, and accordingly

$$\boldsymbol{e}_{+q}{}^* = \boldsymbol{e}_{-q}. \tag{i}$$

for unit vectors. In this case, the amplitude σ_o is arbitrary, and the normalization condition in the reciprocal space can be written as

$$N = \boldsymbol{e}_{+q}{}^* \cdot \boldsymbol{e}_{+q} + \boldsymbol{e}_{-q}{}^* \cdot \boldsymbol{e}_{-q} = 2\boldsymbol{e}_{+q} \cdot \boldsymbol{e}_{-q}. \tag{ii}$$

On the other hand, the vectors $\mathbf{e}_n$ are also normalized in the crystal space, where the normalization can be expressed as

$$N = \sum_n \boldsymbol{e}_n{}^* \cdot \boldsymbol{e}_n = 2\boldsymbol{e}_{+q} \cdot \boldsymbol{e}_{-q} + \boldsymbol{e}_{+q}^2 \exp(2i\boldsymbol{q} \cdot \boldsymbol{r}_n) + \boldsymbol{e}_{-q}^2 \exp(-2\boldsymbol{q} \cdot \boldsymbol{r}_n). \tag{ii$'$}$$

These two equations, (ii) and (ii′), should be identical regardless of n, for which either

$$\exp(2i\boldsymbol{q} \cdot \boldsymbol{r}_n) = \exp(-2i\boldsymbol{q} \cdot \boldsymbol{r}_n) = 0 \tag{iii}$$

or

$$\boldsymbol{e}_{+q}^2 = \boldsymbol{e}_{-q}^2 = 0 \tag{iv}$$

needs to be satisfied. If (iii) is independent of n, we have either $\boldsymbol{q} = 0$ or $\frac{\boldsymbol{G}}{2}$, corresponding to *ferrodistortive* or *antiferrodistortive* arrangements of $\boldsymbol{\sigma}_n$ along these directions of $\boldsymbol{k}$, whereas (iv) indicates that $(e_q)_x^2 + (e_q)_y^2 + (e_q)_z^2 = 0$, for which the vector $\boldsymbol{q}$ can be an arbitrary vector, independent of the lattice periodicity. If taking $\left(e_{q_x}\right)^2 = 1$, for instance, we obtain $\left(e_{q_y}\right)^2 + \left(e_{q_z}\right)^2 = -1$. Writing $e_{q_x} = \exp i\varphi$ and $e_{q_y} \pm ie_{q_z} = i \exp i\varphi$ in this relation, we have a phase angle $\varphi = qx + const.$ in the x-direction. It is noted that such an arrangement of classical pseudospins $\boldsymbol{\sigma}_n$ in the direction of $\boldsymbol{k}$ is *incommensurate* with the lattice period. In this case, the discrete phase $\phi_n = \boldsymbol{q} \cdot \boldsymbol{r}_n - \omega t_n$ is practically continuous in the range $0 \leq \phi_n \leq 2\pi$, so that discrete angles ϕ_n can be replaced by a continuous angle ϕ in the same range $0 \leq \phi \leq 2\pi$.

5.5 Examples of Pseudospin Clusters

Calculating short-range correlations by the formula (5.18a–5.18c) is a familiar method for magnetic crystals, resulting in magnetic spin arrangements of various types [6, 7]. Using Equations (5.18a–5.18c), symmetry changes in structural transformations can be interpreted similarly. In this section, we show similar examples that can be applied to some structural changes.

5.5.1 *Cubic to Tetragonal Transition in $SrTiO_3$*

Figure 5.3 shows a view of $SrTiO_3$ structure, illustrating short-range correlations in the vicinity of TiO_6^{2-} at the center. For the cubic phase, three lattice constants along the symmetry axes are denoted by a, b, and c for convenience, and correlation strengths J_{mn} with the nearest- and next-nearest neighbors are expressed by J and J', respectively. In this model of a pseudospin cluster, we calculate the correlation energy by (5.18a), where the parameter $J_n(\boldsymbol{q})$ determined by (5.18b) is expressed as

$$J(\boldsymbol{q}) = 2J\{\cos(q_a a) + \cos(q_b b) + \cos(q_c c)\} + 4J'\{\cos(q_b b)\cos(q_c c) + \cos(q_c c)\cos(q_a a) + \cos(q_a a)\cos(q_b b)\},$$

omitting the index n, for simplicity. Putting this expression into (5.18c), we can obtain the specific wavevector $\boldsymbol{q} = \boldsymbol{k}$ by solving the equations

$$\sin(k_a a)\{J + 2J'\cos(k_b b) + 2J'\cos(k_c c)\} = 0,$$

$$\sin(k_b b)\{J + 2J'\cos(k_c c) + 2J'\cos(k_a a)\} = 0, \qquad \text{(i)}$$

and

$$\sin(k_c c)\{J + 2J'\cos(k_a a) + 2J'\cos(k_b b)\} = 0$$

for $\boldsymbol{k} = (k_a,\ k_b,\ k_c)$.

We can find that either one of the following wavevectors can satisfy (i). Namely, $\boldsymbol{k}_1 = (k_{1a},\ k_{1b},\ k_{1c})$ for

$$\sin(k_{1a} a) = \sin(k_{1b} b) = \sin(k_{1c} c) = 0. \qquad \text{(ii)}$$

$$\boldsymbol{k}_2 = (k_{2a},\ k_{2b},\ k_{2c})$$

for

$$\sin(k_{2a} a) = 0,\quad \cos(k_{2b} b) = \cos(k_{2c} c) = -1 + \tfrac{J}{2J'},$$

$$\sin(k_{2b} b) = 0,\quad \cos(k_{2c} c) = \cos(k_{2a} a) = -1 + \tfrac{J}{2J'}, \qquad \text{(iii)}$$

$$\sin(k_{2c} c) = 0,\quad \cos(k_{2a} a) = \cos(k_{2b} b) = -1 + \tfrac{J}{2J'}.$$

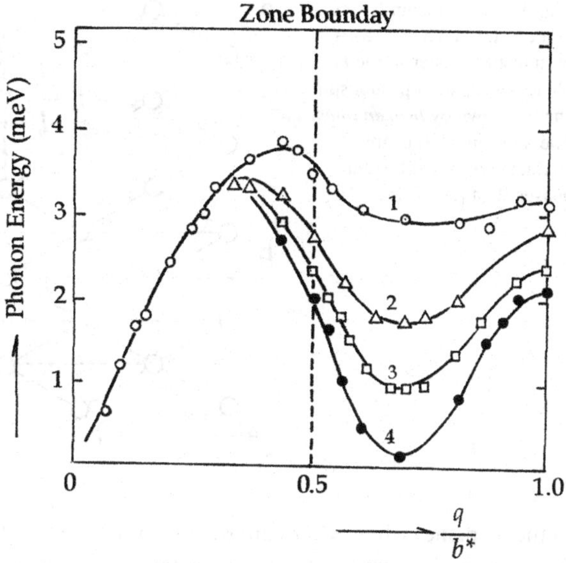

Fig. 5.4 Phonon dispersion curves in K_2SeO_4 near the Brillouin zone boundary obtained by neutron inelastic scatterings. Curves 1, 2, 3, and 4 were determined at 250, 175, 145, and 130 K, respectively. From M. Iizumi, J. D. Axe, G. Shirane, and K. Shimaoka, Phys. Rev. **B15**, 4392 (1977).

And $\boldsymbol{k}_3 = (k_{3a}, k_{3b}, k_{3c})$ for

$$\begin{aligned}\cos(k_{3a}a) &= 0, \quad \cos(k_{3b}b) = \cos(k_{3c}c) = -\tfrac{J}{2J'},\\ \cos(k_{3b}b) &= 0, \quad \cos(k_{3c}c) = \cos(k_{3a}a) = -\tfrac{J}{2J'},\\ \cos(k_{3c}c) &= 0, \quad \cos(k_{3a}a) = \cos(k_{3b}b) = -\tfrac{J}{2J'}.\end{aligned} \qquad \text{(iv)}$$

The solution (ii) gives a wavevector $\boldsymbol{k}_1 = \left(\frac{\pi l}{a}, \frac{\pi m}{b}, \frac{\pi n}{c}\right)$, where l, m, and n are 0 or $\pm$ integers, which gives commensurate arrangements with $J(\boldsymbol{k}_1) = 6J + 12J'$ due to 6 nearest- and 12 next-nearest-neighbors in the cluster. On the other hand, for $\boldsymbol{k}_2$, a component k_{2a} is commensurate along the a-axis, whereas the other k_{2b} and k_{2c} are incommensurate along the b- and c-axes in two dimensions, provided that $\left|1 - \frac{J}{2J'}\right| \leq 1$. The second and third ones in (iii) show a similar two-dimensional incommensurability in the ca and ab planes.

The solution (iv) provides a similar result to (iii); the first set gives an incommensurate arrangement in the bc plane, if $\left|\frac{J}{2J'}\right| \leq 1$, and commensurate along the a-axis, and so on.

Applying these results to $SrTiO_3$, it is clear that a tetragonal phase below T_c occurs at a Brillouin zone boundary incommensurate in two dimensions, which is actually confirmed by the neutron inelastic scattering experiments, showing a dip in the phonon energy at about $T_c \approx 130\,\text{K}$. As shown in Fig. 5.4 [8], the lowest dip appeared at $\frac{q}{b^*} = \frac{1}{2} + \delta_b$, where the shift δ_b was attributed to phase fluctuations at the zone boundary. However, in the magnetic resonance experiment by Müller et al., The incommensurate wavevector can be expressed as

$$k_a = l\left(\frac{1}{2} - \delta_a\right)a^* \quad \text{and} \quad k_c = n\left(\frac{1}{2} - \delta_c\right)c^*,$$

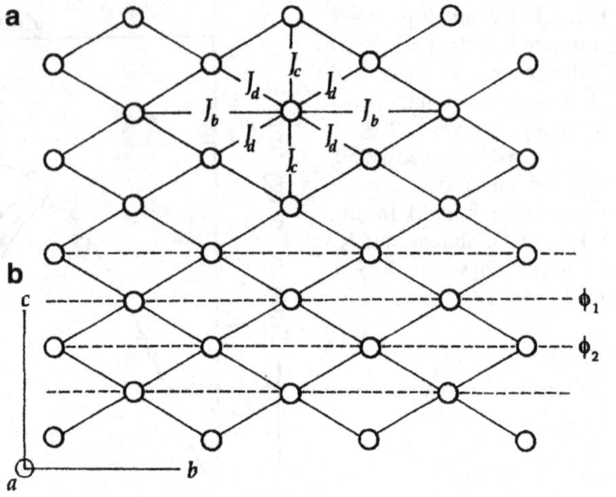

Fig. 5.5 Pseudospin lattice in a crystal of TSCC. (**a**) A pseudospin cluster in the *bc* plane. (**b**) Parallel pseudospin chains shown by *broken lines*. The adjacent chains are characterized by different phases $0<\phi_1, \phi_2<2\pi$.

in the *ac* plane, where a^*, c^* are reciprocal lattice constants. Small shifts δ_a, δ_b, and δ_c are called *incommensurate parameters*, arising from fluctuations at the zone boundary. It is noted that the foregoing argument is only qualitative, due to unknown J and J'. However, the calculation predicts at least the symmetry properties of condensates in perovskites.

5.5.2 Monoclinic Crystals of Tris-Sarcosine Calcium Chloride (TSCC)

A TSCC crystal, where *sarcosine* is an *amino acid* $H_3C–NH_2–CO_2H$, can be *twin* as crystallized from aqueous solutions at room temperature. A singledomain crystal is optically uniform and obtained from a naturally grown crystal, which is spontaneously strained along the a-axis and slightly monoclinic along the b-direction. The molecular arrangement is sketched in Fig. 3.4a. We already discussed the pseudospin model with Fig. 3.4b, where a Ca^{2+} ion is surrounded octahedrally by six O^- of carboxyl ions $–CO_2H$ of sarcosine molecules. In Fig. 5.5a shown is a model for a pseudospin cluster in TSCC, which is composed of pseudospins in nearest-neighbor distances on the *a*-, *b*-, and *c*-axes as well as in the diagonal directions in the *bc* plane.

Denoting these correlation parameters by J_a, J_b, J_c and J_d, (5.18b) can be written as

$$J(\boldsymbol{q}) = 2J_a(q_a a) + 2J_b\cos(q_b b) + 2J_c\cos(q_c c) + 4J_d\cos\frac{q_b b}{2}\cos\frac{q_c c}{2}.$$

From (5.18c) applied to this $J(\boldsymbol{q})$, we can determine the wavevector $\boldsymbol{k}$ for minimal strains. By similar calculation to perovskites in Sect. 5.5.1, the following specific $\boldsymbol{k}$-vectors can be considered for cluster energies. That is,

$$\boldsymbol{k}_1 = \left(\frac{\pi l}{a}, \frac{\pi m}{b}, \frac{\pi n}{c}\right),$$

and

$$\boldsymbol{k}_2 = \left(\frac{\pi l}{a}, k_{2b}, \frac{\pi n}{c}\right) \quad \text{where} \quad \cos\frac{k_{2b}b}{2} = -\frac{J_d}{2J_b},$$

$$\boldsymbol{k}_3 = \left(\frac{\pi l}{a}, \frac{\pi m}{b}, k_{3c}\right) \quad \text{where} \quad \cos\frac{k_{3c}c}{2} = -\frac{J_d}{2J_c},$$

Here, *l, m,* and *n* are zero or integers. Among these, $\boldsymbol{k}_1$ is a commensurate vector with the lattice, for which we have $J(\boldsymbol{k}_1) = 2J_a + 2J_b + 2J_c$.

In contrast, $\boldsymbol{k}_2$ and $\boldsymbol{k}_3$ are incommensurate vectors on the b- and c-axes, provided that $\left|\frac{J_d}{2J_b}\right| \leq 1$ and $\left|\frac{J_d}{2J_c}\right| \leq 1$, respectively. The thermodynamic phase of a TSCC crystal below 120 K is known as ferroelectric and characterized by spontaneous polarization along the b-axis, for which the vector $\boldsymbol{k}_2$ is therefore considered as responsible. For $\boldsymbol{q} = \boldsymbol{k}_2$, we have

$$J(\boldsymbol{k}_2) = 2J_a + 2J_b + 2J_c - \frac{2J_d^2}{J_b} = 2J_a + 2J_c + 2J_b\left(1 - \frac{J_d^2}{J_b^2}\right),$$

this should be positive for stable arrangement. If $J_d = 0$, we have $J(\boldsymbol{k}_2) = J(\boldsymbol{k}_1)$. On the other hand, for $J_d = -2J_b$ we have $J(\boldsymbol{k}_2) = 2J_a - 6J_b + 2J_c < J(\boldsymbol{k}_1)$, which gives a lower correlation energy than the commensurate phase of $\boldsymbol{k}_1$. Writing $\varphi = \frac{k_b b}{2}$ in the relation $\cos\frac{k_b b}{2} = -\frac{J_d}{2J_b}$, the above $J(\boldsymbol{k}_2)$ can be reexpressed as

$$J(\boldsymbol{k}_2) = 2J_a + 2J_c + 2J_b\cos 2\varphi + 4J_d\cos\varphi. \tag{5.19}$$

This is a well-known formula in the theory of magnetism for a spiral arrangement of spins in one dimension [9].

In TSCC crystals, interactions between adjacent pseudospin chains are significant for longitudinal propagation along the b-axis, as shown in Fig. 5.5b, whereas the correlations J_d between chains 1 and 2 appear to be responsible for transverse motion. For a similar situation in charge-density-wave systems, Rice [10] suggested such an interchain potential as $J_d \propto \cos(\phi_1 - \phi_2)$, where two chains are signified by phases ϕ_1 and ϕ_2, as illustrated in Fig. 5.5b, c. In this model, $J_d \to 0$, if assuming $\Delta\phi = \phi_1 - \phi_2 \to 0$; such a phasing mechanism leads all chains in parallel to the b-axis to a domain wall on the ac plane.

Exercise 5

1. Order variables $\boldsymbol{\sigma}_n$ at lattice sites n are subjects of practical observation such as X-ray diffraction. Hence a collision between $\boldsymbol{\sigma}_n$ and an X-ray photon occurs at a space-time (x_n, t_n), and for a collimated X-ray beam these coordinates are distributed over the collision area. Therefore, we have randomly distributed phases in the diffracted beam, which can be regarded as *uncertainty* of observation. Landau's criterion (5.5) is clearly concerned about such uncertainty. Review his argument, referring to the quantum-mechanical uncertainly principle. Also, why is the energy uncertainty can be justified with the equipartition theorem of statistical mechanics?
2. Why should we consider a group of ordered pseudospins as the cluster at the threshold of a structural change? Is it originated from the Born-Huang principle? Discuss this question in qualitative manner.
3. Clustering by short-range correlations discussed in Sect. 5.4 can be characterized as phasing of pseudospins. Justify this process in terms of the Born-Huang principle. Can the process be considered as isothermal in thermodynamic environment? (*Hint*: See if phonon scatterings can be inelastic, when two phonons are considered with correlated pseudospins as proportional to $\boldsymbol{\sigma}_n \cdot \boldsymbol{\sigma}_{n'}$).
4. In a perovskite crystal, either of conditions (iii) and (iv) in Sect. 5.5.1 was considered for a two-dimensional order, which was actually substantiated experimentally. Discuss the theoretical reason for supporting two-dimensional order.

References

1. R. Comes, R. Currat, F. Desnoyer, M. Lambert, A.M. Quittet, Ferroelectrics **12**, 3 (1976)
2. L.D. Landau, E.M. Lifshitz, *Statistical Physics* (Pergamon Press, London, 1958)
3. R. Blinc, B. Zeks, *Soft Modes in Ferroelectrics and Antiferroelectrics* (North-Holland, Amsterdam, 1974)
4. M. Born, K. Huang, *Dynamical Theory of Crystal Lattices* (Oxford Univ. Press, 1968)
5. F.C. Nix, W. Shockley, Rev. Mod. Phys. **10**, 1 (1938)
6. T. Nagamiya, *Solid State Physics*, vol. 20 (Acad. Press, New York, 1963)
7. L.R. Walker, *Magnetism I* (Acad. Press, New York, 1963)
8. M. Iizumi, J.D. Axe, G. Shirane, K. Shimaoka, Phys. Rev. **B15**, 4392 (1977)
9. C. Kittel, *Introduction to Solid State Physics*, 6th edn. (Wiley, New York, 1986)
10. M.J. Rice, Charge density wave systems, in *Solitons and Condensed Matter Physics*, ed. by A.R. Bishop, T. Schneider (Springer, Berlin, 1978)

Chapter 6
Critical Fluctuations

Incorporated with the condensate model, Landau's Gibbs function expanded into power series can be interpreted with an adiabatic potential arising from correlations. On the other hand, a *pinning potential* is mandatory for observing critical anomalies due to fluctuating order variables. In a stable crystalline phase, *pseudospin* modes with inversion symmetry should become in phase with the adiabatic potential, for which Landau's expansion is a logical approach to take Born-Huang's principle into account.

In this chapter, we discuss critical anomalies from condensates in random phases, leaving interpretation of Landau's expansion to Chap. 7. Fluctuating order variables are essential for second-order phase transitions, for which quantum-mechanical uncertainties at all lattice sites are responsible. Thermodynamically, such fluctuations are primarily adiabatic, but diminishing on lowering temperature, implying that pinned condensates are thermally stabilized by transferring excess energies to the lattice.

6.1 The Landau Theory of Binary Transitions

In Landau's thermodynamic theory [1] for binary order, the Gibbs potential is defined as a function of external p, T, and an order parameter η. As a macroscopic variable, η is defined in the range $0 \le \eta \le 1$, representing the degree of order between disordered and ordered states. Representing the mean-field average of $\boldsymbol{\sigma}_n$ overall lattice sites, we assume η as a continuous function of temperature T under a constant p. The critical temperature T_o determined by the Landau theory was found always higher than the observed temperature T_c, indicating that the mean-field theory is an inadequate assumption. Nevertheless, the theory assumes a *singular* behavior of $G(\eta)$ at T_o, signifying binary phase transitions in the first approximation, considering an observed difference between T_o and T_c as the critical anomaly.

Landau assumed that the Gibbs potential below T_o is determined by a power expansion of $G(\eta)$, i.e.,

M. Fujimoto, *Thermodynamics of Crystalline States*,
DOI 10.1007/978-1-4614-5085-6_6,

$$G(\eta) = G(0) + \frac{1}{2}A\eta^2 + \frac{1}{4}B\eta^4 + \frac{1}{6}C\eta^6 + \ldots\ldots\ldots, \tag{6.1}$$

where $G(0)$ is the Gibbs potential at $\eta = 0$, and higher-order terms are significant if η is in finite magnitude. The second-term proportional to η^2 represents a harmonic potential, whereas higher-than-second-order terms can be attributed to nonlinear correlations related to anharmonicity of the lattice. Therefore, a significant role played by the lattice cannot be ignored for $T < T_c$, so that the Gibbs function can be written as $G_L + G(\eta)$, although Landau's original theory disregards G_L as insignificant.

In (6.1), odd-power terms are excluded in the expansion, because of *inversion symmetry*, $\eta \leftrightarrow -\eta$, in a binary system. Thus, (6.1) is consistent with the hypothesis

$$G(\eta) = G(-\eta), \tag{6.2}$$

which is necessary for the order parameter to represent a classical displacement vector in a deformed structure, which is however assumed as a scalar in Landau's theory.

Using the variation principle, the function $G(\eta)$ in an equilibrium crystal is determined by its minimum $\eta = \eta_o$ at T_o, around which we consider any variation $\Delta\eta = \eta - \eta_o$ mathematically. Therefore, a thermodynamic variation of the Gibbs function is expressed as

$$\Delta G = G(\eta) - G(\eta_o) \geq 0,$$

which is determined by $\left(\frac{\partial \Delta G}{\partial \eta}\right)_{p.T} = 0$ in the vicinity of $\eta_o = 0$. The expansion (6.1) can therefore be *truncated* at the fourth-order term η^4 for a small deviation η; hence we have

$$\Delta G(\eta) = \frac{1}{2}A\eta^2 + \frac{1}{4}B\eta^4.$$

Differentiating $\Delta G(\eta)$ with respect to η, we obtain $A\eta_o + B\eta_o^3 = 0$, so that the equilibrium can be determined by either

$$\eta_o = 0 \quad \text{or} \quad \eta_o^2 = -\frac{A}{B} \tag{6.3a}$$

In the former case, $\eta_o = 0$ represents the disordered state. For the latter case, Landau assumed that the factor A is temperature dependent and expressed by

$$A = A'(T - T_o) \quad \text{for} \quad T > T_o \tag{6.4a}$$

and

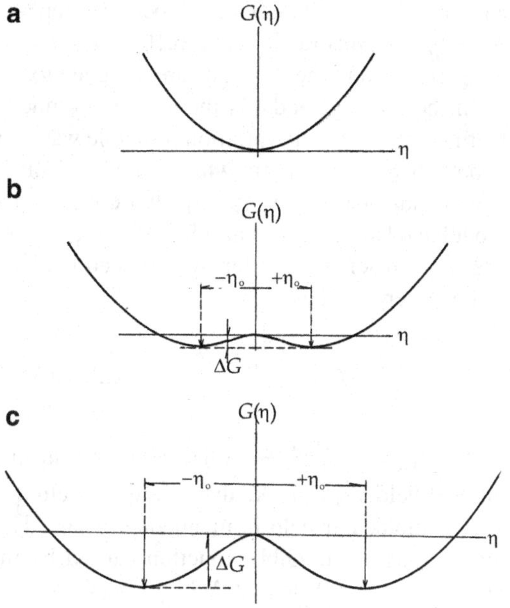

Fig. 6.1 Landau's quartic potentials. (**a**) $T = T_0$; (**b**) and (**c**) $T < T_0$. Binary equilibrium is indicated by two-way shifts and increasing depths of minima with decreasing temperature.

$$A = A'(T - T_0) \quad \text{for} \quad T < T_0 \tag{6.4b}$$

where A' is a positive constant and B is assumed as positive. From the second solution of (6.3a), we then obtain

$$\eta_0^2 = \frac{A'}{B}(T_0 - T) \quad \text{for} \quad T < T_0 \tag{6.3b}$$

The transition can therefore be determined by the factor A that reverses its sign at $T = T_0$, and the positive B emerging at T_0, thereby shifting equilibrium from $\eta_0 = 0$ to $\pm\,\eta_0 = \pm\sqrt{\frac{A'}{B}(T_0 - T)}$ with decreasing temperature. Gibbs functions can thus be written as

$$G(\eta_0) = G(0) \quad \text{at} \quad \eta_0 = 0 \quad \text{for} \quad T \geq T_0$$

and

$$G(\eta_0) = G(0) - \frac{3A'^2}{4B} \quad \text{at} \quad \eta_0 = \pm\sqrt{\frac{A'}{B}} \quad \text{for} \quad T < T_0.$$

Figure 6.1 shows schematically such a change of the Gibbs function with varying temperature. For $T > T_0$, $\eta_0 = 0$ for equilibrium at all temperatures as shown by Fig. 6.1a. For $T < T_0$, $\eta_0 = \pm\sqrt{\frac{A'}{B}(T_0 - T)}$ from (6.3b), shifting equilibrium with a *parabolic* temperature dependence, and the minimum is lowered by $-\frac{3A'^2}{4B}$, as illustrated in Fig. 6.1b, c. The ordering starts at $T = T_o$, with lowering temperature, and the $\pm$ signs are assigned to two domains in the opposite order.

In the above, Landau's theory was outlined in the framework of equilibrium thermodynamics. The variation principle was employed to minimize $G(\eta)$ against a hypothetical variation $\delta\eta$, which can physically be assigned to a dynamic variation in a crystal. A variation $\eta - \eta_o$ at a constant p and T condition is associated with modulated lattice, and a variation $\Delta G(\eta)$ is caused by adiabatic potentials ΔU_n at all sites n. Replacing the latter by the mean field average $\langle \Delta U_n \rangle$, Landau's expansion can be interpreted as

$$\Delta G = \frac{A}{2}\eta^2 + \frac{B}{4}\eta^4 + \ldots\ldots = \langle \Delta U_n \rangle = -(\eta - \eta_o)X_{\text{int}}, \tag{6.5}$$

where $X_{\text{int}} = -\frac{\partial \langle \Delta U_n \rangle}{\partial(\eta - \eta_o)}$ is the *internal field* in the mean-field accuracy. Such an internal field X_{int} is an adiabatic force or field as if applied externally, analogous to Weiss' molecular field in magnetic crystals. Thus, ΔG represents thermodynamic fluctuations in the Gibbs function caused by the internal fluctuations $\Delta\eta$ in the vicinity of $\eta_o = 0$, so that $\Delta G = -\Delta\eta X_{\text{int}}$.

In dielectric and magnetic crystals, the order parameter η represents microscopic dipole and magnetic moments, respectively. In these cases, the change $\Delta\eta$ is the response to applying electric or magnetic fields. Denoting the applied field by X, we can include an additional term $-\eta X$ in the Gibbs function. Assuming a small variation of η in the vicinity of T_o, we can write

$$\Delta G_> = \Delta G(\eta) - \eta X = \frac{1}{2}A\eta^2 - \eta X$$

for $T > T_o$, and the equilibrium can be determined from $\left(\frac{\partial G_>}{\partial \eta}\right)_{p,T} = 0$, namely, $A\eta - X = 0$, from which the *susceptibility* is expressed by

$$\chi_{T>T_o} = \frac{\eta_o}{X} = \frac{1}{A} = \frac{1}{A'(T - T_o)}, \tag{6.6a}$$

where $A = A'(T - T_o)$ from (6.4a). The susceptibility (6.6a) indicates that the system can be ordered by X to such an extent as determined by the term of A in the Gibbs potential at temperatures above T_o.

For $T < T_o$, on the other hand, since $\Delta G(\eta)$ is contributed by $\frac{1}{4}B\eta^4$, the equilibrium is determined by minimizing

$$\Delta G_< = \frac{1}{2}A\eta^2 + \frac{1}{4}B\eta^4 - \eta X,$$

and hence $A\eta_o + B\eta_o^3 - X = 0$, in which $\eta_o^2 = -\frac{A}{B}$ gives equilibrium at $T < T_o$. Accordingly, we have

$$\chi_{T<T_0} = \frac{\eta_0}{2A} = \frac{1}{2A'(T_0 - T)}, \tag{6.6b}$$

where $A = A'(T_0 - T)$ as assumed in (6.4b).

The formulas (6.6a) and (6.6b) comprise the Curie-Weiss law, showing a singularity at T_0, which can be approached with varying temperatures at a rate $\frac{1}{A'}$ or $\frac{1}{2A'}$ for $T > T_0$ or $T < T_0$, respectively. It is noted however that such expressions are valid, only if the magnitude of X is sufficiently small. To study these properties of a crystal experimentally, it is necessary to apply X to detect the response susceptibilities $\chi_{T>T_c}$ and $\chi_{T<T_c}$. The Curie-Weiss laws (6.6a) and (6.4b) were substantiated at temperatures near T_c in many *ferroelectric crystals*. Below T_c however, the temperature dependence of susceptibility $\chi_{T<T_c}$ is not simply proportional to $(T_c - T)^{-1}$, indicating that the effective field X is predominantly internal X_{int}. Qualitatively, the susceptibility for $T < T_c$ shows sufficient evidence for significant correlations in the lattice.

6.2 Internal Pinning of Adiabatic Fluctuations

We consider that the Gibbs potential fluctuates dynamically with the order parameter under the critical condition, as internal correlations fluctuate. In this case, the crystal volume cannot be constant, changing adiabatically under a constant p–T condition. At the threshold of condensate formation, fluctuations are *sinusoidal* in random phases at lattice sites, decreasing with local strains diminishing to another state of equilibrium.

Corresponding to $\Delta G = \frac{A'}{2}(T_0 - T)\eta^2 + \frac{B}{4}\eta^4$ in Landau's theory for $T < T_0$, we can assume changes in local adiabatic potentials at lattice sites in the similar form, namely, $\Delta U_n = \frac{1}{2}a\boldsymbol{\sigma}_n^2 + \frac{1}{4}b\boldsymbol{\sigma}_n^4$ at the outset of fluctuations that occur in a random phase at a temperature $T_c{}^*$ close to T_c. Below $T_c{}^*$, we consider $a < 0$ and $b > 0$, so that the pseudospin vector $\boldsymbol{\sigma}_n$ behaves like a classical displacement, as discussed in Chap. 5. Due to the positive term $\frac{1}{4}b\boldsymbol{\sigma}_n^4$, the equilibrium position shifts closer toward the neighboring pseudospin $\boldsymbol{\sigma}_m (m \neq n)$, where their mutual correlations become significant.

Below $T_c{}^*$, the order variable $\boldsymbol{\sigma}_n$ of a condensate can be expressed primarily as $\boldsymbol{\sigma}_n = \boldsymbol{\sigma}_k \exp i(\boldsymbol{k}\cdot\boldsymbol{r}_n - \omega t_n)$, where $\boldsymbol{k}$ and ω are distributed among lattice sites, which is logical if predominantly $\Delta U_n \approx \frac{1}{2}a\boldsymbol{\sigma}_n^2$ near $T_c{}^*$. In this case, the amplitude $\boldsymbol{\sigma}_n$ is finite, as expressed by the Fourier transform, $\boldsymbol{\sigma}_k = \boldsymbol{\sigma}_n \exp\{-i(\boldsymbol{k}\cdot\boldsymbol{r}_n - \omega t_n)\}$, where the phase $\phi_n = \boldsymbol{k}\cdot\boldsymbol{r}_n - \omega t_n$ are distributed in the range $0 \le \phi_n \le 2\pi$, using the periodic boundary condition in a crystal. Such phases are accompanied with uncertainties $\Delta\phi_n$ related to $\Delta\boldsymbol{r}_n$ and Δt_n, so that the critical region is dominated by distributed phases $\phi_n + \Delta\phi_n$, which we shall call adiabatic phase fluctuations or simply fluctuations in the lattice.

In a continuum-like lattice for a small $|\boldsymbol{k}|$, distributed phases ϕ_n can be replaced by a single continuous phase $\phi = \boldsymbol{k}\cdot\boldsymbol{r} - \omega t$, where $0 \le \phi \le 2\pi$, which represents fluctuations without referring to lattice site n. With this definition, continuously distributed phases $\phi + \Delta\phi$ can signify fluctuations in terms of the whole angle. We can therefore write $\boldsymbol{\sigma}_k = \boldsymbol{\sigma}_0 f(\phi + \Delta\phi)$ to express correlated pseudospins, where $|\boldsymbol{k}|^{-1}$ determines an effective measure of the correlation length in the direction of propagation. Further notable is that $\boldsymbol{\sigma}_k$ becomes nonsinusoidal in the presence of a quartic potential. Expressed by an *elliptic function*, $\boldsymbol{\sigma}_k$ exhibits *nonlinear* character that originates from distant correlations. In general, the amplitude σ_0 depends also on $\boldsymbol{k}$-dependent velocity of propagation, which is typical in nonlinear propagation (See Chap. 8).

Using Born-Huang's principle, we assume that distributed phases $\phi_n + \Delta\phi_n$ in the critical region can become *in phase* with increasing lattice excitations. The process for $\Delta\phi_n \to 0$ is primarily adiabatic and forced to occur by the lattice excitation to minimize lattice strains. It is significant that all the nonlinear $\boldsymbol{\sigma}_k$ and the corresponding adiabatic potentials ΔU_k become in phase after $\Delta\phi_n \to 0$ in the propagating frame of reference. Considering the space inversion $\boldsymbol{r} \to -\boldsymbol{r}$ in a binary crystal, two phase variables $\phi_\pm = \pm(\boldsymbol{k}\cdot\boldsymbol{r} - \omega t)$ are active for such fluctuations. The amplitude should be a real quantity, for which the inversion relation can be written as

$$\boldsymbol{\sigma}_0(+\boldsymbol{k}) = -\boldsymbol{\sigma}_0(-\boldsymbol{k}), \tag{6.7}$$

representing reflection at the inversion center $\phi = 0$ of the potential $\Delta U_k(\phi)$ that is moving in phase with $\boldsymbol{\sigma}_k(\phi)$. In this context, we can say that $\boldsymbol{\sigma}_{\pm k}(\phi)$ are *pinned* by *intrinsic* potentials $\Delta U_{\pm k}(\phi)$ in equilibrium, characterizing *intrinsic pinning* in stress-free crystals. Translational symmetry of an idealized structure can be signified by invariance for $\phi \to \phi + 2n\pi$; hence we may consider that $\phi = 0$ express fluctuations pinned by a moving coordinate system. In practical crystals, there are some unavoidable defects, which can also pin $\boldsymbol{\sigma}_{\pm k}$ *extrinsically* by defect potential V at $\phi = 0$, as expressed by defect pinning with a *pinning energy* $-\boldsymbol{\sigma}_{\pm k}(0)V$. Although categorically similar, intrinsic pinning accompanies long-range order, whereas extrinsic pinning remains local.

We realize that phasing processes considered above should occur thermally, exchanging energy at the boundaries of the heat reservoir. We consider that such phasing should always be observed as a *soft mode* or *thermal relaxation* to gain thermal stability of lattice structure (See Chap. 9).

6.3 Critical Anomalies

Experimentally, the threshold of a second-order transition is determined by the observed critical temperature T_c. We assume theoretically that it is determined for clustered pseudospins $\boldsymbol{\sigma} = \boldsymbol{\sigma}_0 \exp\{\pm i(\boldsymbol{k}\cdot\boldsymbol{r} - \omega t)\}$ to be pinned by the corresponding potential $\Delta U_k = \frac{a}{2}\boldsymbol{\sigma}_k^2 + \frac{b}{4}\boldsymbol{\sigma}_k^4$. At T_c, both $\boldsymbol{\sigma}_k$ and ΔU_k should be stationary in the moving coordinate system, where $\boldsymbol{\sigma}_k$ is pinned by ΔU_k at $\phi = \boldsymbol{k}\cdot\boldsymbol{r} - \omega t = 0$.

Considering Gibbs functions fluctuating between $G(\boldsymbol{\sigma}_{+k})$ and $G(\boldsymbol{\sigma}_{-k})$, we write $\boldsymbol{\sigma}_{+k} = \boldsymbol{\sigma}_0 \exp i\phi$ and $\boldsymbol{\sigma}_{-k} = \boldsymbol{\sigma}_0 \exp(-i\phi)$. In-between, there is a positive

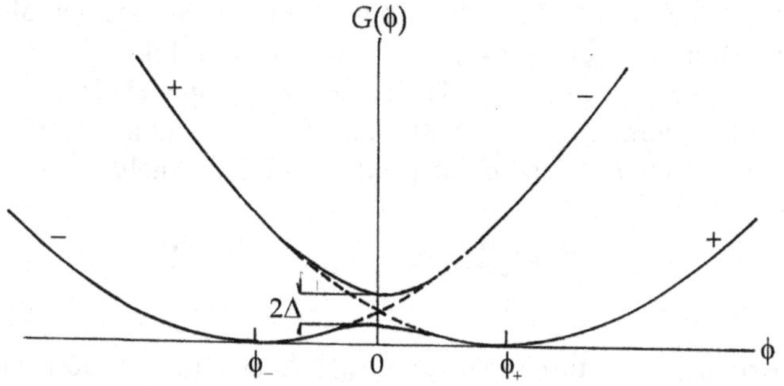

Fig. 6.2 Phase fluctuations of the Gibbs potential at the critical condition.

potential barrier $\Delta U_{\boldsymbol{k}}(\phi) = \Delta U_{\boldsymbol{k}} + \Delta U_{-\boldsymbol{k}} = a\boldsymbol{\sigma}_0^2$ in the vicinity of $\phi = 0$. The potential $\Delta U_{\boldsymbol{k}}(\phi)$ is dominated by the positive term $a\boldsymbol{\sigma}_{\boldsymbol{k}}^2$ at temperatures close to T_c and is symmetric with respect to $\phi = 0$, hence expressed as proportional to cos (2ϕ). Assuming the amplitude $\boldsymbol{\sigma}_0$ as constant, the Gibbs potential fluctuates between $\phi_+ = +\boldsymbol{k}\cdot\boldsymbol{r} - \omega t$ and $\phi_- = -\boldsymbol{k}\cdot\boldsymbol{r} + \omega t$, as illustrated in Fig. 6.2. For two parabolic curves for $G(+\phi)$ and $G(-\phi)$, minimum positions are ϕ_+ and ϕ_-, respectively, where phase fluctuations can occur in the range $\phi_- < \phi < \phi_+$ through the barrier $\Delta U_{\boldsymbol{k}}(\phi) = V\cos 2\phi$, where $V > 0$. As a *level-crossing* problem familiar in quantum mechanics, we can solve this problem by considering $\boldsymbol{\sigma}_{\pm\boldsymbol{k}}(\phi)$ as the wavefunctions of the unperturbed equation $\mathsf{H}\boldsymbol{\sigma}_{\pm\boldsymbol{k}} = \varepsilon_\pm\boldsymbol{\sigma}_{\pm\boldsymbol{k}}$. The degenerate eigenvalues, $\varepsilon_+ = \varepsilon_-$, represent kinetic energies of phase fluctuations, which are then perturbed by the potential $\Delta U_{\boldsymbol{k}}(\phi)$.

In this case, the perturbed function can be expressed as a linear combination of $\boldsymbol{\sigma}_+$and $\boldsymbol{\sigma}_-$, so that, omitting $\boldsymbol{k}$ for brevity, we write

$$\boldsymbol{\sigma} = c_+\boldsymbol{\sigma}_+ + c_-\boldsymbol{\sigma}_- \quad \text{and} \quad c_+^2 + c_-^2 = 1, \tag{6.8}$$

where c_+ and c_- are normalizing coefficients. The perturbed equation, $(\mathsf{H} + V\cos 2\phi)\boldsymbol{\sigma} = \varepsilon\boldsymbol{\sigma}$, can then be solved for ε, if

$$\begin{vmatrix} \varepsilon_+ - \varepsilon & \Delta \\ \Delta & \varepsilon_- - \varepsilon \end{vmatrix} = 0, \tag{6.9}$$

where

$$\begin{aligned} \Delta &= \int_0^{2\pi} \boldsymbol{\sigma}_+{}^*(V\cos 2\phi)\boldsymbol{\sigma}_- \mathrm{d}\phi \Big/ \int_0^{2\pi} \mathrm{d}\phi = \frac{V\boldsymbol{\sigma}_0^2}{2\pi}\int_0^{2\pi} \exp(-2i\phi)\cos(2\phi)\mathrm{d}\phi \\ &= \frac{V\boldsymbol{\sigma}_0^2}{\pi}\int_0^{\pi} \cos^2(2\phi)\mathrm{d}\phi = \frac{V\boldsymbol{\sigma}_0^2}{2\pi}. \end{aligned} \tag{6.10}$$

Solving (6.9) for the perturbed energy ε, we obtain $\varepsilon = \frac{\varepsilon_+ + \varepsilon_-}{2} \pm \sqrt{\left(\frac{\varepsilon_+ - \varepsilon_-}{2}\right)^2 + \Delta^2}$, which is simplified as $\varepsilon = \varepsilon_0 \pm \Delta$, by writing the unperturbed energies as $\varepsilon_+ = \varepsilon_- = \varepsilon_o$. Thus, lifting the degeneracy of the fluctuation energy, we have an energy gap 2Δ, as indicated in Fig. 6.2.

In this case, the coefficients c_+ and c_- should satisfy the relations $c_+^2 = c_-^2 = \frac{1}{2}$, because of the normalization (6.8); hence we obtain symmetrical and antisymmetrical functions of these independent modes, namely,

$$\boldsymbol{\sigma}_\mathrm{A} = \frac{\boldsymbol{\sigma}_+ + \boldsymbol{\sigma}_-}{\sqrt{2}} \quad \text{and} \quad \boldsymbol{\sigma}_\mathrm{P} = \frac{\boldsymbol{\sigma}_+ - \boldsymbol{\sigma}_-}{\sqrt{2}} \tag{6.11}$$

which are assigned to perturbed energies $\varepsilon_o + \Delta$ and $\varepsilon_o - \Delta$, respectively. The perturbed states are therefore characterized by $\boldsymbol{\sigma}_\mathrm{A} = \sqrt{2}\boldsymbol{\sigma}_o \cos\phi$ and $\boldsymbol{\sigma}_\mathrm{P} = \sqrt{2}i\boldsymbol{\sigma}_o \sin\phi$, which are called the *amplitude* and *phase modes*, respectively. The critical region signified by these two modes (6.11) was actually detected in neutron inelastic scattering and magnetic resonance experiments, as described in Chap. 10.

In thermodynamics, such fluctuations are observed as thermal relaxation mode diminishing with lowering temperature. In practice, the amplitude mode exhibits one sharp absorption at $\phi = 0$, whereas the phase mode characterized by two peaks separated with decreasing temperature. It is noted that the Gibbs potential for the phase mode consists of quadratic and quartic potentials, shifting equilibrium from $\phi = 0$ to $\phi = \phi_\pm$, representing domains in a crystal. The presence of these two modes is substantiated by the *central peak* and *soft mode* from observed time-dependent fluctuation spectra. (See Chaps. 9 and 10 for the detail).

6.4 Observing Anomalies

In the transition region, the fluctuations are distributed in different shapes from random fluctuations, as revealed by observed *critical anomalies*. In fact, symmetrical and asymmetrical modes, $\boldsymbol{\sigma}_\mathrm{A}$ and $\boldsymbol{\sigma}_\mathrm{P}$, exhibit anomalies that are signified by distributed phases ϕ in the range $0 \le \phi \le 2\pi$. In experiments, we use probes that can be associated with $\boldsymbol{\sigma}$ to sample fluctuations. In this case, the timescale of observation t_o should be sufficiently shorter than the time $t = 2\pi/\omega$; otherwise, fluctuations are averaged out. Leaving experimental details to Chap. 10, here we discuss the principle of sampling.

In sampling experiments, the time averages $\langle\boldsymbol{\sigma}_\mathrm{A}\rangle_t$ and $\langle\boldsymbol{\sigma}_\mathrm{P}\rangle_t$ can be analyzed from observed quantities. Considering $\boldsymbol{\sigma}_o$ as constant, we have

$$\langle\boldsymbol{\sigma}_\mathrm{A}\rangle_\mathrm{t} = \frac{\sqrt{2}\boldsymbol{\sigma}_o}{t_o}\int_0^{t_o} \cos(\boldsymbol{k}\cdot\boldsymbol{r} - \omega t)\mathrm{d}t = 2\sqrt{2}\boldsymbol{\sigma}_o \frac{\sin\omega t_o}{\omega t_o}\cos\boldsymbol{k}\cdot\boldsymbol{r}$$

and

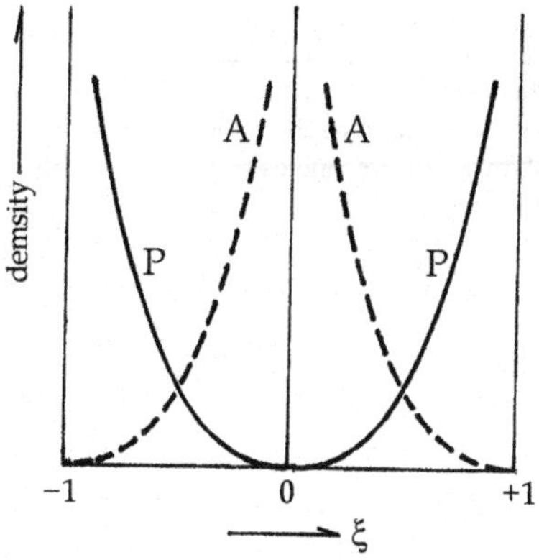

Fig. 6.3 Intensity distributions of the amplitude mode (A) and phase mode (P).

$$\langle \boldsymbol{\sigma}_\mathrm{P} \rangle_\mathrm{t} = \frac{2i\sqrt{2}\boldsymbol{\sigma}_\mathrm{o}}{t_\mathrm{o}} \int_0^{t_\mathrm{o}} \sin(\boldsymbol{k}\cdot\boldsymbol{r} - \omega t)\mathrm{d}t = 2i\sqrt{2}\boldsymbol{\sigma}_\mathrm{o} \frac{\sin \omega t_\mathrm{o}}{\omega t_\mathrm{o}} \sin \boldsymbol{k}\cdot\boldsymbol{r}.$$

These mode amplitudes are reduced by the factor $\Gamma = \frac{\sin \omega t_o}{\omega t_o}$, determining the width of fluctuations, which depends on t_o; for example, if $\omega t_o \to 0$, we have $\Gamma \to 1$. Further $\boldsymbol{k}\cdot\boldsymbol{r}$ represents the *spatial phase* ϕ_s that determines the measured fluctuations in the effective width $\Gamma|\boldsymbol{\sigma}_o|$ in the timescale t_o of observation.

Owing to the classical character, these $\langle \boldsymbol{\sigma}_\mathrm{A} \rangle_\mathrm{t}$ and $\langle \boldsymbol{\sigma}_\mathrm{P} \rangle_\mathrm{t}$ may be considered as longitudinal and transverse components of a pseudospin vector $\boldsymbol{\sigma} = \boldsymbol{\sigma}_\mathrm{A} + \boldsymbol{\sigma}_\mathrm{P}$. We can write the vector σ in a complex form $\sigma = \sigma_\mathrm{A} + i\sigma_\mathrm{P}$, whose real and imaginary parts are as follows:

$$\sigma_\mathrm{A} = \sigma_\mathrm{o} \cos \phi_\mathrm{s} \quad \text{and} \quad \sigma_\mathrm{P} = \sigma_\mathrm{o} \sin \phi_\mathrm{s}. \tag{6.12}$$

Here the amplitude σ_o substitutes for $2\sqrt{2}|\boldsymbol{\sigma}_o|\Gamma$. In practice, these components are measured from a sample crystal oriented in the laboratory reference. Therefore, the quantities proportional to $\boldsymbol{\sigma}_{\mathrm{A,P}}$ can be expressed as $\int_0^{2\pi} f_{\mathrm{A,P}}(\phi_s)\sigma_{\mathrm{A,P}}\mathrm{d}\phi_s$, where $f_{\mathrm{A,P}}(\phi_s)$ are densities of fluctuations distributed between $\phi_s$and $\phi_s + \mathrm{d}\phi_s$. In practice, a linear variable ξ, defined by $\xi = \cos \phi_s$, and hence $\pm\sqrt{1-\xi^2} = \sin \phi_s$, is more convenient than the angular variable ϕ_s. So converting variables from ϕ_s to ξ, we obtain the following quantities:

$$\int_{-1}^{+1} f_\mathrm{A} \frac{\sigma_\mathrm{o}}{\xi} \mathrm{d}\xi \quad \text{and} \quad \int_{-1}^{+1} f_\mathrm{P} \frac{\sigma_\mathrm{o}}{\sqrt{1-\xi^2}} \mathrm{d}\xi,$$

for σ_A and σ_P modes, respectively. Figure 6.3 shows curves of these density functions, where A and P modes exhibit clearly distinctive shapes. The latter is distributed between $+\sigma_o$ and $-\sigma_o$ with the width $2\sigma_o \propto \Gamma$, whereas the former is a

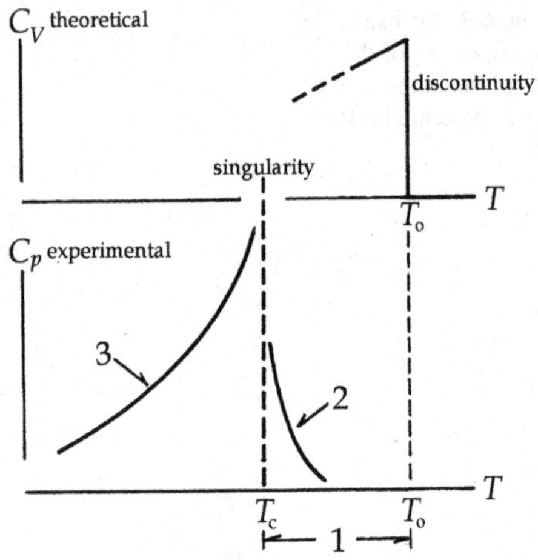

Fig. 6.4 Transition anomalies in C_V(mean-field theory) compared with experimental C_p(schematic), showing discrepancies 1, 2, and 3.

single line at the center $\phi = 0$. Practical examples of observed anomalies [2] will be discussed in Chap. 10.

In contrast, early thermal experiments on crystals as a whole were not logical for studying $\boldsymbol{\sigma}_k$. It is realized that the crystalline state at and below $T_c{}^*$ is heterogeneous, where specific heats C_p and C_V defined for homogeneous materials are invalid concepts. Therefore, such experimental results for C_p as shown in Fig. 5.2 do not logically represent thermodynamic properties. Figure 6.4 shows typical temperature dependence of experimental C_p, as compared with the mean-field results. Nevertheless, we notice three kinds of discrepancies in the comparison: (1) anomalous difference between T_c and T_o, (2) a steep rise of the singularity at T_c vs. discontinuity at T_o, and (3) a gradual decay for $T < T_c$ vs. a linear decay for $T < T_o$. We will analyze the anomaly (3) in Chap. 8.

6.5 Extrinsic Pinning

An ideal crystal is characterized by translational symmetry of a periodic structure. In contrast, practical crystals are by no means perfect, because of the presence of imperfections of many types, such as lattice defects, dislocations, impurities, and surfaces. Aside from surfaces that determine thermal properties of crystals, the amount of imperfections in crystals can technically be reduced by the careful preparation, making the sample as nearly perfect. Among practical imperfections, *point defects* are most significant in idealized crystals, disrupting locally the lattice periodicity. On the other hand, some defects are located at sites arranged in different periodicities from the lattice, constituting *pseudo-symmetry*. If order variables are pinned by such defect potentials, we may have a phase transition characterized by such pseudo-symmetry. Nonetheless, it is important experimentally to distinguish intrinsic pining from external pinning.

6.5.1 Pinning by Point Defects

A practical crystal contains some unavoidable imperfections that are mostly missing constituents or impurity ions at lattice sites. Translational space symmetry is disrupted by such *point defects*, by which collective pseudospin modes can be pinned, exhibiting standing waves in a crystal.

A point defect can be represented by a potential function of distance $\boldsymbol{r} - \boldsymbol{r}_\text{i}$ from a defective lattice site $\boldsymbol{r}_\text{i}$. Such a defect potential is usually anharmonic in a crystal, where $\boldsymbol{\sigma}_n$ is confined symmetrically to the vicinity of the defect site. Nevertheless, in a crystalline phase with a low defect density, propagation of pseudospin modes $\boldsymbol{\sigma}_\text{o}\exp(\pm i\phi_n)$ can be perturbed by a defect potential at site n, offering a problem that is mathematically the same as in Sect. 6.3.

At such an *extrinsic* defect in a crystal, fluctuations can be either symmetric or antisymmetric with respect to the defect center: the former is for defects at a regular lattice site, whereas the latter can be applied at surface sites. Considering the symmetric combination, $\frac{(\boldsymbol{\sigma}_+ + \boldsymbol{\sigma}_-)_n}{\sqrt{2}} = \sqrt{2}\boldsymbol{\sigma}_\text{o}\cos\phi_n$, we can define the pinning potential as

$$V(\phi) = -V_\text{o}\cos\phi, \tag{6.13}$$

where ϕ is the phase variable in the range $0 \le \phi \le 2\pi$; the minimum is located at $\phi = 0$. Here, the magnitude V_o should be redefined to include $\sigma_{\pm k}(0)$, as discussed in Sect. 6.2; nevertheless, (6.13) is a meaningful potential as a function of ϕ.

Considering surfaces as imperfections, we take antisymmetric function $\frac{(\boldsymbol{\sigma}_+ - \boldsymbol{\sigma}_-)_n}{\sqrt{2}}$ to express the pinning potential $-V_\text{o}\sin\phi$, which however does not correspond to $\phi = 0$. Therefore, we may consider such antisymmetric pinning at $\phi = \frac{\pi}{2}$, for which the pinning potential is defined by the same expression as (6.13) considering ϕ is an angular deviation from $\frac{\pi}{2}$. Point defects, including surface defects as well, can thus be considered as extrinsic perturbations. The phase mode at the threshold of transition can also be regarded as antisymmetric phase pinning. In any case, the magnitude V_o is a significant factor for pinning to occur.

6.5.2 Pinning by an Electric Field

If the order variables are electrically dipolar, the arrangement of $\boldsymbol{\sigma}_n$ can be perturbed by an electric field $\boldsymbol{E}$ applied externally. In this case, the crystal is stabilized with an additional potential energy $-\left(\sum_n \boldsymbol{\sigma}_n\right) \cdot \boldsymbol{E}$. Here, the field $\boldsymbol{E}$, if uniform, is *asymmetrical* at all sites n; hence those $\boldsymbol{\sigma}_n$ modes interacting with the field are expressed by an asymmetric combination $\frac{\boldsymbol{\sigma}_{n,+k}-\boldsymbol{\sigma}_{n,-k}}{\sqrt{2}}$ that is proportional to $\sin\phi_{n,k}$. In this case, the pinning potential can be expressed as

$$V_E(\phi) = -\sigma_o E \sin\phi. \tag{6.14a}$$

However, $V_E(0) = 0$ at the pinning center is $\phi = 0$. The value of this potential can be lowered, if the phase ϕ shifts by $\frac{\pi}{2}$, namely,

$$V_E\left(\phi + \frac{\pi}{2}\right) = -\sigma_o E \sin\left(\phi + \frac{\pi}{2}\right) = -\sigma_o E \cos\phi, \tag{6.14b}$$

giving minimum at $\phi = \frac{\pi}{2}$.

In magnetic resonance studies on the ferroelectric phase transition of TSCC crystals, two fluctuation modes σ_A and σ_P were identified [2]. In the presence of a weak electric field, it was observed that the spectrum of σ_A splits into two lines with increasing E at a temperature below T_c, which was interpreted by a change in the pinning potential from (6.14a) to (6.14b) with shifting ϕ by $\frac{\pi}{2}$ in a thermal relaxation.

6.5.3 Surface Pinning

Surfaces are not defects in lattice, but a pseudospin mode $\boldsymbol{\sigma}_s$ can interact with an antisymmetric field A similar to an electric field, assuming that surface point behaves like a reflecting wall. Assuming perfect reflection, the mode should behave like a standing wave $\frac{(\boldsymbol{\sigma}_{-k}-\boldsymbol{\sigma}_{+k})_{\text{surface}}}{\sqrt{2}}$ inside of a crystal, whereas $(\boldsymbol{\sigma}_+)_{\text{surface}} = 0$ outside, respectively. Therefore, we can write the pinning potential as

$$V_S = -\sigma_o A \sin\phi \quad \text{for} \quad -\pi < \phi \leq 0 \tag{6.15}$$

Thermodynamically, heat exchange with the surroundings can be considered for V_S to associate with a lattice mode. The corresponding lattice displacement $\boldsymbol{u}$ at $\phi = 0$ is responsible for the heat transfer as perturbed usually by scatterings of air particles.

Exercise 6

1. In the Landau theory, the order parameter was assumed to be a scalar quantity. Considering the theory for a vector variable $\boldsymbol{\sigma} = \boldsymbol{\sigma}_o \exp i\phi$, the Gibbs function (6.1) is a function of the magnitude square $\boldsymbol{\sigma}_o^2$. In this case, the theory can disregard the phase ϕ and all space-time uncertainties. Is this theory compatible with the mean-field approximation?

2. Critical fluctuations are observed in two modes, $\boldsymbol{\sigma}_A$ and $\boldsymbol{\sigma}_P$. Experimentally, we know that $\boldsymbol{\sigma}_A$ mode is steady at $\phi = 0$ and T_c with respect to the moving coordinate system at a speed ω/k, whereas the phase of $\boldsymbol{\sigma}_P$ shifts with lowering temperature for $T < T_c$. What is the reason for that? Explain it, using a quartic potential $(b/4)\boldsymbol{\sigma}^4$.
3. Critical fluctuations should be observed in two modes, $\boldsymbol{\sigma}_A$ and $\boldsymbol{\sigma}_P$ of clustered pseudospins, if pinned by the adiabatic potential. Extrinsically pinned pseudospins are either symmetric or antisymmetric, but hence not distinguishable experimentally from $\boldsymbol{\sigma}_A$. Further, $\boldsymbol{\sigma}_P$ is characterized by phase shifting on lowering temperature, while $\boldsymbol{\sigma}_A$ and defect-pinned σ remain unchanged. Discuss these features of pinned pseudospins, regarding the difference between intrinsic and extrinsic pinning.

References

1. L.D. Landau, E.M. Lifshitz, *Statistical physics* (Pergamon Press, London, 1958)
2. M. Fujimoto, S. Jerzak, W. Windsch, Phys. Rev. **B34**, 1668 (1986)

Chapter 7
Pseudospin Correlations

Landau's expansion (6.1) can be interpreted for the adiabatic potential due to pseudospin correlations in a mesoscopic state. Using a truncated potential in one dimension, we obtained longitudinal propagation of classical vectors $\boldsymbol{\sigma}_k$ along the direction of $\boldsymbol{k}$ in crystals. In this chapter, verifying that $\boldsymbol{\sigma}_k$ can be expressed by an elliptic wave, such an expansion is compatible with the condensate model. In this approach, the traditional *long-range order* is an embedded concept in the expansion. For classical displacements in the field of $\boldsymbol{\sigma}_k$, transverse components $\sigma_{k\perp}$ can be responsible for energy dissipation in the thermodynamic environment.

7.1 Propagation of a Collective Pseudospin Mode

Pseudospins $\boldsymbol{\sigma}_n$ and their Fourier transform $\boldsymbol{\sigma}_k$ in a periodic lattice constitute a *vector field*, which is convenient to describe propagation along the direction of $\boldsymbol{k}$. In such a field, a classical propagation can be discussed for a small $|\boldsymbol{k}|$, similar to phonon field.

A collective order variable mode $\boldsymbol{\sigma}_k$ can be in free propagation in unperturbed crystals but modulated when perturbed by an adiabatic potential $\Delta U(\boldsymbol{\sigma}_k) = \frac{1}{2}a\boldsymbol{\sigma}_k^2 + \frac{1}{4}b\boldsymbol{\sigma}_k^4$. Setting the transverse component $\sigma_{k\perp}$ aside, here we consider the longitudinal component σ_k, so denoted for convenience, in the direction of $\boldsymbol{k}$, for propagation. We write the one-dimensional wave equation for σ_k as

$$m\left(\frac{\partial^2}{\partial t^2} - v_0^2\frac{\partial^2}{\partial x^2}\right)\sigma_k(x,t) = -\frac{\partial U_k}{\partial x} = -\sigma_{ko}k\left(a\sigma_k + b\sigma_k^3\right), \qquad (7.1a)$$

M. Fujimoto, *Thermodynamics of Crystalline States*,
DOI 10.1007/978-1-4614-5085-6_7,

where $\sigma_k(x,t)=\sigma(kx-\omega t)$, $v_0=\frac{\omega}{k}$ the speed of propagation in a crystal and the amplitude σ_{ko} is infinitesimal at a small k. Krumhansl and Schrieffer [1, 2] have shown that (7.1a) can be solved by reexpressing it in a simplified form.

$$\frac{d^2Y}{d\phi^2}+Y-Y^3=0, \tag{7.1b}$$

for which we defined rescaled variables

$$Y=\frac{\sigma_k}{\sigma_{ko}} \quad \text{and} \quad \phi=k(x-vt), \tag{7.1c}$$

where

$$\sigma_{ko}=\sqrt{\frac{|a|}{b}},\quad k^2=\frac{|a|}{m(v_0^2-v^2)}=\frac{k_0^2}{1-\frac{v^2}{v_0^2}} \quad \text{and} \quad k_0^2=\frac{|a|}{mv_0^2}.$$

Corresponding to k, the frequency can be written with the general relation $\omega=vk$ and is expressed as

$$\omega^2=v_0^2(k^2-k_0^2), \tag{7.2}$$

which indicates that the propagation is *dispersive*. In (7.2), we notice that $\omega=0$ if $k=k_0$, but for a finite ω, we should have $v<v_o$, for which we have $k>k_o$. Considering the former case for $T>T_c$, the latter can be assigned to the region $T<T_c$. The dispersion relation (7.2) is then attributed to the nonlinear potential proportional to $\sigma_{\mathbf{k}}^3$ in (7.1a).

For a small σ_k, the term Y^3 in (7.1b) can be ignored, in which case (7.1b) is a linear equation whose solution is sinusoidal, that is, $Y=Y_0\sin(\phi+\phi_0)$. Here, Y_0 is infinitesimal, and $\phi_0=0$ can be chosen for the reference phase. The nonlinear equation (7.1b) can then be solved analytically, as shown below.

Integrating (7.1b) once, we obtain

$$2\left(\frac{dY}{d\phi}\right)^2=(\lambda^2-Y^2)(\mu^2-Y^2), \tag{7.3}$$

where

$$\lambda^2=1-\sqrt{1-\alpha^2} \quad \text{and} \quad \mu^2=1+\sqrt{1-\alpha^2}.$$

Here $\alpha=\left(\frac{dY}{d\phi}\right)_{\phi=0}$ is a constant of integration. Writing $\xi=\frac{Y}{\lambda}$ for convenience, (7.3) can be reexpressed in integral form as

$$\frac{\phi_1}{\sqrt{2}\kappa} = \int_0^{\xi_1} \frac{d\xi}{\sqrt{(1-\xi^2)(1-\kappa^2\xi^2)}}, \tag{7.4a}$$

where ξ_1 is the upper limit of the variable ξ , and the corresponding phase ϕ is specified as ϕ_1. The ratio defined by $\kappa = \frac{\lambda}{\mu}$ is called the *modulus* of the *elliptic integral of the first kind* (7.4a). It is noted that the constants λ and μ can be written in terms of the modulus κ, that is,

$$\lambda = \frac{\sqrt{2}\kappa}{\sqrt{1+\kappa^2}} \quad \text{and} \quad \mu = \frac{\sqrt{2}}{\sqrt{1+\kappa^2}}.$$

The inverse function of the integral (7.4a) can be written as

$$\xi_1 = \text{sn}\frac{\phi_1}{\sqrt{2}\kappa}, \tag{7.4b}$$

which is known as *Jacobi's elliptic sn-funcion.* In previous notations, the nonlinear mode σ_1 can be expressed as

$$\sigma_1 = \lambda\sigma_0 \text{sn}\frac{\phi_1}{\sqrt{2}\kappa}, \tag{7.5}$$

which is an elliptical wave modified from a sinusoidal wave by λ, μ, κ, and α. Also significant is that the amplitude $\lambda\sigma_0$ and phase ϕ_1 are both determined by the integral in (7.4a) that is typical for nonlinear waves.

Using an angular variable Θ defined by $\xi = \sin\Theta$, (7.4a) is expressed as

$$\frac{\phi_1}{\sqrt{2}\kappa} = \int_0^{\Theta_1} \frac{d\Theta}{\sqrt{1-\kappa^2\sin^2\Theta}}, \tag{7.6a}$$

where Θ_1 represents an *effective sinusoidal phase* of ξ_1. This ξ_1 is a useful variable to express the propagation in comparison of a sinusoidal function, that is,

$$\text{sn}\frac{\phi_1}{\sqrt{2}\kappa} = \sin\Theta_1. \tag{7.6b}$$

In this way, σ_1 can be regarded as the longitudinal component of a classical vector, that is, $\sigma_1 = \lambda\sigma_0 \sin\Theta_1$, with effective amplitude $\lambda\sigma_0$. Jacobi called the angle Θ_1 as the *amplitude function* defined as $\Theta_1 = \text{am}\left(\int_0^{\Theta_1} \frac{d\Theta}{\sqrt{1-\kappa^2\sin^2\Theta}}\right)$; we, nevertheless, use the definition that signifies *effective phase angle*. Figure 7.1 shows the relation

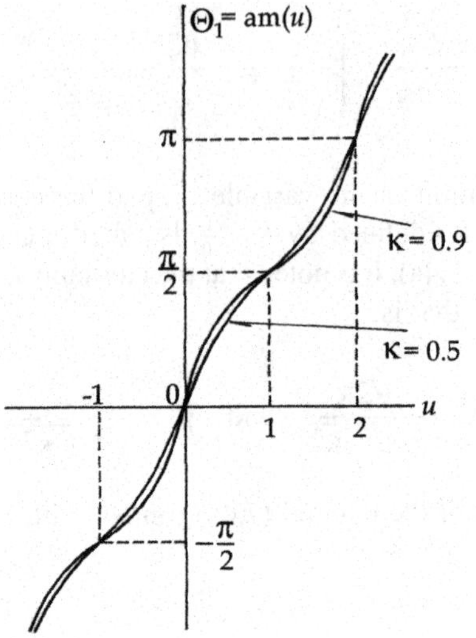

Fig. 7.1 Effective phase $\Theta_1(u)$ (Jacobi's amplitude function) versus u.

between the elliptic integral $u(\kappa) = \int_0^{\Theta_1} \frac{d\Theta}{\sqrt{1-\kappa^2\sin^2\Theta}}$ and Jacobi's $\Theta_1 = \mathrm{am}\, u$, plotted for $\kappa = 0.5$ and 0.9 in particular.

Except for $\kappa = 1$, the sn-function (7.6b) is periodic, whose periodicity is expressed by the repeating unit determined by the difference between $\Theta_1 = 0$ and $\Theta_1 = 2\pi$. The integral

$$K(\kappa) = \int_0^{\frac{\pi}{2}} \frac{d\Theta}{\sqrt{1-\kappa^2\sin^2\Theta}} \tag{7.7}$$

is called the *complete elliptic integral*. We consider that long-range correlations are included implicitly in (7.5) for $0 < \kappa < 1$. The range of $0 < \Theta < \pi$ is defined as the *period* of sn-function, that is, $4K(\kappa)$.

For $\kappa = 1$ or $\lambda = \mu = 1$, (7.6a) can specifically be integrated as

$$\sigma_1 = \sigma_0 \tanh \frac{\phi_1}{\sqrt{2}}, \tag{7.8}$$

showing $\sigma_1 \rightarrow \sigma_0$ in the limit of $\phi_1 \rightarrow \infty$. Figure 7.2 illustrates the curves of (7.5) and (7.8) for representative values of the modulus κ.

Although not immediately clear, such σ_1 as expressed by (7.8) corresponds to a completely ordered state, as inferred from $\kappa = 1$, for which $\sigma_1 = \sigma_0$ at all angles, except for $\phi_1 = 0$. Accordingly, we may consider the parameter in the range $0 \leq \kappa \leq 1$ to express the degree of *long-range order* in crystals.

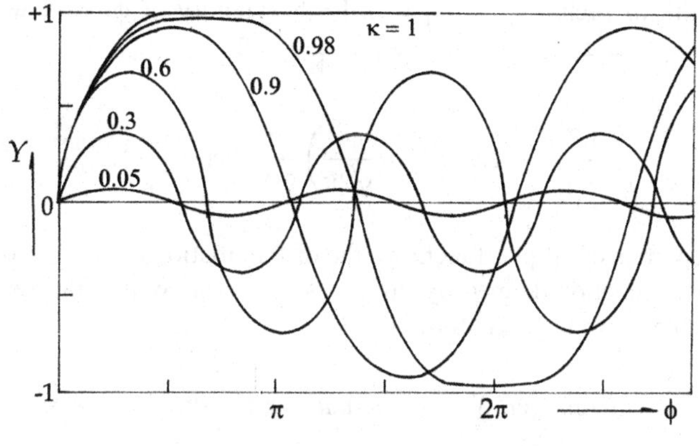

Fig. 7.2 Elliptic functions $\mathrm{sn}(\phi/\sqrt{2}\kappa)$ plotted against ϕ for various values of the modulus κ.

7.2 Transverse Components and the Cnoidal Potential

In the above, the amplitude σ_o is left undetermined, which should, however, take a finite value, if considering thermal interactions with the surroundings at a given temperature. Leaving the temperature dependence to later discussions, for the periodic solution (7.5) for $0<\kappa<1$, we have $\sigma_1 = \lambda\sigma_o \sin\Theta_1$, representing the longitudinal component of a vector $\lambda\boldsymbol{\sigma}_o$. Therefore, we can logically consider the transverse component $\sigma_{1\perp}$ that can be written as

$$\sigma_{1\perp} = \lambda\sigma_o \cos\Theta_1 = \lambda\sigma_o \mathrm{cn}\frac{\phi_1}{\sqrt{2}\kappa} \quad \text{for} \quad 0<\kappa<1, \tag{7.9a}$$

where we have the amplitude relation $\sigma_1^2 + \sigma_{1\perp}^2 = \lambda^2\sigma_o^2$.

Equation (7.8) for $\kappa = 1$, the transverse component can be expressed as

$$\sigma_{1\perp} = \sigma_o \mathrm{sech}\frac{\phi_1}{\sqrt{2}} \quad \text{for} \quad \kappa = 1, \tag{7.9b}$$

and we have $\sigma_1^2 + \sigma_{1\perp}^2 = \sigma_o^2$.

Owing to the transverse component of longitudinally correlated pseudospins, we can consider an adiabatic potential $\Delta U(\sigma_{1\perp}) = \frac{1}{2}a\sigma_{1\perp}^2$, assuming that there is no *cross correlations* between adjacent chains in a crystal. On the other hand, reversing σ_1 across a perpendicular plane for $\phi_1 = 0$ and $n\pi$ (n is integers), a work by a force $F = -\frac{\partial\Delta U}{\partial x} = -k\frac{\partial\Delta U}{\partial\phi_1}$ is required, which can be calculated as $W = -\int_{-\delta}^{+\delta} \sigma_1 \frac{\partial\Delta U}{\partial\phi_1} d\phi_1$ over the region $-\delta<\phi_1<+\delta$, where 2δ is the effective width of sech h ϕ_1 at $\phi_1 = 0$. In the vicinity of $\phi_1 = 0$, we consider $\sigma_1(\phi_1) = \sigma_{1\perp}(\phi_1 + \frac{\pi}{2}) \approx \lambda\sigma_o$ and $\frac{\partial \Delta U}{\partial \phi_1} \propto \frac{d}{d\phi_1} \mathrm{cn}^2 \frac{\phi_1}{\sqrt{2}\kappa}$, so that

$$\frac{dW}{d\phi_1} = -2\lambda^2\sigma_o^2 \left(\mathrm{cn}\frac{\phi_1}{\sqrt{2}\kappa}\right) \frac{d}{d\phi_1}\left(\mathrm{cn}\frac{\phi_1}{\sqrt{2}\kappa}\right).$$

Using properties of elliptic functions, the differentiation can be performed with another elliptic function defined by $\mathrm{dn}^2 u = 1 - \kappa^2 \mathrm{sn}^2 u$. With a dn-function for a general variable u, we have the formula

$$(\mathrm{sn}u)' = \mathrm{cn}u\ \mathrm{dn}u, (\mathrm{cn}u)' = -\mathrm{sn}u\ \mathrm{dn}u \quad \text{and} \quad (\mathrm{dn}u)' = -\kappa^2 \mathrm{sn}u\ \mathrm{cn}u$$

as listed in Appendix. Letting $u = \frac{\phi_1}{\sqrt{2}\kappa}$, with these formula the above dW can be written as

$$\frac{dW}{d\phi_1} = \frac{1}{\sqrt{2}\kappa}\frac{dW}{du} = \frac{2\lambda^2\sigma_o^2}{\sqrt{2}\kappa}\mathrm{cn}u\ \mathrm{sn}u\ \mathrm{dn}u = -\frac{2\lambda^2\sigma_o^2}{\sqrt{2}\kappa^3}(\mathrm{dn}u)(\mathrm{dn}u)'$$
$$= -\frac{\lambda^2\sigma_o^2}{\kappa^2}\frac{d}{d\phi_1}\mathrm{dn}^2\frac{\phi_1}{\sqrt{2}\kappa}.$$

Expressing W for the adiabatic potential, we have

$$\Delta U(\kappa) = \mu^2\sigma_o^2 \mathrm{dn}^2 \frac{\phi_1}{\sqrt{2}\kappa} \quad \text{for} \quad 0 < \kappa < 1. \tag{7.10a}$$

A potential proportional $\mathrm{dn}^2 u$ shows the same wave form as $\mathrm{sn}^2 u$ and $\mathrm{cn}^2 u$, so that (7.10a) is generally referred to as a *cnoidal potential*. On the other hand, for $\kappa = 1$, $\mathrm{dn}u = \mathrm{sech}u$.

$$\Delta U(\kappa = 1) = \sigma_o^2 \mathrm{sech}^2 \frac{\phi_1}{\sqrt{2}} \quad \text{for} \quad \kappa = 1. \tag{7.10b}$$

In this case, $\sigma_{1\perp}$approaches $\pm\ \sigma_o \times \infty$ in the limit of $\phi_1 \to \infty$, which can be assigned to a *domain boundary* where the pseudospin direction is reversed. These boundaries at $\phi_1 = 0$ and ∞ are identical, as indicated by inversion symmetry. In Figure 7.3a–c, the curves of $\boldsymbol{\sigma}_1, \sigma_1,$ and $\sigma_{1\perp}$ near the boundary are plotted, respectively, against ϕ to illustrate their behaviors.

From the foregoing, the variable $\boldsymbol{\sigma}_k(\phi)$ is clearly a classical vector, characterized by longitudinal and transverse components, where the amplitude and phase are specified by the parameter κ.

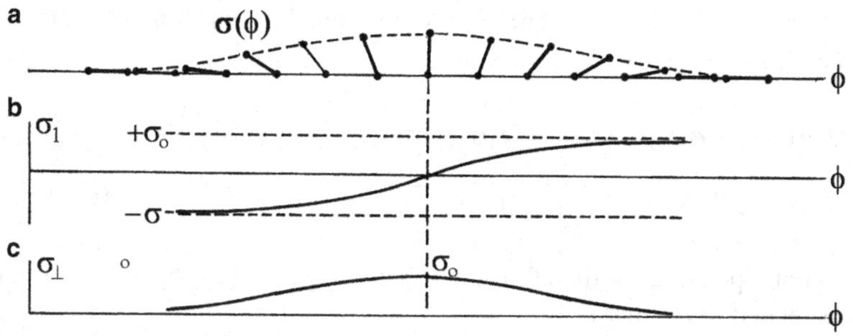

Fig. 7.3 A pseudospin mode along x direction for $\kappa = 1$. (**a**) Distributed mode vectors in the x–z plane. (**b**) The longitudinal component $\sigma_1(\phi)$. (**c**) The transverse component $\sigma_{1\perp}(\phi)$.

7.3 The Lifshitz Condition for Incommensurability

The elliptic function for $0 < \kappa < 1$ is periodic, but its period is not the same as lattice periodicity. Such a pseudospin mode is generally *incommensurate* with the lattice period, although commensurate in some cases. In an equilibrium crystal, the mode is pinned in phase with the adiabatic potential; however, at pinning, we should consider fluctuations due to phase uncertainties. Assuming fluctuations, Lifshitz derived the incommensurability condition for thermodynamic quantities, providing a significant criterion for mesoscopic states [3].

For practical analysis, such mesoscopic pseudospin modes as expressed by elliptic and hyperbolic functions can be interpreted conveniently by components

$$\sigma_1 = \sigma_o \sin\Theta_1 \quad \text{and} \quad \sigma_\perp = \sigma_o \cos\Theta_1$$

of a classical vector $\boldsymbol{\sigma} = (\sigma_1, \sigma_\perp)$, where σ_o and $\Theta_1(x,t)$ are the amplitude and phase. For condensates, these variables are observed as temperature dependent because the corresponding lattice displacements $(u_1,\ u_\perp)$ can scatter phonons inelastically. It is noted that the phase variable $\Theta_1(x,\ t)$ is periodic in the lattice with the period $4K(\kappa)$, which is not necessarily in phase with the lattice. In any case, the observed pseudospins can be expressed with temperature-dependent amplitude and phase. The Gibbs potential $G(p,T;\boldsymbol{\sigma})$ can therefore be specified by such a mesoscopic variable $\boldsymbol{\sigma}$, or by σ_o and Θ_1, taking a minimum value in equilibrium states. Assuming the amplitude σ_o as constant, correlations between pseudospins can be specified by different phases.

Lifshitz considered statistically such correlation energies as determined by $J_{ij}\boldsymbol{\sigma}_i.\boldsymbol{\sigma}_j$ to derive the thermodynamic condition for incommensurability. Such time-dependent quantities in thermodynamic experiments are usually averaged out in a long timescale of observation t_o. For sufficiently short t_o, in contrast, we observed them in broadened functions of x, varying between x_i and x_j on the longitudinal axis, as signified by δx

$= x_i - x_j$ and $x = \frac{1}{2}(x_i + x_j)$. The observed correlation function can therefore be written as proportional to

$$\langle \sigma^*(x_i)\sigma(x_j) + \sigma(x_i)\sigma^*(x_j)\rangle_t = 2\langle \sigma^*(x)\sigma(x)\rangle_t + \left\langle \sigma^*(x)\frac{\partial\sigma(x)}{\partial x} - \sigma(x)\frac{\partial\sigma*(x)}{\partial x}\right\rangle_t \delta x$$
$$+\left\langle \sigma^*(x)\frac{\partial\sigma(x)}{\partial x} + \sigma(x)\frac{\partial\sigma^*(x)}{\partial x}\right\rangle_t (\delta x)^2 + \left\langle \frac{\partial\sigma^*(x)}{\partial x}\frac{\partial^2\sigma(x)}{\partial x^2} - \frac{\partial\sigma(x)}{\partial x}\frac{\partial^2\sigma^*(x)}{\partial x^2}\right\rangle_t (\delta x)^3 + \ldots .$$

The Gibbs potential is therefore contributed significantly by such correlation terms as related to δx, that is,

$$G_L = \frac{iD}{2}\int_0^L \left\langle \sigma^*\frac{\partial\sigma}{\partial x} - \sigma\frac{\partial\sigma^*}{\partial x}\right\rangle_t \frac{dx}{L} + \frac{iD}{2}\int_0^L \left\langle \frac{\partial\sigma^*0}{\partial x}\frac{\partial^2\sigma}{\partial x^2} - \frac{\partial\sigma}{\partial x}\frac{\partial^2\sigma^*}{\partial x^2}\right\rangle_t (\delta x)^2\frac{dx}{L} + \ldots, \tag{7.11}$$

where $\frac{iD}{2}$ is defined as a constant proportional to δ_x. We can then write

$$G(\sigma) = G(0) + \int_0^L \left\langle \frac{mv_o^2}{2}\left|\frac{\partial\sigma}{\partial x}\right|^2 + \frac{a}{2}|\sigma|^2 + \frac{b}{4}|\sigma|^4\right\rangle_t \frac{dx}{L} + G_L, \tag{7.12}$$

where the upper limit L represents a sampling length of the mesoscopic σ as related to practical experiments.

Writing $\sigma = \sigma_o \exp i\phi, G(\sigma)$ as a function of σ_0 and ϕ, we have (7.12) reexpressed as

$$G(\sigma_0,\phi) = G(0) + \int_0^L \left\langle \frac{a\sigma_o^2}{2} + \frac{b\sigma_o^4}{4} + \frac{mv_o^2}{2}\left(\frac{d\sigma_o}{dx}\right)^2 + \frac{mv_o^2\sigma_o^2}{2}\left(\frac{d\phi}{dx}\right)^2\right.$$
$$\left. + D\sigma_o^2\frac{d\phi}{dx}\left(1 + \delta x^2\frac{d^2\phi}{dx^2}\right)\right\rangle_t \frac{dx}{L},$$

which is minimized for equilibrium by setting $\frac{\partial G}{\partial \sigma_o} = 0$ and $\frac{\partial G}{\partial \phi} = 0$ simultaneously. Carrying out these partial differentiations of $G(\sigma_o,\ \phi)$, we obtain the equations

$$a\sigma_o + b\sigma_o^3 + mv_o^2\frac{d^2\sigma_o}{dx^2} + mv_o^2\left(\frac{d\sigma_o}{dx}\right)^2 + 2D\sigma_o\frac{d\phi}{dx}\left(1 + \delta x^2\frac{d^2\phi}{dx^2}\right) = 0$$

and

$$\left(mv_o^2\frac{d\phi}{dx} + D\sigma_o^2\right)\frac{d}{d\phi}\left\{\frac{d\phi}{dx}\left(1 + \delta x^2\frac{d^2\phi}{dx^2}\right)\right\} = 0.$$

We see immediately that the second equation is fulfilled if

$$\frac{d\phi}{dx} = -\frac{D}{mv_o^2}. \tag{7.13}$$

Therefore, setting $-\frac{D}{mv_o^2} = q$, (7.13) represents the wavenumber of fluctuations defined independently from the lattice, because both D and mv_o^2 are unrelated constants with the lattice, giving an incommensurate wavevector. As σ_o is assumed as constant, the first relation can be solved for σ_o^2, that is,

$$\sigma_o^2 = -\frac{a - \dfrac{D}{mv_o^2}}{b}.$$

The wave $\sigma_o \exp iqx$ is therefore incommensurate if $D \neq 0$, which is known as the Lifshitz theorem. Incommensurability originates, in fact, from a nonvanishing displacement $\langle \delta x \rangle_t$ of lattice points; hence, $D \neq 0$ is an obvious consequence.

7.4 Pseudopotentials

The space group can be modified with additional translational symmetry, signifying a periodic rotation of constituents along a symmetry axis, for instance, rotations over m-times of the lattice constant [4]. On a so-called *screw axis* of m-fold rotation, we can consider a commensurate potential V_m^L, where correlated pseudospins $\sigma(\phi)$ can be pinned, if the spatial phase of ϕ matches such lattice spacing. Assuming that a lattice with such *pseudosymmetry* is signified by m-times rotations at lattice points

$$u = u_o \exp(\pm i\theta_p), \text{where} \quad \theta_p = \frac{2\pi}{m}p \quad \text{and} \quad p = 0, 1, 2, \ldots, m-1 \tag{7.14}$$

along the m-fold screw axis, we can consider the potential

$$V_m^L(\theta_0, \theta_1, \ldots, \theta_{m-1}) \propto \sum_p u_p = u_o \sum_p \{\exp i\theta_p + \exp(-i\theta_p)\} = 2u_o \sum_p \cos\theta_p.$$

Here, these angles θ_p can be reexpressed by the lattice coordinates $x_p = pa_o$ combined with (7.14); $\theta_p = \frac{2\pi}{m}\frac{x_p}{a_o} = G_m x_p$, where $G_m = \frac{2\pi}{ma_o}$, and therefore, we have $V_m^L \propto \sum_p \cos(G_m x_p)$. If a pseudospin mode $\sigma = \sigma_o \exp i\phi$ is pinned by the adiabatic potential $V_m(\phi)$ corresponding to V_m^L, we should have a phase matching $G_m x_p = m\phi$, so that

$$\cos(\mathrm{m}\phi) = \frac{1}{2}\{\exp(i\mathrm{m}\phi) + \exp(-i\mathrm{m}\phi)\} = \frac{1}{2\sigma_\mathrm{o}^\mathrm{m}}(\sigma^\mathrm{m} + \sigma^{-\mathrm{m}}).$$

Therefore, the potential $V_\mathrm{m}(\phi)$ is characteristically a function of $\sigma(\phi)$ and expressed as

$$V_\mathrm{m}(\phi) = \frac{2\rho}{\mathrm{m}}(\sigma^\mathrm{m} + \sigma^{-\mathrm{m}}) = \frac{2\rho\sigma_\mathrm{o}^\mathrm{m}}{\mathrm{m}}\cos(\mathrm{m}\phi), \tag{7.15}$$

where ρ is the proportionality constant.

The Gibbs function can then be written as

$$G(\sigma) = \int_0^L \left\{\frac{a\sigma^*\sigma}{2} + \frac{b(\sigma^*\sigma)^2}{4} + \frac{mv_\mathrm{o}^2}{2}\frac{\partial\sigma^*}{\partial x}\frac{\partial\sigma}{\partial x} + V_\mathrm{m}(\phi)\right\}\frac{\mathrm{d}x}{L}.$$

Replacing the integrand by its time average $\langle.....\rangle_t$, we have

$$G(\sigma_\mathrm{o},\phi) = \int_0^L \left\langle\frac{a\sigma_\mathrm{o}^2}{2} + \frac{b\sigma_\mathrm{o}^4}{4} + \frac{mv_\mathrm{o}^2}{2}\left(\frac{\partial\sigma_\mathrm{o}}{\partial x}\right)^2 + \frac{mv_\mathrm{o}^2\sigma_\mathrm{o}^2}{2}\left(\frac{\partial\phi}{\partial x}\right)^2 + \frac{2\rho\sigma_\mathrm{o}^\mathrm{m}}{\mathrm{m}}\cos(\mathrm{m}\phi)\right\rangle_\mathrm{t}\frac{\mathrm{d}x}{L}.$$

Setting $\frac{\partial G(\sigma_\mathrm{o},\phi)}{\partial\sigma_\mathrm{o}} = 0$ and $\frac{\partial G(\sigma_\mathrm{o},\phi)}{\partial\phi} = 0$ for minimum of $G(\sigma_\mathrm{o},\phi)$, we obtain

$$a\sigma_\mathrm{o} + b\sigma_\mathrm{o}^3 + 2\rho\sigma_\mathrm{o}^{\mathrm{m}-1}\cos(\mathrm{m}\phi) + mv_\mathrm{o}^2\left\{\sigma_\mathrm{o}\left(\frac{\mathrm{d}\phi}{\mathrm{d}x}\right)^2 + \frac{\mathrm{d}^2\sigma_\mathrm{o}}{\mathrm{d}x^2}\right\} = 0 \tag{i}$$

and

$$mv_\mathrm{o}^2\sigma_\mathrm{o}^2\frac{\mathrm{d}^2\phi}{\mathrm{d}x^2} - 2\rho\sigma_\mathrm{o}^\mathrm{m}\sin(\mathrm{m}\phi) = 0. \tag{ii}$$

Integrating the second equation, (ii) can be modified as $\frac{1}{2}mv_\mathrm{o}^2\sigma_\mathrm{o}^2\left(\frac{\mathrm{d}\phi}{\mathrm{d}x}\right)^2 + V_\mathrm{m}(\phi)$ = const., expressing a conservation law that is analogous to a simple pendulum with a finite amplitude. Using the abbreviations $\Psi = \mathrm{m}\phi$ and $\zeta = \frac{2\mathrm{m}\rho\sigma_\mathrm{o}^{\mathrm{m}-2}}{mv_\mathrm{o}^2}$, (ii) be simplified as

$$\frac{\mathrm{d}^2\Psi}{\mathrm{d}x^2} - \zeta\sin\Psi = 0. \tag{7.16a}$$

This is standard form of the *sine–Gordon equation*. For the conservation law, (7.16a) can be integrated as

$$\frac{1}{2}\left(\frac{d\Psi}{dx}\right)^2 - \zeta\cos\Psi = E, \tag{7.16b}$$

where the constant E represents the energy of an equivalent pendulum. On the other hand, the equation (i) determines the amplitude σ_0, which is not constant as in a simple pendulum, depending on the phase Ψ. Such a phase-related amplitude is typical for a nonlinear oscillation.

Equation (7.16b) can be written in an integrated form

$$x - x_0 = \int_0^{\Psi_1} \frac{d\Psi}{\sqrt{2(E + \zeta\cos\Psi)}},$$

where the upper and lower limit Ψ_1 and 0 correspond to coordinates x and x_0. This integral can be expressed by the elliptic integral of the first kind if the modulus κ is defined by $\kappa^2 = \frac{2\zeta}{E+\zeta}$, that is, if $\kappa^2 < 1$, then

$$x - x_0 = \frac{\kappa}{\sqrt{\zeta}} \int_0^{\Theta_1} \frac{d\Theta}{\sqrt{1 - \kappa^2 \sin^2\Theta}},$$

where $\Theta = \frac{\Psi}{2}$. Therefore,

$$\sin\Theta_1 = \mathrm{sn}\frac{\sqrt{\zeta}(x - x_0)}{\kappa} \quad \text{for} \quad 0 < \kappa < 1,$$

and

$$\sin\Theta_1 = \tanh\sqrt{\zeta}(x - x_0) \quad \text{for} \quad \kappa = 1.$$

To discuss the phase-matching condition, it is convenient to define $x - x_0 = \Lambda(\kappa)$ corresponding to the phase difference between $\Theta_1 = 0$ and $\Theta_1 = \pi$, that is,

$$\Gamma(\kappa) = \frac{\kappa}{\sqrt{\zeta}} \int_0^{\pi} \frac{d\Theta}{\sqrt{1 - \kappa^2 \sin^2\Theta}} = \frac{2\kappa K(\kappa)}{\sqrt{\zeta}}, \tag{7.17}$$

where $K(\kappa)$ is the complete elliptic integral of the first kind as defined in (7.7). Phase matching with the pseudopotential $V_m^L = V_0 \sum_p \cos(G_m x_p)$ may be stated by $G_m x_p = \Theta_p = \frac{2\pi p}{m}$, assuming that the pseudospin wave reflects perfectly on the potential barrier V_0. Given by (7.15), V_0 behaves not only as a simple *resistive* reflector but also as *inductive*, resulting in a phase shift. At the transition

Fig. 7.4 Phase variables of a collective mode pinned by a pseudopotential $V_m(\phi)$.

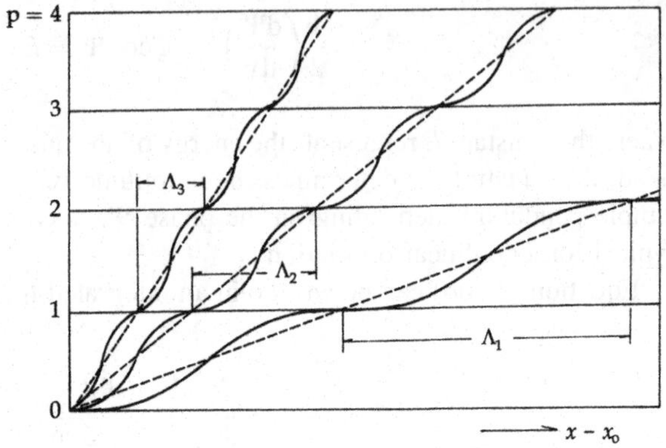

temperature, quantum-mechanical uncertainties at lattice sites can be considered as responsible for such a shift, and the pseudopotential can be expressed as

$$V_m^L = -V_o \sum_p \cos\{\Theta_p(1-\delta_p)\}, \tag{7.18}$$

where δ_p is called the *incommensurate parameter*, exhibiting transition anomalies due to uncertain phase fluctuations at p.

Figure 7.4 illustrates the behavior of Θ_p, as described by $\sin\frac{2\pi p}{m} = \mathrm{sn}\frac{\sqrt{\zeta}(x-x_o)}{\kappa}$, against $x-x_o$, showing characteristic lengths $\Gamma(\kappa)$ that determined by (7.17). In the above, we assumed that $\kappa^2 < 1$, which means $E < -\zeta$ by definition. On the other hand, if $E > -\zeta$, we have a case for $\kappa^2 > 1$. If ζ can be ignored, the sine–Gordon equation can be reduced to an equation for free motion; no pinning occurs if $V_m(\phi)$ is disregarded. In this context, the change from $\kappa > 1$ to $\kappa < 1$ with lowering temperature signifies a transition to a mesoscopic phase in commensurate structure.

Although one-dimensional correlations are a valid assumption for anisotropic crystals, experimentally such a model should be evaluated on sample crystals of high quality characterized by a small defect density. For such phase-locking phase transitions in K_2ZnCl_4 and Rb_2ZnCl_4 crystals, Pan and Unruh [5] reported laminar patterns of such *discommensurattion* lines perpendicular to the x direction after transitions recorded by transmission electron microscopy (TEM). Although there were additional "splittings" and "vortex-like" patterns, the dominant laminar structure in the dark fields is undeniable evidence for pseudopotential as discussed above.

Exercise 7

1. Derive (7.1a) from the wave equation

$$m\left(\frac{\partial^2}{\partial t^2} - v_o^2\frac{\partial^2}{\partial x^2}\right)\sigma_k(x,t) = -k\left(a\sigma_k + b\sigma_k^3\right)$$

using rescaled variables (7.1b).

2. The Lifshitz' theory is based on statistical correlations whose energy is minimized by the variation principle, for which arbitrary variations can be considered as related to thermal fluctuations. If Lifshitz' calculation can be repeated with this interpretation, the incommensurability should be consistent with observed results obtained with varying temperature. Discuss the validity of Lifshitz' theory for thermal fluctuations.
3. Pinning by a pseudosymmetry potential (7.18) is regarded as a transition from incommensurate to commensurate phases. Discuss the reason why such a pinning can be regarded for a thermodynamic phase transition.

References

1. J.A. Krumshansl, J.R. Schrieffer, Phys. Rev. **B11**, 3535 (1975)
2. S. Aubry, J. Chem. Phys. **64**, 3392 (1976)
3. R. Blinc, A.P. Levanyak, *Incommensurate Phases in Dielectrics* (North-Holland, Amsterdam, 1986)
4. H.D. Megaw, *Crystal Structures: A Working Approach* (Saunders, Philadelphia, 1973)
5. X. Pan, H.-G. Unruh, J. Phys. Condens. Matter **2**, 323 (1990)

Chapter 8
Soliton Theory of Long-Range Order

In the foregoing, we considered for the adiabatic potential to originate from pseudospin correlations in the crystal space. In this chapter, the correlation field is discussed mathematically, where pseudospin waves are expressed by elliptic functions determined by solutions of the Korteweg–deVries equation. The concept of *long-range order* in crystalline states is embedded in the solutions, which can be signified as dissipative in thermodynamic environments. In addition, the longitudinal solution of collective pseudospins is dispersive, implying interactions with neighboring modes. In this chapter, we discuss long-range order of cluster arrangements, as inferred from soliton solution subjected to phonon scatterings. For mathematical theories in this chapter, I owe greatly to Professor G. L. Lamb Jr. [1].

8.1 A Longitudinal Dispersive Mode of Collective Pseudospins

In the expression $\boldsymbol{\sigma}_k = \boldsymbol{\sigma}_o \exp i\phi$ for a collective cluster mode at a small $|k|$, the phase ϕ is considered as independent of the amplitude $\boldsymbol{\sigma}_o$. However for a finite amplitude $\boldsymbol{\sigma}_o$, the propagation of the classical vector $\boldsymbol{\sigma}_k$ is *dispersive* and *dissipative* in a crystal, where the amplitude depends on the speed of propagation v that is not simply proportional to $|k|$. In addition, there should be an internal *force* to drive it, where the transversal component $\sigma_\perp$ cannot be zero.

We consider such mesoscopic variables as $\boldsymbol{\sigma}_k$ are represented by a classical *vector field* in crystal space, whose propagation at long wavelengths can be described by analogy of a continuous fluid. In this approach, we can consider the pseudospin density $\rho = |\boldsymbol{\sigma}_k|^2$ along the direction of propagation x and its conjugated current density $j = \rho v$, representing a continuous flow at a constant speed v under a constant pressure p. On the other hand, if there is a pressure gradient $-\frac{\partial p}{\partial x}$ along the direction x, the speed v cannot be constant, resulting in speed-dependent amplitude and phase. For a condensate, the gradient $-\frac{\partial p}{\partial x}$ can be assumed to arise

M. Fujimoto, *Thermodynamics of Crystalline States*,
DOI 10.1007/978-1-4614-5085-6_8,

from the force $-\frac{\partial \Delta U}{\partial x}$ of the adiabatic potential ΔU, in which case we can write the equation

$$\frac{\partial v}{\partial t}+v\frac{\partial v}{\partial x}=-\frac{1}{\rho_o}\frac{\partial p}{\partial x} \tag{i}$$

for the variable speed v and constant $\rho_o = \sigma_o^2$, if the nonlinear term $v\frac{\partial v}{\partial x}$ in (i) can be ignored. In general, the density ρ is not confined to the x-axis, spreading transversally as well, so we can write $\rho_o = \rho_x + \rho_z$, assuming ρ_o as a function of x and z. Disregarding the y-direction for simplicity, the equations of continuity can be written as

$$\frac{\partial \rho_x}{\partial t}+\rho_o\frac{\partial v}{\partial x}=0 \quad \text{and} \quad \frac{\partial \rho_z}{\partial t}=0. \tag{ii}$$

The density ρ_x represents a flow along the x-axis, assuming that ρ_z does not propagate in the z-direction. We assume however that the density ρ_z remains in the vicinity of the x-axis, satisfying the equation

$$\frac{\partial^2 \rho_z}{\partial t^2}+\alpha(\rho_z-\rho_o)=p, \tag{iii}$$

where α is a restoring constant.

Equation (i) is nonlinear because of the term $v\frac{\partial v}{\partial x}$, whereas (ii) and (iii) are linear equations. Equation (i) can be linearized, if $v\frac{\partial v}{\partial x}$ is considered as a perturbation. In this case, the components ρ_x and ρ_z are both functions of x and z, but we assume that the total density is unchanged, i.e., $|\rho_x| = |\rho_z| = \rho_o$.

Introducing a set of reduced variables $x' = \sqrt{\alpha}x$, $t' = \sqrt{\alpha}t$, and $\rho' = \frac{\rho-\rho_o}{\rho_o}$, (i)–(iii) can be linearized as

$$\frac{\partial v}{\partial t'}+\frac{\partial p'}{\partial x'}=0, \quad \frac{\partial \rho'}{\partial t'}-\frac{\partial v}{\partial x'}=0 \quad \text{and} \quad \frac{\partial^2 \rho'}{\partial t^2}+\rho'=p', \tag{iv}$$

where we wrote as $p' = \frac{p}{\rho_o}$ for simplicity. Assuming that these variables ρ', v, p' are all proportional to $\exp i(kx' - \omega t')$ in the first approximation for $\omega = vk$, we can obtain from (iv) an equation for their amplitudes. From this, we can derive the dispersion relation

$$\omega^2 = \frac{k^2}{1+k^2}$$

which can be approximated as $\omega \approx k - \frac{1}{2}k^3$ for a small k, indicating that the speed $v = \omega/k$ is no longer constant. Writing $k(x' - t') = \xi$ and $\frac{1}{2}k^3 t' = \tau$ in the expression

$\exp i(kx' - \omega t') = \exp i\{k(x' - t') + \frac{1}{2}k^3 t'\}$, we perform a coordinate transformation from (x', t') to (ξ, τ) by differential relations:

$$\frac{\partial}{\partial x'} = k\frac{\partial}{\partial \xi} \quad \text{and} \quad \frac{\partial}{\partial t'} = -k\frac{\partial}{\partial \xi} + \frac{1}{2}k^3\frac{\partial}{\partial \tau}. \tag{v}$$

Expressing (i) and (ii) in terms of x', t' and p', we obtain

$$\frac{\partial v}{\partial t'} + v\frac{\partial v}{\partial x'} + \frac{\partial p'}{\partial x'} = 0 \quad \text{and} \quad \frac{\partial \rho'}{\partial t'} + \frac{\partial v}{\partial x'} + \frac{\partial(\rho' v)}{\partial x'} = 0.$$

Combining the second relation with (iv), these equations can be transformed by (v) to express with coordinates ξ and τ. That is,

$$-\frac{\partial v}{\partial \xi} + \frac{1}{2}k^2\frac{\partial v}{\partial \tau} + v\frac{\partial v}{\partial \xi} + \frac{\partial p'}{\partial \xi} = 0$$

$$-\frac{\partial \rho'}{\partial \xi} + \frac{1}{2}k^2\frac{\partial \rho'}{\partial \tau} + \frac{\partial v}{\partial \xi} + \frac{\partial(\rho' v)}{\partial \xi} = 0$$

$$p' = \rho' + k^2\frac{\partial^2 \rho'}{\partial \xi^2} - k^4\frac{\partial^2 \rho'}{\partial \xi \partial \tau} + \frac{1}{4}k^6\frac{\partial^2 \rho'}{\partial \tau^2}, \tag{vi}$$

and

$$p' = \rho' + k^2\frac{\partial^2 \rho'}{\partial \xi^2} - k^4\frac{\partial^2 \rho'}{\partial \xi \partial \tau} + \frac{1}{4}k^6\frac{\partial^2 \rho'}{\partial \tau^2}.$$

Here, quantities ρ', p', and $v - v_0$ emerging at T_c can be expressed in power series with respect to small k^2 in *asymptotic* form:

$$\rho' = k^2 \rho_1{}' + k^4 \rho_2{}' + \cdots,$$

$$p' = k^2 p_1{}' + k^4 p_2{}' + \cdots,$$

and

$$v - v_0 = k^2 v_1 + k^4 v_2 + \cdots$$

Substituting these expansions for ρ', p' and v' in (vi), we compare terms of $k^2, k^4, \ldots$ separately. From the factors proportional to k^2, we obtain

$$-\frac{\partial \rho_1{}'}{\partial \xi}+\frac{\partial v_1}{\partial \xi}=0, \quad -\frac{\partial v_1}{\partial \xi}+\frac{\partial p_1{}'}{\partial \xi}=0 \quad \text{and} \quad p_1{}'=\rho_1{}'.$$

Combining the first and second relations, the third one can be expressed as

$$p_1{}'=\rho_1{}'=v_1+\varphi(\tau), \tag{vii}$$

where $\varphi(\tau)$ is an arbitrary function of τ.

Next, comparing coefficients of the k^4 terms, we obtain

$$-\frac{\partial \rho_2{}'}{\partial \xi}+\frac{1}{2}\frac{\partial \rho_1{}'}{\partial \tau}+\frac{\partial v_2}{\partial \xi}+\frac{\partial(\rho_1{}'v_1)}{\partial \xi}=0,$$

$$-\frac{\partial v_2}{\partial \xi}+\frac{1}{2}\frac{\partial v_1}{\partial \tau}+v_1\frac{\partial v_1}{\partial \xi}+\frac{\partial p_2{}'}{\partial \xi}=0,$$

and

$$p_2{}'=\rho_2{}'+\frac{\partial^2 \rho_1{}'}{\partial \xi^2}.$$

Eliminating $\rho_2{}'$, v_2, and $p_2{}'$ from these equations, we arrive at the relations for $\rho_1{}'$, v_1, and $p_1{}'$, i.e.,

$$\frac{\partial v_1}{\partial \tau}+3v_1\frac{\partial v_1}{\partial \xi}+\frac{\partial^3 v_1}{\partial \xi^3}+\varphi\frac{\partial v_1}{\partial \xi}+\frac{\partial \varphi}{\partial \tau}=0$$

and

$$\frac{\partial \rho_1{}'}{\partial \tau}+3\rho_1{}'\frac{\partial \rho_1{}'}{\partial \xi}+\frac{\partial^3 \rho_1{}'}{\partial \xi^3}-\varphi\frac{\partial \rho_1{}'}{\partial \xi}-\frac{\partial \varphi}{\partial \tau}=0.$$

Noted that these equations for v_1 and $\rho_1{}'$ become identical, if the function $\varphi(\tau)$ is chosen to satisfy the relation

$$\varphi\frac{\partial(\rho_1{}', v_1)}{\partial \xi}+\frac{\partial \varphi}{\partial \tau}=0.$$

Due to the relation (vii), we notice that $p_1{}'$ also satisfies this equation for v_1 or $\rho_1{}'$. We can therefore write an equation in general form:

$$\frac{\partial V_1}{\partial \tau}+3V_1\frac{\partial V_1}{\partial \xi}+\frac{\partial^3 V_1}{\partial \xi^3}=0, \tag{8.1}$$

where V_1 represents $v_1{}'$, $\rho_1{}'$ and $p_1{}'$ that are functions of ξ and τ. Equation (8.1) is known as the Korteweg–deVries equation, where ξ is a phase of propagation, and τ is a parameter for evolving nonlinearity. It is significant that τ is not necessarily the *real time*, but representing such a variable as temperature, if we deal with V_1 in thermodynamic environment. The quantity k^2V_1 is called a *soliton* because of its particle-like behavior, as proven by the soliton theory. In this approximation, the mesoscopic pseudospins are driven in phase by the adiabatic field to increase nonlinearity.

In the above, we discussed a one-dimensional chain of classical pseudospins in analogy of fluid, where the pressure gradient $-\frac{\partial p}{\partial x}$ can be interpreted as related with $\frac{\partial p'}{\partial \xi} = k^2 \frac{\partial^3 \rho_1{}'}{\partial \xi^3}$ in the accuracy of k^4. In such a *hydrodynamic* approach, the pressure p can be interpreted as "viscous correlations" among pseudospins.

8.2 The Korteweg–deVries Equation

We continue to discuss the function $\boldsymbol{\sigma}(\phi)$ in one dimension, taking only a longitudinal component, as the first step. We can disregard the time t to write the function $\sigma(x,t)$ simply as $\sigma(x)$, which is legitimate in idealized crystals that are in invariant structure for the translation $x - v_0 t \rightarrow x$, allowing the origin of space–time coordinates to be at an arbitrary lattice site. This is a valid assumption, if the crystal is idealized by using Born–von Kármán's boundary conditions. With respect to such a stationary reference, $\sigma(x)$ represents collective pseudospins, for which we have a differential equation

$$\mathcal{D}^2\sigma(x) = \varepsilon_0\sigma(x) \tag{8.2}$$

for basic propagation in idealized crystal space, where $\mathcal{D} = \frac{\partial}{\partial x}$ is a *differential operator*, and the eigenvalue ε_0 represents a kinetic energy of propagation. However, when ordering is in progress, an adiabatic potential $V(x)$ should hinder moving $\boldsymbol{\sigma}(x)$, so (8.2) is modified as

$$\mathcal{L}\sigma = \left(\mathcal{D}^2 + V\right)\sigma = \varepsilon\sigma. \tag{8.3}$$

However, we are interested in such a wave as $\sigma(x)$ whose eigenvalue ε remains unchanged in thermodynamic environment. Therefore, expressing developing order by a parameter τ, we apply the condition $\frac{d\varepsilon}{d\tau} = 0$, $\varepsilon = \varepsilon_0$; otherwise $V = 0$ for $\tau = 0$, meaning no development. Thermodynamically, the parameter τ can be either the temperature change ΔT or pressure change while representing dynamically the time t.

The nonlinearity can evolve with progressing order, as described by the equation

$$\frac{\partial}{\partial \tau}\sigma(x,\tau) = \mathcal{B}\sigma(x,\tau), \tag{8.4}$$

where $\mathcal{B}$ is the developing operator for spatial order by means of $\mathcal{D}$, and τ the parameter for developing order thermodynamically. Hence, we call (8.4) the *developing equation.*

Suppose $\mathcal{B}$ is determined in a simple form $\mathcal{B}_1 = c\mathcal{D}$, where c is constant, we obtain $\sigma_\tau = c\sigma_x$ from (8.4), signifying that σ is a function of $x - c\tau$ that is nothing but a phase translation. Also noted is that a $\mathcal{B}_2 \propto \mathcal{D}^2$ operator cannot be responsible for a nonlinear progress, so ε remains constant and $V(x,\tau) = \text{const}$. In contrast, if $\mathcal{B}$ consists of a third-order derivative $\mathcal{D}^3$, we can verify that a potential $V(x,\tau)$ should be significant in (8.3), as shown in the following.

Considering the presence of $V(x,\tau)$, the wave equation can be written as

$$\mathcal{L}\sigma(x,\tau) = \{\mathcal{D}^2 + V(x,\tau)\}\sigma(x,\tau) = \varepsilon\sigma(x,\tau) = \varepsilon_0\sigma(x,\tau). \tag{8.5}$$

For convenience, differentiations are shorthanded in the following arguments by such suffixes in $\frac{\partial V}{\partial x} = V_x$, $\frac{\partial^2 V}{\partial x^2} = V_{xx}$, etc. Differentiating (8.5) with respect to τ, we have

$$\frac{\partial}{\partial\tau}(\mathcal{L}\sigma) = \mathcal{L}_\tau\sigma + \mathcal{L}\sigma_\tau = -V_\tau\sigma + \mathcal{L}\mathcal{B}\sigma$$

and

$$\frac{\partial}{\partial\tau}(\varepsilon\sigma) = \varepsilon_\tau\sigma + \varepsilon\sigma_\tau = \varepsilon_\tau\sigma + \varepsilon(\mathcal{B}\sigma) = \varepsilon_\tau\sigma + \mathcal{B}\mathcal{L}\sigma.$$

Therefore

$$(-V_\tau + [\mathcal{L},\mathcal{B}])\sigma = \varepsilon_\tau\sigma, \quad \text{where} \quad [\mathcal{L},\mathcal{B}] = \mathcal{L}\mathcal{B} - \mathcal{B}\mathcal{L}.$$

If the condition $\varepsilon_\tau = 0$ is fulfilled, the equation to be solved for V is given by

$$(-V_\tau + [\mathcal{L},\mathcal{B}])\sigma(x,\tau) = 0. \tag{8.6}$$

Assuming $\mathcal{B} = \mathcal{B}_3 = a\mathcal{D}^3 + c\mathcal{D} + b$ with another constant a, and variable terms b and c that are functions of x and τ, we have

$$[\mathcal{L},\mathcal{B}_3]\sigma = (2c_x + 3aV_x)\mathcal{D}^2\sigma + (c_{xx} + 2b_x + 3aV_{xx})\mathcal{D}\sigma + (b_{xx} + aV_{xxx} + cV_x)\sigma.$$

To obtain a differential equation for $V(x - c\tau)$ from (8.5), the coefficients of $\mathcal{D}^2\sigma$ and $\mathcal{D}\sigma$ should be vanished, so we write

$$2c_x + 3aV_x = 0 \quad \text{and} \quad c_{xx} + 2b_x + 3aV_{xx} = 0.$$

Integrating these, the relations can be obtained for the coefficients c and b to be determined. Hence, we then obtain

$$c = -\tfrac{3}{2}aV + C \quad \text{and} \quad b = -\tfrac{3}{4}aV_x + B,$$

where C and B are integration constants.

Consequently, we have $[\mathcal{L}, \mathcal{B}_3]\sigma = \left\{\frac{a}{4}(V_{xxx} - 6VV_x) + CV_x\right\}\sigma$, and (8.6) can be expressed as

$$\frac{a}{4}(V_{xxx} - 6VV_x)\sigma + (CV_x - V_\tau)\sigma = 0.$$

Transforming variables (x, τ) back to $(x - c\tau, \tau)$, and setting $C = 0$ and $a = -4$, this equation can be expressed as

$$V_\tau - 6VV_x + V_{xxx} = 0. \tag{8.7}$$

This is a Korteweg–deVries' equation for the potential $V = V(x - c\tau)$, for which the corresponding pseudospin variable $\boldsymbol{\sigma}(x - c\tau)$ is in phase with $V = V(x - c\tau)$, sharing a common equation of propagation at the same eigenvalue ε. In this case, the developing equation is

$$\sigma_\tau = \left(-4\mathcal{D}^3 + 6V\mathcal{D} + 3V_x + \mathcal{B}\right)\sigma,$$

where $\sigma = \sigma(x - c\tau)$ and c represents the speed of development. If the variable τ is considered for the rate of energy transfer to the heat reservoir, we consider that $\tau \propto \Delta T$.

8.3 Solutions of the Korteweg–deVries Equation

The equation (8.7) is analytically soluble for the potential $V = V(x - c\tau)$. Since such a potential is stationary in the reference frame moving at a constant phase, we require the condition $\frac{\partial V}{\partial(x-c\tau)} = 0$, and hence $V_\tau = cV_\mathbf{x}$. Therefore (8.7) can be written as $cV_x - 6VV_x + V_{xxx} = 0$, that is,

$$cV_x - 3\frac{\mathrm{d}V^2}{\mathrm{d}x} + \frac{\mathrm{d}V_{xx}}{\mathrm{d}x} = 0.$$

This equation can be integrated as

$$V_{xx} = 3V^2 - cV + a,$$

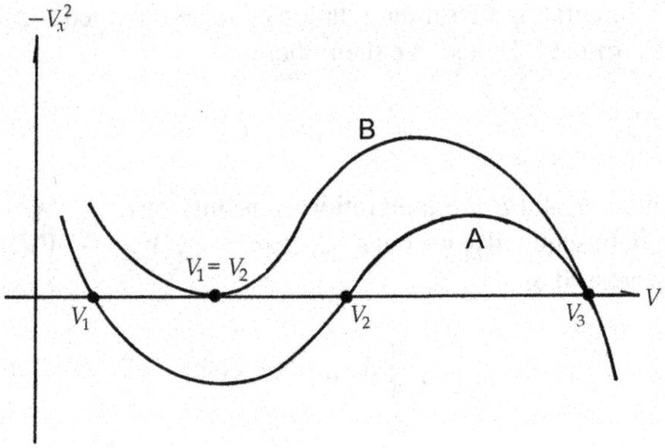

Fig. 8.1 Solving the Korteweg–deVries equation. Oscillatory and solitary solutions are represented by the curves A and B, respectively.

where a is a constant of integration. Multiplying by V_x, we integrate this equation once more, and obtain

$$V_x^2 = 3V^3 - cV^2 + 2aV + b,$$

where b is another constant. The right side is an algebraic expression of third-order with respect to V, and hence written as

$$V_x^2 = -2(V - V_1)(V - V_2)(V - V_3).$$

Here, V_1, V_2, and V_3 are three roots of the third-order equation $V_x^2(V) = 0$. Illustrated in Fig. 8.1 are curves of the equation $-V_x^2 = 0$ plotted against V, which can cross the horizontal axis at three points V_1, V_2 and V_3, if these are all real (case A), otherwise crossing at two real roots (case B) or only at one point (not shown). Since our interest is only in a real V_x, we consider the region $V_2 < V < V_3$ of a curve A as significant for our purpose, which is characterized by $-V_x^2 > 0$.

We define the variable V by writing $V - V_3 = -g$ for the above algebraic equation, which can then be expressed as

$$g_x^2 = 2g(V_3 - V_1 - g)(V_3 - V_2 - g).$$

Further, introducing another variable ξ by $g = (V_3 - V_2)\xi^2$, this equation can be written for ξ as

$$\xi_x^2 = \frac{1}{2}(V_3 - V_1)(1 - \xi^2)(1 - \kappa^2\xi^2), \quad \text{where} \quad \kappa^2 = \frac{V_3 - V_2}{V_3 - V_1}.$$

Expressing the phase by $\phi = \sqrt{V_3 - V_1}x$, the equation can be further modified as

$$2\xi_\phi^2 = (1 - \xi^2)(1 - \kappa^2\xi^2).$$

Integrating this expression, we obtain

$$\frac{\phi_1}{\sqrt{2}} = \int_0^{\xi_1} \frac{d\xi}{\sqrt{(1-\xi^2)(1-\kappa^2\xi^2)}}, \tag{8.8a}$$

where the phase ϕ_1 is determined by the upper limit ξ_1 of this integral. The reverse function of ϕ_1, known as Jacobi's *sn-function*, can be written as

$$\xi_1 = \mathrm{sn}\left(\frac{\phi_1}{\sqrt{2}};\ \kappa\right). \tag{8.8b}$$

The potential $V(\phi)$ can then be expressed by

$$V(\phi) = V_3 - (V_3 - V_2)\mathrm{sn}^2\left(\sqrt{V_3 - V_1}\phi;\ \kappa\right) \quad \text{for} \quad 0<\kappa<1, \tag{8.9a}$$

where the sn^2-function is periodic with the period

$$2K(\kappa) = 2\int_0^1 \frac{d\xi}{\sqrt{(1-\xi^2)(1-\kappa^2\xi^2)}}.$$

On the other hand, the potential $V(\phi)$ in (8.9a) has a periodic interval of $\frac{2K(\kappa)}{\sqrt{V_3-V_2}}$, corresponding a finite amplitude $V_3 - V_2$. If $V_2 \to V_1$, we have the curve B in Fig. 8.1, for which $\kappa \to 1$ and $K(1) = \infty$. In this case the potential is

$$V(\phi) = V_3 + (V_3 - V_2)\mathrm{sech}^2\left(\sqrt{V_3 - V_1}\phi\right). \tag{8.9b}$$

This shows a pulse-shaped potential with height $V_3 - V_2$, propagating with effective phase $\sqrt{V_3 - V_1}\phi$. Specifically if $V_3 - V_2$ is infinitesimal, the modulus κ is almost zero; (8.9a) exhibits a sinusoidal variation:

$$V(\phi) = V_3 + (V_3 - V_2)\sin^2\left(\sqrt{V_3 - V_1}\phi\right).$$

It is noted that the coefficients in (8.9a) and (8.9b) are related with the modulus κ. By definition, we have $V_3 - V_2 = \kappa^2(V_3 - V_1)$, and hence the amplitude $V_3 - V_2$ is proportional to κ^2. Adjusting values of $V_1, V_2,$ and V_3, the potential $V(\phi)$ can be made as consistent with adiabatic potentials derived in Chap. 7.

Thus, solutions of the Korteweg–deVries equation are expressed by elliptic functions of the propagating phase, which is either oscillatory or in a pulse shape, depending on the modulus κ; a solitary potential of (8.9b) is known as the *Eckart potential.* The latter potential behaves like an independent particle, which can be proven rigorously, and hence called a soliton, but we shall not go into the detail.

Physically, the transverse density ρ_z can be responsible for dispersion of the longitudinal σ_1 mode; mathematically the relation between $\rho_z(\phi)$ and the soliton potential $V(\phi)$ is a direct solution of the Korteweg–deVries equation. In a crystal, the adiabatic potential ΔU should reflect symmetry of the three-dimensional lattice. In addition to $V(\phi)$ of (8.9a) that corresponds to the longitudinal density $\rho_x(\phi)$, we should have a transverse potential:

$$V_\perp(\phi') = V_2 + (V_3 - V_2)\mathrm{cn}^2\left(\sqrt{V_3 - V_1}\phi';\ \kappa'\right) \tag{8.10}$$

can be associated with the transverse density $\rho(\phi')$, where $\phi' = \phi - K$ and $\kappa' = \kappa$. Judging from what we discussed in Sect. 8.1, this is an acceptable postulate. The total adiabatic potential for the density $\boldsymbol{\sigma}_\mathrm{o} * \boldsymbol{\sigma}_\mathrm{o}$ should therefore be given by

$$\Delta U = V(\phi) + V_\perp(\phi - K), \tag{8.11}$$

which acts on the total density $\sigma_x^2 + \sigma_z^2$.

Jacobi's elliptic functions can be expanded into power series as in Landau's expansion, which can be truncated at a quartic term σ^4 (see Appendix). It was therefore logical to consider such expansions in the Landau theory. The corresponding potential can be expressed as $\Delta U = \frac{a}{2}\sigma^2 + \frac{b}{4}\sigma^4$ in the adiabatic approximation. The above interpretation of Landau's expansion owes the developing operator $\mathcal{B}_3$ that signifies the dispersive nature of the developing (8.4), as supplemented by Born–Huang's principle in thermodynamic environment. Consequently, elliptic and quartic potentials are both useful approximations, for practical analysis of mesoscopic variables.

It is notable that the soliton potential (8.9b) can describe a domain wall, in which the reversing energy is proportional to $\sigma_\perp^2$ that shows singularity at $\phi = 0$, after a thermal process for $\kappa \to 1$. The adiabatic potential $\Delta U(\sigma_\perp)$ in (7.9b) represents therefore the *domain-wall energy* in the limit of $\kappa \to 1$.

8.4 Cnoidal Theorem and the Eckart Potential

It is significant that adiabatic potentials are *conformable* with the order variables density in the Korteweg–deVries equation; both are signified by the phase $\phi = x - v\tau$, as they are in thermal equilibrium. Furthermore, by the coordinate transformation $x - v\tau \to x'$, the potential $V(x')$ becomes always in phase with the order density $\sigma(x')^2$ in the system of x'. Hence disregarding τ, these $V(x,\kappa)$ and $\sigma(x,\kappa)^2$ can be expressed in the limit $\kappa \to 1$ as proportional to $\mathrm{sech}^2 x$.

The Eckart potential (8.9b) characterized by a solitary peak can be written in such a form as

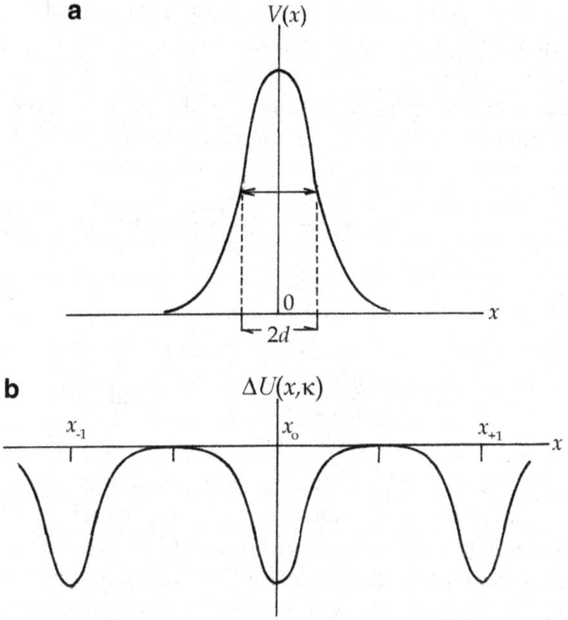

Fig. 8.2 (**a**) Eckart's potential of the half-width $2d$. (**b**) A sketch of a cnoidal potential $\mathrm{cn}^2(\phi/\sqrt{2}\kappa)$ with periodic peaks.

$$V(x) = -V_o \mathrm{sech}^2\left(\frac{x}{d}\right) \quad \text{for} \quad \kappa = 1, \tag{8.12}$$

excluding the constant term. Here, d is introduced to indicate the width $2d$ of the symmetric peak around $x = 0$, as shown in Fig. 8.2a. It is noted that (8.12) is related to σ_1 in (7.8), whose transverse component can be written as $\sigma_\perp\left(\frac{x}{d}\right)$ that corresponds to $V(x)$.

In contrast, the potential (8.9a) is oscillatory and periodic in space. Although expressed as sn^2 for $0 < \kappa < 1$, such a potential is practically the same as cn^2-function because of the relation $\mathrm{sn}^2 + \mathrm{cn}^2 = 1$. In Chap. 7, we have shown that the transverse component $\sigma_\perp(x,\ \kappa)$ is in motion with an adiabatic cnoidal potential

$$V(x,\kappa) = -V_o \mathrm{dn}^2 u(x,\kappa), \tag{8.13a}$$

where V_o is a proportionality factor. Because of the relation $\mathrm{dn}^2 u = 1 - \kappa^2 \mathrm{sn}^2 u$, (8.13a) gives an identical potential

$$V(x,\kappa) = -V_o \kappa^2 \mathrm{sn}^2 u(x,\kappa) \tag{8.13b}$$

ignoring the additive constant. In any case, these squared elliptic functions represent a periodic potential, called a *cnoidal* curve, as sketched in Fig. 8.2b. For a physical application, omitting a trivial additive constant, the adiabatic potential $\Delta U(\sigma)$ is expressed by (8.13a) or (8.13b).

At this point, we pay attention to a mathematical theorem for such a sn^2 function, stating that peaks in the cnoidal expansion can be replaced by sech^2 peaks. The rigorous proof is available in the advanced theory of elliptic functions, which is however too tedious to follow at the present level of discussion. Quoting the resulting formula, here we use the theorem

$$2\kappa^2 \mathrm{sn}^2 x = -2a \sum_{\mathrm{m}=-\infty}^{\mathrm{m}=+\infty} \mathrm{sech}^2\left(\sqrt{a}x - c\mathrm{m}\right) + \text{const.}, \tag{8.14}$$

where

$$a = \frac{\pi^2}{4K'(\kappa)^2} \quad c = \frac{\pi K(\kappa)}{K'(\kappa)} \quad \text{and} \quad K'(\kappa) = K\left(\sqrt{1-\kappa^2}\right).$$

Here, the index m specifies the peak positions at x_{m} for $-\infty < \mathrm{m} < +\infty$ along the x-axis; $K(\kappa)$ is the complete elliptic integral (7.7) and $K'(\kappa)$ is defined in the above. Nevertheless, those readers who are interested in the derivation of (8.14) can consult a standard reference book on elliptic functions. Owing to (8.14), the cnoidal potential can be replaced by a periodic array of Eckart potentials located at $x = \frac{c}{\sqrt{a}}\mathrm{m}$, where m are $\pm$ integers. By virtue of this theorem, problems for the oscillatory cnoidal potential can be replaced by series of Eckart's sech^2-peaks.

On the other hand, physically such a transition to sech^2 potentials should be interpreted by Born–Huang principle for $\boldsymbol{\sigma}(x,\tau)$ to be placed in thermodynamic environment; taking $\kappa \to 1$, we use $K(1)$and $K'(1) = K(0)$ for (8.14). In this interpretation, pseudospins are ordered in small oppositely polarized domains of $4K(1)$ in succession in the mesoscopic state.

8.5 Condensate Pinning by the Eckart Potentials

In this section we discuss the eigenvalue problem of $\sigma(x,\kappa)$ in the Eckerd potential field $V(x, \kappa)$. Denoting the eigenvalue by ε, the perturbed wave equation is written as

$$\frac{\mathrm{d}^2\sigma(x)}{\mathrm{d}x^2} + \left(\varepsilon + V_{\mathrm{o}}\ \mathrm{sech}^2\frac{x}{d}\right)\sigma(x) = 0.$$

For stable pinning, the eigenvalue should be negative, so we write $\varepsilon = -\mu^2$. Replacing $\frac{x}{d}$ by x, and setting $V_{\mathrm{o}}d^2 = v_{\mathrm{o}}$ and $\mu^2 d^2 = \beta^2$, this equation can be expressed as

$$\frac{\mathrm{d}^2\sigma(x)}{\mathrm{d}x^2} + \left(-\beta^2 + v_{\mathrm{o}}\ \mathrm{sech}^2 x\right)\sigma(x) = 0. \tag{8.15}$$

This is a differential equation familiar in Mathematical Physics, whose solution is expressed by *hypergeometric series*. Following Morse and Feshbach [2], we transform (8.15) to the hypergeometric equation in standard form.

By defining the relation $\sigma(x) = Af(x)\ \mathrm{sech}^{\beta}x$, the differential equation for the function $f(x)$ can be obtained as

$$\frac{d^2f}{dx^2} - 2\beta(\tanh x)\frac{df}{dx} + \left(v_0 - \beta^2 - \beta\right)\left(\mathrm{sech}^2 x\right)f(x) = 0.$$

Further, this equation can be modified by another variable $\zeta = \frac{1}{2}(1 - \tanh x)$ as

$$\zeta(1-\zeta)\frac{d^2f}{d\zeta^2} + (1+\beta)(1-2\zeta)\frac{df}{d\zeta} + \left(v_0 - \beta^2 - \beta\right)f(x) = 0.$$

We define such parameters **a**, **b**, and **c** that satisfy the relations

$$\mathbf{a} + \mathbf{b} = 2\mathbf{c} - 1,$$

where

$$\mathrm{c} = 1 + \beta \quad \text{and} \quad \mathbf{ab} = -\mathbf{v}_0 + \beta^2 - \beta,$$

that is,

$$\mathbf{a}, \mathbf{b} = \frac{1}{2} + \beta \pm \sqrt{\mathbf{v}_0 + \frac{1}{4}}.$$

Using **a**, **b**, and **c**, we obtain the hypergeometric equation expressed in standard form, i.e.,

$$\zeta(1-\zeta)\frac{d^2f}{d\zeta^2} + \{\mathrm{c} - (\mathbf{a} + \mathbf{b} + 1)\zeta\}\frac{df}{d\zeta} - \mathbf{ab}f = 0. \tag{8.16}$$

Expressing as $f(\zeta) = F(\mathbf{a},\ \mathbf{b},\ \mathbf{c};\ \zeta)$, the solution of (8.16) is called a hypergeometric function.

The mesoscopic pseudospin variable $\sigma(x, \kappa)$ can then be expressed as

$$\sigma(x) = A\ \mathrm{sech}^{\beta}xF(\mathbf{a},\ \mathbf{b}, \mathbf{c};\ \ \zeta). \tag{8.17}$$

However, we are only interested in extreme cases that can be specified by $\zeta \to 0$ and $\zeta \to 1$, in order to use it physically at distant points $x \to \pm\infty$, for which we expand $F(\mathbf{a}, \mathbf{b}, \mathbf{c}; \zeta)$ as

$$F(\mathbf{a},\ \mathbf{b},\ \mathbf{c}; \zeta) = 1 + \frac{\mathbf{ab}}{1!\mathbf{c}}\zeta + \frac{\mathbf{a}(\mathbf{a}+1)\mathbf{b}(\mathbf{b}+1)}{2!\mathbf{c}(\mathbf{c}+1)}\zeta^2 + \cdots. \tag{8.18}$$

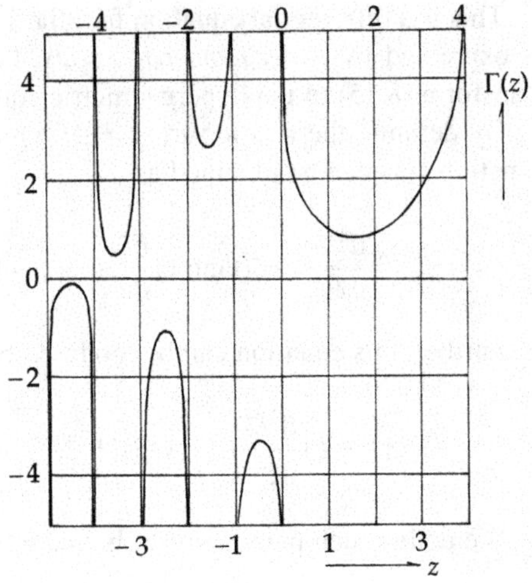

Fig. 8.3 Gamma function Γ (z).

Corresponding to $\zeta \to 0$, for $x \to +\infty$ we have

$$\sigma(x)_{x\to\infty} \to A2^{\beta}\exp(-\beta x),$$

whereas for $\zeta \to 1$, we consider $x \to -\infty$.

For calculating $\sigma(x)$ in these cases, it is convenient to use the following formula [3]:

$$F(\mathbf{a},\ \mathbf{b},\ \mathbf{c};\ \zeta) = \frac{\Gamma(\mathbf{c})\Gamma(\mathbf{c}-\mathbf{a}-\mathbf{b})}{\Gamma(\mathbf{c}-\mathbf{a})\Gamma(\mathbf{c}-\mathbf{b})}F(\mathbf{a},\ \mathbf{b},\ \mathbf{a}+\mathbf{b}-\mathbf{c}+1;\ \zeta)$$
$$+ (1-\zeta)^{\mathbf{c}-\mathbf{a}-\mathbf{b}}\frac{\Gamma(\mathbf{c})\Gamma(\mathbf{a}+\mathbf{b}-\mathbf{c})}{\Gamma(\mathbf{a})\Gamma(\mathbf{b})}F(\mathbf{c}-\mathbf{a},\ \mathbf{c}-\mathbf{b},\ \mathbf{c}-\mathbf{a}-\mathbf{b}+1;\ 1-\zeta),$$

where $\Gamma(\ldots)$ are *gamma functions*. It is noted from (8.15) that the first term dominates $F(\mathbf{a},\ \mathbf{b},\ \mathbf{c};\ \zeta)$, if $\zeta \to 0$. On the other hand, if $\zeta \to 1$ we have $(1-\zeta)^{\mathbf{c}-\mathbf{a}-\mathbf{b}} \approx 2^{\beta}\exp(-\beta x)$ and $F(\ldots;\ 1-\zeta) \to 1$ in the second term, dominating $F(\mathbf{a},\ \mathbf{b},\ \mathbf{c};\ \zeta)$. Considering a small β, $\mathrm{sech}^{\beta}x \approx 2^{\beta}\exp(+\beta x)$ and $F(\mathbf{a},\ \mathbf{b}, 1+\beta;\ \zeta) \to 1$ in the first term. In this case, writing $\beta = -ik$ to express propagating waves, we can write that

$$\sigma(x)_{x\to\infty} \propto \frac{\Gamma(\mathbf{c})\Gamma(\mathbf{a}+\mathbf{b}-\mathbf{c})}{\Gamma(\mathbf{a})\Gamma(\mathbf{b})}\exp(+ikx) + \frac{\Gamma(\mathbf{c})\Gamma(\mathbf{c}-\mathbf{a}-\mathbf{b})}{\Gamma(\mathbf{c}-\mathbf{a})\Gamma(\mathbf{c}-\mathbf{b})}\exp(-ikx). \quad (8.19)$$

Using definitions of **a**, **b**, and **c**, we notice that the numerators of these coefficients are $\Gamma(1+\beta)\Gamma(-\beta)$ and $\Gamma(1+\beta)\Gamma(\beta)$, respectively, whereas the denominators have singularities, depending on the value of v_0. In Fig. 8.3 the curve of a gamma function $\Gamma(z)$ plotted against the variable z is shown, where there are a number of poles at $z = 0, -1, -2, \ldots$. Hence, the first coefficient has singularities at

$$\mathbf{a}, \mathbf{b} = \frac{1}{2} + \beta \pm \sqrt{v_0 + \frac{1}{4}} = -\mathrm{m}, \quad \mathrm{m} = 0, -1, -2, \ldots \tag{i}$$

whereas the second coefficient can be specified by singularities for the denominator to become infinity, i.e.,

$$\Gamma(\mathbf{c}-\mathbf{a})\Gamma(\mathbf{c}-\mathbf{b}) = \Gamma\left(\frac{1}{2} + \sqrt{v_0 + \frac{1}{4}}\right)\Gamma\left(\frac{1}{2} - \sqrt{v_0 + \frac{1}{4}}\right) = \frac{\pi}{\cos\left(\pi\sqrt{v_0 + \frac{1}{4}}\right)} = \infty;$$

hence

$$\sqrt{v_0 + \frac{1}{4}} = \mathrm{n} + \frac{1}{2} \quad \text{or} \quad \mathrm{v_0} = \mathrm{n}(\mathrm{n}+1), \tag{ii}$$

where n can be any integer. Combining (i) and (ii), we obtain

$$\beta = \left(\mathrm{n} + \frac{1}{2}\right) - \mathrm{m} - \frac{1}{2} = \mathrm{n} - \mathrm{m}, \tag{iii}$$

where $\mathrm{m} = 0, 1, 2, \ldots, \mathrm{n}-1$. Therefore the eigenvalues for $\beta > 0$ can be represented as $\beta_\mathrm{p} = \mathrm{n} - \mathrm{m} = \mathrm{n}, \mathrm{n}-1, \mathrm{n}-2, \ldots, 1$. Imposing conditions for no transmission and reflection on (8.19), the pseudospin wave $\sigma(x)$ can be in phase with the soliton potential, if $\rho(x)$ and $V(x)$ are conformal with integers specified by (i) and (ii). Therefore, by expressing the potential by eigenvalues p, the steady-state equation (8.15) for longitudinal wave functions $\sigma_\mathrm{p}(x)$ can be written as

$$\frac{\mathrm{d}^2\sigma_\mathrm{p}(x)}{\mathrm{d}x^2} + \{-\beta^2 + \mathrm{p}(\mathrm{p}+1)\,\mathrm{sech}^2 x\}\sigma_\mathrm{p}(x) = 0, \tag{iv}$$

where discrete eigenvalues $\varepsilon_\mathrm{p} = -\beta_\mathrm{p}^2$ can be determined by $\beta_\mathrm{p}^2 = \mathrm{p}^2$ that are indexed by $\mathrm{p} = \mathrm{n}, \mathrm{n}-1, \mathrm{n}-2, \ldots 2, 1$. Writing the (iv) for two adjacent levels at n and $\mathrm{n}-1$, we have

$$\frac{\mathrm{d}^2\sigma_\mathrm{n}}{\mathrm{d}x^2} = \{-\mathrm{n}^2 + \mathrm{n}(\mathrm{n}+1)\,\mathrm{sech}^2 x\}\sigma_\mathrm{n} = 0$$

and

$$\frac{\mathrm{d}^2\sigma_{\mathrm{n}-1}}{\mathrm{d}x^2} + \left\{-(\mathrm{n}-1)^2 + (\mathrm{n}-1)\mathrm{n}\,\mathrm{sech}^2 x\right\}\sigma_{\mathrm{n}-1} = 0.$$

Hence, for a change $\sigma_{\mathrm{n}-1} \rightarrow \sigma_\mathrm{n}$ the potential value decreases by $\Delta V_{\mathrm{n}-1,\mathrm{n}} = -2\mathrm{n}\,\mathrm{sech}^2 x$. We consider that such transitions can take place as a

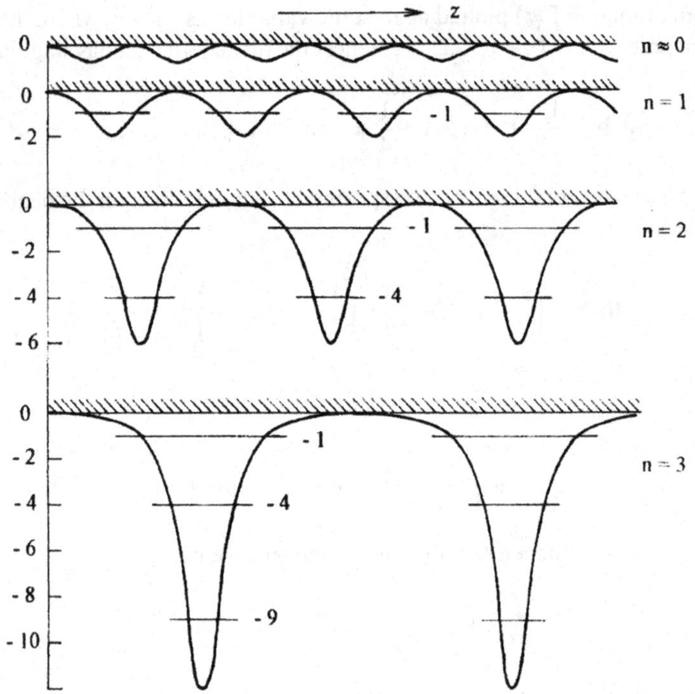

Fig. 8.4 Soliton levels in cnoidal potentials.

thermal relaxation process with varying temperature. Dynamically inaccessible, but such transitions should be observed thermally as a consequence of phonon scatterings. Corresponding lattice displacements u_n should occur in a relaxation process between strain energies related to $u_{n-1}^2 \rightarrow u_n^2$, corresponding to $T_{n-1} \rightarrow T_n$ in virtue of the equipartition theorem.

Figure 8.4 shows a change of the periodic potential (8.13b) with lowering temperature, where eigenvalues of the pseudospin density are indicated in each of periodic Eckart's potentials. Not shown in the figure, but the density levels are practically in band structure whose widths are significantly broader at smaller n. Therefore, it is reasonable that stepwise thermal relaxation processes are responsible for these transitions toward the lowest level determined by n^2 in each potential well.

8.6 Born–Huang's Transitions

For two transverse components $\sigma_1(x)$ and $\sigma_2(x)$ for the same eigenvalue $-\beta^2$, there is generally a finite potential difference between $V_1(x)$ and $V_2(x)$, which can be shown in the following.

Writing the wave equations as

$$\mathcal{D}^2\sigma_1(x) = \{-\beta^2 + V_1(x)\}\sigma_1(x) \quad \text{and} \quad \mathcal{D}^2\sigma_2(x) = \{-\beta^2 + V_2(x)\}\sigma_2(x) \quad \text{(i)}$$

we consider that these components are related by

$$\sigma_2(x) = A(x,\beta)\sigma_1(x) + B\,\mathcal{D}\sigma_1(x) \quad \text{(ii)}$$

which is adequate for the present analysis, although B is assumed as constant. Substituting (ii) into (i), and setting the front factors of σ_1 and $\mathcal{D}\sigma_1$ equal to zero, we obtain the relations

$$A_{xx} + V_{1x} + A(V_1 - V_2) = 0 \text{ and } 2A_x + (V_1 - V_2) = 0, \quad \text{(iii)}$$

respectively. Eliminating $V_1 - V_2$ in (iii), we derive the relation $A_x A_{xx} - A_x A - V_{1x}$ $A_x = 0$, which is integrated as

$$A^2 - A_x - V_1 = -\tilde{\beta}^2,$$

where $-\tilde{\beta}^2$ is a constant of integration. This equation, known as *Riccati's equation*, can be linearized by transforming A to $\tilde{\sigma}(x)$, i.e.,

$$A = -\frac{\mathcal{D}\tilde{\sigma}}{\tilde{\sigma}} = -\mathcal{D}(\ln\tilde{\sigma}), \quad \text{(iv)}$$

resulting in

$$\mathcal{D}^2\tilde{\sigma}(x) = \{-\beta^2 + V_1(x)\}\tilde{\sigma}_1(x).$$

This is identical to equations (i), indicating that the function $\tilde{\sigma}(x)$ satisfies (i), if $\tilde{\beta} = \beta$. In this case, using (iv) in the second equation of (iii), we can write

$$\mathcal{D}^2\sigma_2(x) = \{-\beta^2 + V_1(x) - 2(\ln\tilde{\sigma})\}\sigma_2(x) \quad \text{(v)}$$

and hence

$$V_2(x) - V_1(x) = -2(\ln\tilde{\sigma}), \quad \text{(vi)}$$

representing a significant property of the adiabatic potential. Although consistent with equations (iv) in Sect. 8.5, the physical implication of $\tilde{\sigma}(x)$ is not immediately clear.

In the above analysis, the variable x represents the phase of propagation, and the functions $\sigma(x)$ and $V(x)$ are conformal with the transformation $x \to x - \alpha\tau$, where α is constant, and τ is an arbitrary parameter. For thermodynamic applications, we consider that the temperature T can be assigned to τ for changing $V(x)$. Therefore, writing

$$\Delta V(x-\alpha\tau)=\left(\frac{\partial V}{\partial x}\right)_{\tau}\Delta x-\alpha\left(\frac{\partial V}{\partial \tau}\right)_{x}\Delta\tau, \qquad \text{(vii)}$$

we can assume $\Delta\tau \propto \Delta T$. In this case, an *isothermal change* can be specified by $\Delta T=0$, where

$$\Delta V(x-\alpha\tau)=\left(\frac{\partial V}{\partial x}\right)_{\tau}\Delta x=2\frac{\tilde{\sigma}'}{\tilde{\sigma}}\Delta x.$$

For an adiabatic change, we have

$$\Delta V(x-\alpha\tau)=-\alpha\left(\frac{\partial V}{\partial \tau}\right)_{x}\Delta\tau=-\frac{2\alpha\tilde{\sigma}'}{\tilde{\sigma}}\Delta\tau.$$

Subjecting to phonon scatterings, such ΔV can be calculated as proportional to $\langle \boldsymbol{K}+\boldsymbol{k}|\Delta V|\boldsymbol{K}\rangle^2$, where $\boldsymbol{K}$ is the phonon vector in a crystal, and then evaluated by phonon densities at $\boldsymbol{K}+\boldsymbol{k}$ and $\boldsymbol{K}$ states. Using the equipartition theorem in statistical mechanics, both of these rates, $\left(\frac{\partial V}{\partial x}\right)_{\tau}$ and $\left(\frac{\partial V}{\partial \tau}\right)_{x}$ for phonon scatterings, can be expressed as proportional to $\Delta T=T_2-T_1$, except for $\Delta\boldsymbol{K}=0$ in the isothermal process. Therefore, the latter process cannot be totally adiabatic while involved in thermal relaxation. The expression (vii) may therefore be called *Born–Huang transitions* in thermodynamic environment.

In real crystals, excited states of correlations can be attributed to soliton excitations of order variables, where the adiabatic potential is composed of sech^2-potentials, emerging at a critical temperature T_c, after the crystal is stabilized by Born–Huang's transition. Accordingly, $\mathrm{n}=0$ is logically for the ground state of $\tilde{\sigma}(x)=\cosh x$, which is symmetrical combination of $\exp(\pm x)$, whereas the mesoscopic state at lower temperatures can be signified by increasing number n, becoming to $\tilde{\sigma}(x)=\cosh^{\mathrm{n}+1}x$. Therefore, we can consider that the number n represents an elemental decrement of the adiabatic potential. In this interpretation, $\cosh^{\mathrm{n}+1}x$ is a multi-soliton function in the correlation field.

Assuming that the transformation (iv) considered for $\sigma_1 \rightarrow \sigma_2$ represents a Born–Huang transition, we can explain experimental results. On the other hand, subject to phonon scatterings in crystals, a change $\Delta\sigma=\sigma_1-\sigma_2$ should accompany a lattice displacement $\Delta u=u_1-u_2$, which is proportional to a temperature change ΔT. The temperature-dependence in long tail of the specific heat anomaly below T_c, the so-called λ-transitions, can therefore be analyzed as illustrated in Fig. 8.5. It is instructive to see that the tail for $T<T_c$ can be schematically interpreted as stepwise isothermal and adiabatic changes, characterizing the latter processes by ΔT and $\Delta\mathrm{n}$, as shown in the figure.

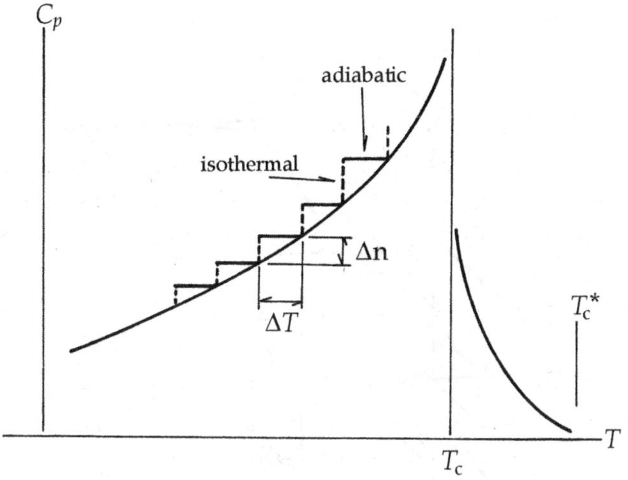

Fig. 8.5 A schematic diagram for $C_p - T$ curve. A fast phasing process of σ_k for $T>T_c$, and a slow phasing of clustered pseudospins $\boldsymbol{\sigma}_k$ for $T<T_c$, which is analyzable as adaibatic-isothermal steps.

In this interpretation, the Gibbs function of a mesoscopic thermodynamic state can be expressed by

$$G(\mathrm{n}, \mathrm{p}, T) = G_\sigma(\mathrm{n}_\sigma, p, T) + G_\mathrm{L}(\mathrm{n}_\mathrm{L}, p, T),$$

where n_σ and n_L are numbers of solitons in the order-variable system and the lattice, respectively, but $\mathrm{n} = \mathrm{n}_\sigma = \mathrm{n}_\mathrm{L}$ in thermodynamic equilibrium. At constant p, we showed in the above that $\mathrm{n} = \mathrm{n}(T) \propto T$ and $\Delta\mathrm{n} \propto \Delta T$. To express the long-range order for G_σ thermodynamically, we may define the chemical potential μ_σ with the standard relation $\Delta G_\sigma = -\mu_\sigma\,\Delta\mathrm{n}_\sigma$, which are nevertheless the same as writing $\Delta G = -\mu\,\Delta\mathrm{n}$.

8.7 Topological Mapping of Mesoscopic Fields

One-dimensional scatterings of σ_p by an Eckart's potential were discussed in the previous section, which are found to be discrete in terms of the parameter p. In Sect. 8.2, for one-dimensional propagation of σ_p along the x-direction, the driving potential $V(x)$ was found as a solution of the Korteweg–deVries equation. The propagation is basically dispersive, as signified by the third-order derivative $\mathcal{D}^3\sigma_\mathrm{p}$. Accordingly, the adiabatic potential cannot be strictly one dimensional, as implied by the transversal component $\sigma_\perp(x)$ that can interact with neighboring $\sigma(x)$ in a crystal. The potential $V(\boldsymbol{r})$ should therefore represent a geometrical surface in crystal space, expressed by the coordinates x, y, and z. Including coordinates y and z, the three-dimensional Eckart potential $V(\boldsymbol{r})$ is free from reflection of a

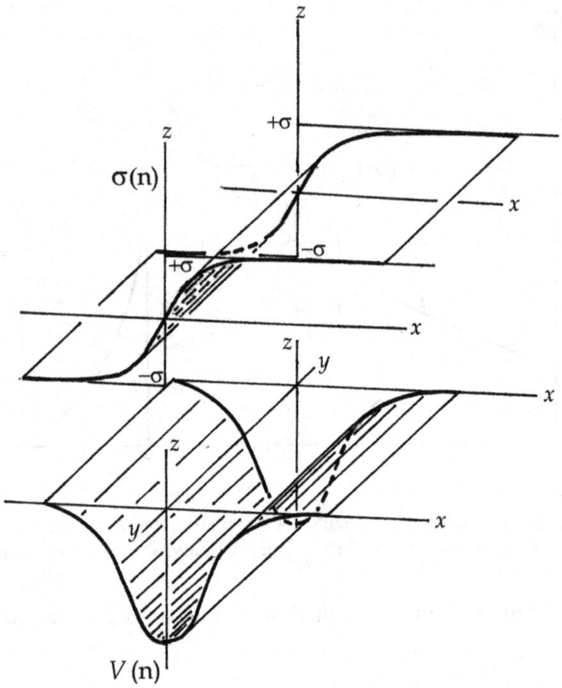

Fig. 8.6 Three-dimensional view of a domain boundary wall; $\boldsymbol{\sigma}(\mathrm{n})$ and $V(\mathrm{n})$, characterized by inversion and in trough shape, respectively.

propagating wave $\boldsymbol{\sigma}(\boldsymbol{r})$. Such an Eckart potential modified for three-dimensional crystals can therefore describe topological features of the correlation field.

Except for the critical region in the vicinity of T_c, the noncritical phase for $T < T_c$ can be represented by the adiabatic potential $V(\mathrm{n})$, where $\kappa = 1$ after Born–Huang's relaxation. Depending on n, the magnitude of $V(\mathrm{n})$ signifies the *long-range order* in a crystal. In the idealized binary order, where the correlation field is primarily one dimensional, $V(\mathrm{n}, \mathrm{x})$ represents a *trough* at $\phi = 0$ along the y-direction with the depth on the z-axis, as illustrated in Fig. 8.6, which constitutes a *domain wall*. In a *hydrodynamic model* for order variables, these quantities $\sigma(x)$ and $V(x)$ can be treated like fluid material, flowing through a crystal. In this view applied to condensates, we can geometrically map the mesoscopic state by $V(\mathrm{n})$, where domain walls and other singularities can be illustrated, analogous to a geographical contour specified by altitudes.

A basic theorem in the classical field theory is that the field consists of laminar (irrotational) and rotational component fields that are characterized by div $\boldsymbol{\sigma} \neq 0$ and curl $\boldsymbol{\sigma} \neq 0$, respectively. For example, the pattern shown in Fig. 8.6 represents a singularity in the laminar field; on the other hand, the rotational field can be specified by such an angular velocity as defined by $\boldsymbol{\omega} = r \times \boldsymbol{\sigma}$, representing a field of vector potential $\mathbf{A} \propto \boldsymbol{\omega}$ related by the Lorentz condition $\mathrm{div}\mathbf{A}(\mathrm{n}) \propto \frac{\partial V(\mathrm{n})}{\partial t}$. Such a vector potential $\mathbf{A}(\mathrm{n})$ can therefore represent a *vortex* in the potential field.

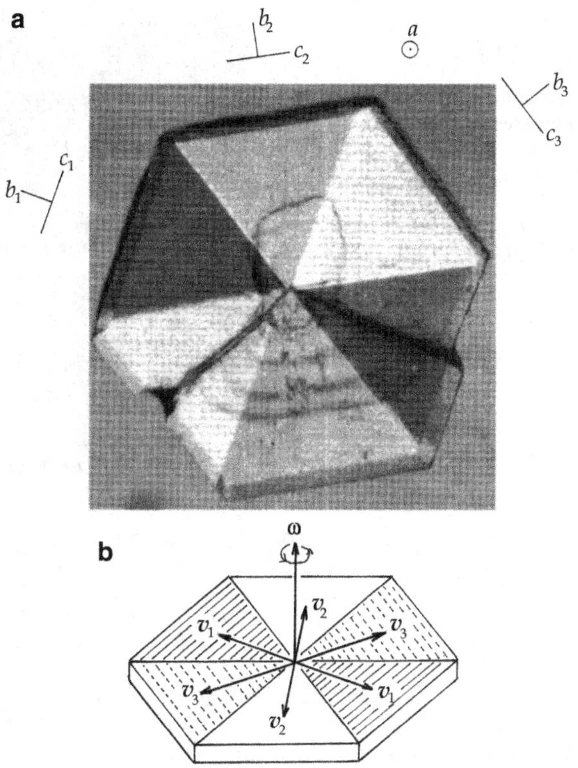

Fig. 8.7 A TSCC crystal photographed under a polarizing microscope perpendicular to the a-axis, showing three ferroelastic domains in near-trigonal arrangement. A vortex of the elastic field can be recognized along the a-axis perpendicular to the page. From M. Fujimoto, Ferroelectrics, **47**, (1983).

We notice that vortexes presumably occur at impurity or defective sites in crystals. Figure 8.7 shows a photograph of a TSCC crystal obtained under polarizing microscope, which was taken from a crystal grown from aqueous solution at room temperature; exhibiting well-identifiable ferroelastic domains in hexagonal shape [4]. Domain boundaries are clearly distinguishable by shaded areas of the crystal, indicating the potential $V(\mathrm{n})$ in *trigonal* symmetry. The central line of the pattern parallel to the a-axis is terminated at some points, indicating a trigonal vortex around the singular a-axis.

Figure 8.8a shows vortexes in K_2ZnCl_4 crystal [5], where many *discommensuration lines* exhibit a clear indication of pseudopotentials of C_3 screw axis in this dark-field image of satellite reflections of electron beams from (100) planes, whereas points at which three lines join together can be interpreted as vortexes. In Fig. 8.7b a schematic interpretation of a vortex is shown, which may occur at defective sites in the lattice.

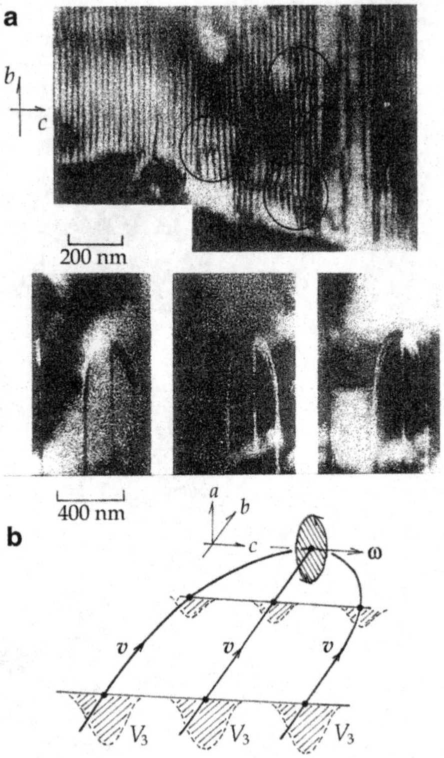

Fig. 8.8 (**a**) An electron-microscopic dark-field image from a K_2ZnCl_4 crystal, showing discommensuration lines along C_3-screw axes terminated at vortexes. (**b**) Schematic illustration of lines at a vortex. From Xiaoquin Pan and H. -G. Unruh, J. Phys. Cond. Matter, **2**, 323 (1990).

Exercise 8

1. Consider a one-dimensional oscillator, where the spring force F is given in a nonharmonic relation with the displacement x. Assuming $F = -ax - bx^3$, the equation of motion can be expressed as $m\ddot{x} = -ax - bx^3$. Replacing $\sqrt{a/m}t$ and $\sqrt{b/a}x$ by t and x, respectively, we write the equation as

$$\frac{\mathrm{d}^2x}{\mathrm{d}t^2} = -x - x^3.$$

Multiplying by $\mathrm{d}x/\mathrm{dt}$ to integrate this equation, we obtain

$$\frac{1}{2}\left(\frac{\mathrm{d}x}{\mathrm{d}t}\right)^2 = E - \frac{x^2}{2} - \frac{x^4}{4},$$

where E is the integration constant. In this case, if $U = \frac{x^2}{2} + \frac{x^4}{4}$ is the potential energy, we have the energy conservation law $E =$ const. Show that the solution of the above nonlinear equation is given by an elliptic function.

Answer: Solving the above for $\dot{x}$, we can write

$$t = \pm \int_0^x \frac{dx}{\sqrt{2E - x^2 - \frac{1}{2}x^4}}.$$

Noting that the quantity inside the square root must be positive for a real solution, we assume that the motion is limited within a range $-a \le x \le +a$, where $E \le U$, as illustrated in Fig. 8.8a. Such limits $\pm a$ are determined by

$$a^4 + 2a^2 - 4E = 0 \quad \text{or} \quad a^2 = -1 + \sqrt{1 = 4E}.$$

We can write

$$2E - x^2 - \tfrac{1}{2}x^4 = \left(a^2 - x^2\right)\left(2 + a^2 + x^2\right).$$

Therefore, setting $x = a \cos \Theta$,

$$t_1 = \pm \frac{1}{\sqrt{1 + a^2}} \int_0^{\Theta_1} \frac{d\Theta}{\sqrt{1 - k^2 \sin^2 \Theta}} \quad \text{where } k^2 = \frac{a^2}{2(1 + a^2)},$$

and the period of oscillation is

$$T = \frac{4\sqrt{2}}{a} kK(k).$$

2. For the potential $U(x)$ in a more general form than in the problem (1), we can repeat the same argument. In a given $U(x)$, the motion is restricted in the range $a_1 \le x \le a_2$ in Fig. 8.5a. In this case, we can express that $2(E - U) = (x - a_1)(a_2 - x)V(x)$. Converting x to an angular variable Θ by

$$x = \frac{a_1 + a_2}{2} - \frac{a_1 - a_2}{2} \cos \Theta,$$

we can obtain

$$t_1 = \pm \int_0^{\Theta_1} \frac{d\Theta}{\sqrt{V(\Theta)}}.$$

Using this result for the potential $U(x) = \frac{1}{2}ax^2 + \frac{1}{4}bx^4$, confirm the previous formula in the problem (1).

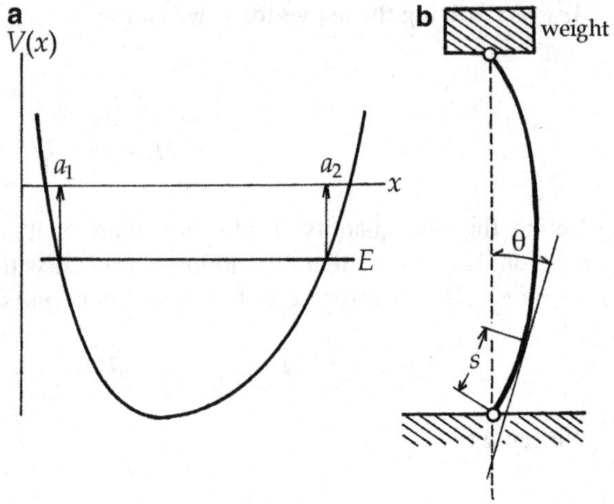

Fig. 8.9 (**a**) The oscillatory region $a_1<x<a_2$ for $E \le V(x)$. (**b**) A model for buckling of an elastic rod.

3. Anharmonic potentials as discussed in problems (1) and (2) represent the property of the spring force. On the other hand, we consider them as arising from condensates, which are temperature dependent in thermodynamic environment. The anharmonic part of the potential is considered as the adiabatic potential, although dynamically we cannot make it distinct from the spring force. While the crystal stability is attained by harmonic interaction, the anharmonicity arises from the lattice as an adiabatic potential. Discuss the physical origin of anharmonicity with respect to pseudospin–lattice interactions.
4. For stability of an elastic body, there is a nonlinear problem known as *buckling*, which is important for its structural stability. Consider a uniform elastic rod. When compressed by external forces F and $-F$ applied at both ends, the rod stays straight until F reaches a critical strength, beyond which it is stable in bent shape, as illustrated in Fig. 8.9.

   The condition for elastic stability can be obtained with a simple analysis. Denoting the radius of curvature by R, mechanical equilibrium can be described as the torque $F \times y$ proportional to R^{-1}, i.e., $Fy = cR^{-1}$ where c is a constant. On the other hand, $\frac{1}{R} = \frac{d\theta}{ds}$, where θ and s are the angle and the length of the curved rod measured from one end, as indicated in the figure. Therefore, using the relation $\frac{dy}{ds} = \sin\theta$, we obtain

$$\frac{d^2\theta}{ds^2} - \zeta \sin\theta = 0 \quad \text{where } \zeta = \frac{F}{c}.$$

   This is a sine-Gordon equation, whose solution was discussed in Chap. 7. The problem can be concluded by stating that the longitudinal length of the bent rod l is given by

$$\Lambda(\kappa) = l,$$

   for which the value of κ can be determined numerically.

It is interesting that the buckling phenomenon can be compared with an elliptic function for $\sigma_1 = \lambda\sigma_0 \mathrm{sn}\frac{\phi_1}{\sqrt{2K}}$; in a continuous crystal, we have such a similarity to buckling.

5. The soliton theory requires a nonlinear wave equation for applications. Physically, an external agent drives, in principle, the nonlinearity, but internal correlations are responsible in crystals. A one-dimensional wave equation represents a free propagation, so the theory makes us to assume the presence of a driving agent. Discuss this issue, regarding a wave propagation in crystalline states.

References

1. G.L. Lamb Jr., *Elements of Soliton Theory* (Wiley, New York, 1980)
2. P.M. Morse, H. Feshbach, *Methods of Theoretical Physics*, (McGraw-Hill, New York, 1953), p. 1651
3. M. Abramovitz, I.A. Stegan, *Handbook of Mathematical Functions*, (National Bureau of Standards Applied Math Series, 1964), p. 253
4. M. Fujimoto, Ferroelectrics **47**, 177 (1983)
5. X. Pan, H.-G. Unruh, J. Phys. Cond. Matter **2**, 323 (1990)

Chapter 9
Soft Modes

Soft modes were once regarded as direct evidence for binary structural changes, as implied by their frequencies diminishing toward the critical temperature to destabilize structure. In dielectric crystals, the condensate can be studied by applying time-dependent electric fields at the Brillouin-zone center. In contrast, neutron inelastic scatterings can be analyzed for soft modes at zone boundaries for cell-doubling transitions. Although the spatially distribution at T_c is implicit, soft modes can be studied for symmetry changes in crystals with varying temperature. Driven by the adiabatic potential, the soft mode exhibits a *thermal response* of condensates in crystals near T_c, as implied by the Born–Huang principle. In this chapter, soft modes are discussed as related to symmetry changes in the lattice.

9.1 The Lyddane–Sachs–Teller Relation

In ionic crystals, the electric polarization $\boldsymbol{P}(\boldsymbol{r},t)$ is associated with charge displacements related with lattice excitations. Induced by an applied electric field $\boldsymbol{E} = \boldsymbol{E}_o \exp(-i\omega t)$, where $\boldsymbol{E}_o$ and ω are the amplitude and frequency, respectively, $\boldsymbol{P}(\boldsymbol{r},t)$ is not only related to ionic mass displacements $\boldsymbol{u}(\boldsymbol{r},t)$ but also contributed by the *ionic polarization* expressed by $\alpha\boldsymbol{E}$, where α is called the ionic polarizability. At low frequencies, we can consider that these functions $\boldsymbol{P}(\boldsymbol{r},t)$ and $\boldsymbol{u}(\boldsymbol{r},t)$ constitute *continuous fields* in space–time, $\boldsymbol{r}$ and t, just as in the vibration field discussed in Chap. 2. Following Elliott and Gibson [1], we write the equation

$$\boldsymbol{P} = a\boldsymbol{u} + b\boldsymbol{E}, \tag{9.1}$$

M. Fujimoto, *Thermodynamics of Crystalline States*,
DOI 10.1007/978-1-4614-5085-6_9, © Springer Science+Business Media New York 2013

and for the polar lattice mode $\boldsymbol{u}$, we have an equation of motion

$$\frac{\partial^2 \boldsymbol{u}}{\partial t^2} = a'\boldsymbol{u} + b'\boldsymbol{E} \tag{9.2}$$

for a sufficiently weak $\boldsymbol{E}_o$; a, b and a', b' in (9.1) and (9.2) are constants.

Mass displacements $\boldsymbol{u}(\boldsymbol{r}, t)$ occur in propagating waves along the direction of an applied field $\boldsymbol{E}(\boldsymbol{r}, t)$, for which we can generally consider two independent modes that are characterized by $\mathrm{curl}\boldsymbol{u} = 0$ and $\mathrm{div}\boldsymbol{u} = 0$, known as *irrotational* and *laminar* component fields, respectively. Therefore, the vector $\boldsymbol{u}$ is generally expressed as $\boldsymbol{u} = \boldsymbol{u}_\mathrm{L} + \boldsymbol{u}_\mathrm{T}$; $\boldsymbol{u}_\mathrm{L}$ is defined by $\mathrm{curl}\boldsymbol{u}_\mathrm{L} = 0$ and $\mathrm{div}\boldsymbol{u}_\mathrm{L} \neq 0$, whereas $\boldsymbol{u}_\mathrm{T}$ is determined by $\mathrm{curl}\boldsymbol{u}_\mathrm{T} \neq 0$ and $\mathrm{div}\boldsymbol{u}_\mathrm{T} = 0$. Here the suffixes L and T indicate *longitudinal* and *transverse modes* with respect to the direction of $\boldsymbol{E}$.

Applying over the crystal, the electric field is written as $\boldsymbol{E} = \boldsymbol{E}_o \exp(-i\omega t)$ with constant amplitude $\boldsymbol{E}_o$. On the other hand, for internal variables $\boldsymbol{P}$ and $\boldsymbol{u}$, we write $\boldsymbol{P} = \boldsymbol{P}_o \exp(-i\omega t)$ and $\boldsymbol{u} = \boldsymbol{u}_o \exp(-i\omega t)$, where their amplitudes $\boldsymbol{P}_o$ and $\boldsymbol{u}_o$ are proportional to $\exp i\boldsymbol{k}\cdot\boldsymbol{r}$. From (9.1) and (9.2), we can relate these space-dependent amplitudes by

$$\boldsymbol{P}_o = a\boldsymbol{u}_o + b\boldsymbol{E}_o \quad \text{and} \quad -\omega^2\boldsymbol{u}_o = a'\boldsymbol{u}_o + b'\boldsymbol{E}_o,$$

where a, b, a', and b' are constants. Eliminating $\boldsymbol{u}_o$ from these relations, we obtain the relation

$$\boldsymbol{P}_o = \left(b + \frac{ab'}{-a' - \omega^2}\right)\boldsymbol{E}_o.$$

In the dielectric response function $\varepsilon(\omega)$ defined by $\boldsymbol{D}_o = \varepsilon(\omega)\boldsymbol{E}_o = \varepsilon_o\boldsymbol{E}_o + \boldsymbol{P}_o$, we consider the susceptibility χ by the relation $\boldsymbol{P}_o = \chi\varepsilon_o\boldsymbol{E}_o$, and the function $\varepsilon(\omega)$ can be expressed as

$$\varepsilon(\omega) - \varepsilon_o = b + \frac{ab'}{-a' - \omega^2},$$

and hence,

$$\varepsilon(\omega) = \varepsilon(\infty) + \frac{\varepsilon(0) - \varepsilon(\infty)}{1 - \dfrac{\omega^2}{\omega_o^2}}, \tag{9.3}$$

where $b = \varepsilon(\infty) - \varepsilon_o$ and $a' = -\omega_o^2$, and hence, $\frac{ab'}{\omega_o^2} = \varepsilon(0) - \varepsilon(\infty)$. Equation (9.3) is the dielectric function at an arbitrary point $\boldsymbol{r}$ at a constant frequency ω_o, while the displacement $\boldsymbol{u}_o$ at $\boldsymbol{r}$ depends on the strength of an applied field $\boldsymbol{E}_o$.

In a continuum crystal, we can decompose the displacement $\boldsymbol{u}(\boldsymbol{r},t)$ into *laminar* and *irrotational* components, as characterized by div$\boldsymbol{u}_{\mathrm{T}}=0$ and curl $\boldsymbol{u}_{\mathrm{L}}=0$, respectively. As amplitudes of these components $\boldsymbol{u}_{\mathrm{L}}$and $\boldsymbol{u}_{\mathrm{T}}$ are proportional to $\exp i\boldsymbol{k}\cdot\boldsymbol{r}$, we have the relations among these amplitudes $\boldsymbol{u}_{\mathrm{oL}}$,$\boldsymbol{u}_{\mathrm{oT}}$, and $\boldsymbol{k}$, namely,

$$i\boldsymbol{k}\cdot\boldsymbol{u}_{\mathrm{oT}}=0 \quad \text{or} \quad \boldsymbol{k}\perp\boldsymbol{u}_{\mathrm{oT}}$$

and

$$i\boldsymbol{k}\times\boldsymbol{u}_{\mathrm{oL}}=0 \quad \text{or} \quad \boldsymbol{k}\parallel\boldsymbol{u}_{\mathrm{oL}}.$$

Hence, the applied field is decomposed as $\boldsymbol{E}=\boldsymbol{E}_{\mathrm{L}}+\boldsymbol{E}_{\mathrm{T}}$ where $\boldsymbol{E}_{\mathrm{T}}\perp\boldsymbol{u}_{\mathrm{L}}$ and $\boldsymbol{E}_{\mathrm{L}}\parallel\boldsymbol{u}_{\mathrm{L}}$, respectively, and dielectric response functions can be separately obtained by $\boldsymbol{E}_{\mathrm{L}}$ and $\boldsymbol{E}_{\mathrm{T}}$.

At sufficiently low frequencies, the electric displacement $\boldsymbol{D}$ should satisfy the basic equation div$\boldsymbol{D}=0$. Hence, we write div$\boldsymbol{P}=-\varepsilon_{\mathrm{o}}div\boldsymbol{E}$, and from (9.1), we have the relation $a$div$\boldsymbol{u}=-(\varepsilon_{\mathrm{o}}+b)div\boldsymbol{E}$. Thus, for a transverse mode, we can write div $\boldsymbol{u}_{\mathrm{T}}=0$ and div$\boldsymbol{E}_{\mathrm{T}}=0$. On the other hand, a longitudinal mode is characterized by div$\boldsymbol{u}_{\mathrm{L}}\neq 0$, so that we have

$$\mathrm{div}\boldsymbol{E}_{\mathrm{L}}=-\frac{a}{\varepsilon_{\mathrm{o}}+b}\mathrm{div}\boldsymbol{u}_{\mathrm{L}} \quad \text{and} \quad -\omega_{\mathrm{L}}^{2}\boldsymbol{u}_{\mathrm{L}}=a'\boldsymbol{u}_{\mathrm{L}}+b'\boldsymbol{E}_{\mathrm{L}}.$$

Therefore, we have separate equations of propagation for $\boldsymbol{u}_{\mathrm{T}}$ and $\boldsymbol{u}_{\mathrm{L}}$, namely,

$$\frac{\partial^{2}\boldsymbol{u}_{\mathrm{T}}}{\partial t^{2}}=a'\boldsymbol{u}_{\mathrm{T}}=-\omega_{\mathrm{o}}^{2}\boldsymbol{u}_{\mathrm{T}}$$

and

$$\frac{\partial^{2}\boldsymbol{u}_{\mathrm{L}}}{\partial t^{2}}=\left(a'+\frac{ab'}{\varepsilon_{\mathrm{o}}+b}\right)\boldsymbol{u}_{\mathrm{L}}=-\frac{\varepsilon(0)}{\varepsilon(\infty)}\omega_{\mathrm{o}}^{2}\boldsymbol{u}_{\mathrm{L}},$$

where

$$\omega_{\mathrm{oT}}=\omega_{\mathrm{o}} \quad \text{and} \quad \omega_{\mathrm{oL}}=\omega_{\mathrm{o}}\sqrt{\frac{\varepsilon(0)}{\varepsilon(\infty)}}$$

are characteristic frequencies of transverse and longitudinal modes. Accordingly, for these two modes, we find the relation between frequencies and dielectric constants, that is,

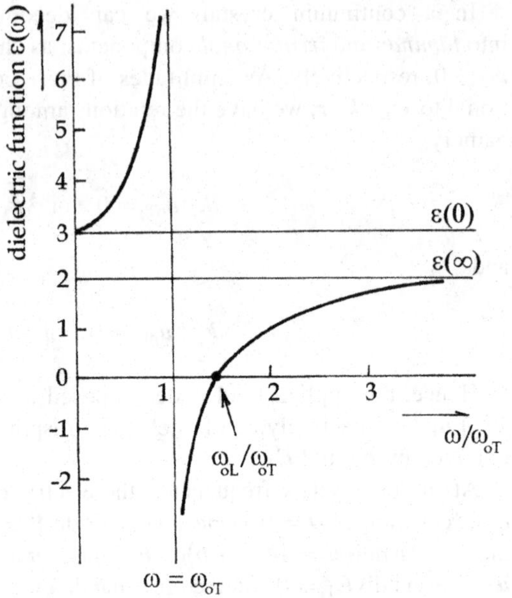

Fig. 9.1 Dispersion curve for a dielectric function $\varepsilon(\omega)$. A forbidden gap between $\varepsilon(0)$ and $\varepsilon(\infty)$; at $\omega = \omega_L$, we have $\varepsilon(\omega_L) = 0$.

$$\frac{\omega_{oL}^2}{\omega_{oT}^2} = \frac{\varepsilon(0)}{\varepsilon(\infty)}. \tag{9.4}$$

This is known as the Lyddane–Sachs–Teller (LST) formula.

The dielectric function (9.3) plotted against ω exhibits a *forbidden gap* $\varepsilon(0) - \varepsilon(\infty)$, as shown in Fig. 9.1, where there is a singularity at $\omega = \omega_{oT}$ and $\varepsilon(\omega_{oL}) = 0$ at $\omega = \omega_{oL}$. The LST relation is therefore consistent with the Landau theory in the mean-field approximation. For dielectric functions above and below the singular temperature T_o, we can write

$$\varepsilon_>(0) = \varepsilon_0(1 + \chi_>) \approx \varepsilon_0\chi_> \quad \text{and} \quad \varepsilon_<(0) = \varepsilon_0(1 + \chi_<) \approx \varepsilon_0\chi_<.$$

Considering that the Curie–Weiss laws (6.6a, b) determine the critical temperature T_o, we can express

$$\omega_{oT}^2(T > T_o) \propto T - T_o \quad \text{and} \quad \omega_{oT}^2(T < T_o) \propto T_o - T, \tag{9.5}$$

implying that frequencies of the transverse modes diminish or *soften*, if T_o is approached from both sides. We may say that soft modes arise from the LST formula, if combined with Curie–Weiss laws.

9.2 Soft Lattice Modes of Condensates

In the above theory, amplitudes $\boldsymbol{P}_o$ and $\boldsymbol{u}_o$ as functions of a position $\boldsymbol{r}$ are distributed in crystal space. As these are proportional to the exponential factor $\exp i\boldsymbol{k}\cdot\boldsymbol{r}$, experimental studies with time-dependent electric field can be performed with no reference to lattice points, if the condition $\exp i\boldsymbol{k}\cdot\boldsymbol{r} = 1$ is fulfilled, namely, either for $\boldsymbol{k} = 0$ at $\boldsymbol{r} = \boldsymbol{R}$ or for $\boldsymbol{k} = \frac{\boldsymbol{G}}{2}$ at $\boldsymbol{r} = 2\boldsymbol{R}$. These cases correspond to the center or boundaries of first Brillouin zone in the reciprocal lattice. In the previous section, we actually discussed the response of a dielectric polarization to an applied field at the zone center $\boldsymbol{G} = 0$. Similar soft-mode studies can be carried out at the zone boundary $\boldsymbol{k} = \frac{\boldsymbol{G}}{2}$ in neutron inelastic scattering experiments, as discussed in Chap. 10. Dielectric properties of crystals originate from displaced charges at the lattice sites $\boldsymbol{R}$; on the other hand, the neutron inelastic scatterings exhibit the response of mass displacements to a structural change associated with combing two unit cells in size $2\boldsymbol{R}$. In both of these cases, soft modes are due to condensates in thermodynamic environment.

9.2.1 *The Lattice Response to Collective Pseudospins*

In the above approximation, harmonic displacements $\boldsymbol{u}_o$ were assumed as given by (9.2). In the presence of correlations however, pseudospins are in collective propagation, driven by the internal adiabatic potential in modulated lattice, shifting frequency and damping energy via phonon scatterings. Accordingly, shifting soft-mode frequencies is significantly related to symmetry and anharmonicity of the adiabatic potential.

Considering longitudinal and transverse displacements $\boldsymbol{u}_L \parallel \boldsymbol{k}$ and $\boldsymbol{u}_T \perp \boldsymbol{k}$ from equilibrium positions, the adiabatic lattice potential in a condensate can be expressed by

$$\begin{aligned}\Delta U = {} & V^{(1)}\left(u_L + \sum_T u_T\right) + V^{(2)}\left(u_L^2 + \sum_T u_T^2\right)\\ & + V^{(3)}\left\{u_L\left(\sum_T u_T^2\right) + u_L^2\left(\sum_T u_T\right)\right\} + V^{(4)}\left\{u_L^4 + u_L^2\left(\sum_T u_T^2\right)\right\}\\ & + \cdots\cdots,\end{aligned} \tag{9.6}$$

where the summation $\sum_T \ldots$ is for transverse directions along the y and z axes, and $V^{(1)}, V^{(2)}, V^{(3)}, V^{(4)}, \ldots..$ are expansion coefficients of the corresponding powers 1, 2, 3, 4, ... Here $u_L$and u_T are finite but small displacements at temperatures T in the vicinity of T_c. The energy ΔU expressed by (9.6) can then be truncated at the fourth order, composed of component terms of $V^{(1)} \dots V^{(4)}$, if u_L and u_L are sufficiently small.

Expressing the phonon wavefunction by $|\boldsymbol{q}, \omega\rangle$, we can calculate the matrix elements $\langle \boldsymbol{q}', \omega'|\Delta U|\boldsymbol{q}, \omega\rangle$ for the isothermal average $\langle \Delta U\rangle_T$ at T. The secular part, written as ΔU_s, for brevity, can be written as

$$\Delta U_s = \Delta U_{sL} + \Delta U_{sT},$$

where the longitudinal component is

$$\Delta U_{sL} = \left(V^{(2)} + V^{(4)}u_{oL}^2\right)u_L^2 + V^{(4)}u_L^4 \tag{9.7a}$$

and the transverse potential can be decomposed as

$$\Delta U_{sT} = \Delta U_{s+} + \Delta U_{s-},$$

where

$$\Delta U_{s+} = \left(V^{(2)} + V^{(4)}u_{oL}^2\right)u_+^2, \quad \dots, \Delta U_{s-} = \left(V^{(2)} + V^{(4)}u_{oL}^2\right)u_-^2, \dots\dots \tag{9.7b}$$

and $\sum_T u_T^2 = u_+^2 + u_-^2$. On the other hand, the time-dependent term arises from matrix elements $\langle \boldsymbol{q}', \omega'|u_T|\boldsymbol{q}, \omega\rangle$ and $\langle \boldsymbol{q}', \omega'|u_T^3|\boldsymbol{q}, \omega\rangle$ for $\boldsymbol{q}' \neq \boldsymbol{q}$, which are subjected to *inelastic phonon scatterings* at T. The temperature-dependent terms consist of these nonzero elements for $\boldsymbol{q}' - \boldsymbol{q} = \boldsymbol{k}$ and $\omega' - \omega = \Delta\omega$, related to *inelastic phonon scatterings*. Taking the term of $V^{(1)}$, for example, we have

$$\langle \boldsymbol{q}',\ \omega'|\Delta U^{(1)}|\boldsymbol{q},\ \omega\rangle = V^{(1)}\langle \boldsymbol{q}'|u_{L,T}|\boldsymbol{q}\rangle \exp i(t\Delta\omega),$$

the scattering probability can thereby be expressed as the time average of

$$P = 2\left|V^{(1)}\right|^2 \langle \boldsymbol{q}'|u_{L,T}|\boldsymbol{q}\rangle^2 \frac{1 - \cos(t\Delta\omega)}{\Delta\omega^2}$$

integrated over phonon collision time t_o. Calculating $\langle P\rangle_t = \frac{1}{2t_o}\int_{-t_o}^{+t_o} P dt$, we obtain

$$\langle P\rangle_t = 2\left|V^{(1)}\right|^2 \langle \boldsymbol{q}'|u_{L,T}|\boldsymbol{q}\rangle^2 \frac{\sin(t_o\Delta\omega)}{t_o\Delta\omega}, \tag{9.8}$$

where the factor $\frac{\sin(t_o\Delta\omega)}{t_o\Delta\omega}$ is nearly equal to 1, if $t_o < \frac{1}{\Delta\omega}$. For modes $u_L = u_{oL}\exp(-i\omega t)$ and $u_T = u_{oT}\exp(-i\omega t)$, the probability at temperature T can be calculated with (9.8) for inelastic phonon scatterings $\boldsymbol{q}' - \boldsymbol{q} = \boldsymbol{k}$, while $\hbar(\omega' - \omega) = \hbar\Delta\omega$ represents the energy transfer to the surroundings, which is considered to be equal to $k_B T$ at a temperature T. Thus, the factor $\left|V^{(1)}\right|^2$ represents energy damping from the u_L mode. Damping is also contributed by the third-order term $V^{(3)}u_T^3$ and higher terms, so that the overall damping can effectively be expressed by an empirical constant γ as in the following analysis. Setting u_L aside, the equations of motion for the transverse mode u_T can be written as

$$m\left(\frac{d^2u_T}{dt^2}+\gamma\frac{du_T}{dt}+\omega_o^2 u_T\right)=-\frac{\partial U_{s\pm}}{\partial x}=F_{s\pm}\exp(-i\omega t), \tag{9.9}$$

where $F_{s\pm}=-\left(V^{(2)}+V^{(4)}u_{oL}^2\right)\frac{\partial u_\pm^2}{\partial x}$ that is the driving force; ω_o is the characteristic frequency and γ the damping constant of the unforced mode.

Equation (9.9) represents a harmonic oscillator forced by an external force, which can be used for a zone-center transition at $\boldsymbol{G}=0$ perturbed by an applied electric field; this also is applicable to a zone-boundary transition at $\frac{1}{2}\boldsymbol{G}$, where the momentum loss of neutrons is equal to $\frac{1}{2}\hbar\boldsymbol{G}$, and $\boldsymbol{q}'-\boldsymbol{q}=\frac{1}{2}\boldsymbol{G}+\boldsymbol{k}$.

Using $u_T=u_{o\pm}\exp(-i\omega t)$in (9.9), we obtain the steady solution of (9.9)

$$m\left(-\omega^2-i\omega\gamma+\omega_o^2\right)u_{o\pm}=F_{s\pm},$$

from which we can define the *complex susceptibility* as

$$\chi(\omega)=\frac{mu_{o\pm}}{F_{s\pm}}=\frac{1}{\omega_o^2-\omega^2-i\omega\gamma}. \tag{9.10}$$

Writing this in a complex form $\chi(\omega)=\chi'(\omega)-i\chi''(\omega)$, the real and imaginary parts of (8.10) can be expressed as

$$\chi'(\omega)=\frac{\omega_o^2-\omega^2}{\left(\omega_o^2-\omega^2\right)^2+\omega^2\gamma^2} \quad\text{and}\quad \chi''(\omega)=\frac{\omega\gamma}{\left(\omega_o^2-\omega^2\right)^2+\omega^2\gamma^2}, \tag{9.11}$$

respectively. Real and imaginary parts of the complex susceptibility are plotted against ω in Fig. 9.2. The imaginary part $\chi''(\omega)$ is a dumbbell-shaped symmetric curve, if $\gamma<\frac{|\omega_o^2-\omega^2|}{\omega^2}\approx 2\Delta\omega$, where $\Delta\omega=\omega-\omega_o$. Otherwise, $\chi''(\omega)\sim\frac{1}{\omega\gamma}$ for $\gamma>\Delta\omega$, showing no peak at $\omega=\omega_o$, which is *overdamped*; otherwise, the case for $\gamma<\Delta\omega$ is called *underdamped*. Neutron experiments have revealed that the energy loss expressed as proportional to $\chi''(\omega_o)$.

Experimentally, the characteristic frequency ω_o of soft modes becomes close to zero in the vicinity of T_c, where the equation (9.9) is dominated by the damping term. Therefore, we write

$$\gamma\frac{du_T}{dt}+\omega_o^2 u_T=\frac{F_{o\pm}}{m}\exp(-i\omega t).$$

Writing $\frac{\omega_o^2}{\gamma}=\frac{1}{\tau}$ and $\frac{F_{o\pm}}{m\gamma}=F'$, this equation is expressed as

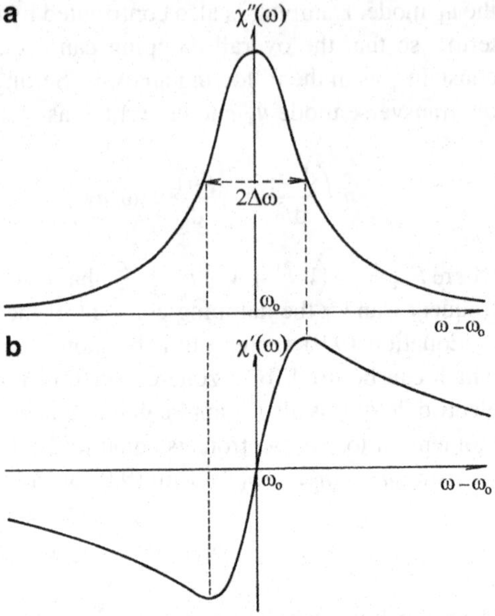

Fig. 9.2 A complex susceptibility $\chi(\omega) = \chi'(\omega) - i\chi''(\omega)$. (**a**) Imaginary part $\chi''(\omega)$ and (**b**) real part $\chi'(\omega)$.

$$\frac{\mathrm{d}u_T}{\mathrm{d}t} + \frac{u_T}{\tau} = F' \exp(-i\omega t), \tag{9.12}$$

where τ is a *relaxation time*. Equation (9.12) is known as Debye relaxation. Setting $u_T = A\exp(-i\omega t)$, we obtain the susceptibility for such a *relaxator*, that is,

$$\chi_D(\omega) = \frac{1}{-i\omega + \dfrac{1}{\tau}} = \frac{\tau}{1+\omega^2\tau^2} + \frac{i\omega\tau^2}{1+\omega^2\tau^2}. \tag{9.13}$$

Equations (9.12) and (9.13) are known as Debye relaxation, where τ and F' are related to γ and $F_{s\pm}$ in the overdamped soft mode. In practical crystals with lattice imperfections, such a relaxation formula are also valid, if $\omega\tau \ll 1$, although ω is not necessarily low. Nevertheless, the soft mode near zero frequencies looks like Debye *relaxator*.

Figure 9.3 shows an example of dielectric functions from ferroelectric crystals of $BaTiO_3$ obtained by Petzelt and his coworkers [2], showing the presence of soft modes in the paraelectric phase, as characterized by dispersive $\varepsilon'(\omega)$ of Debye's type and absorptive $\varepsilon''(\omega)$ of dumbbell shape. In observed squared frequencies ω_o^2 versus $T_c - T$ shown in Fig. 9.4 [3], there is a noticeable deviation from the straight line predicted by the mean-field theory.

9.2.2 Temperature Dependence of Soft-Mode Frequencies

Equation (9.9) represents a dynamical oscillator in thermodynamic environment. Experimentally, we know that the damping constant γ and characteristic frequency

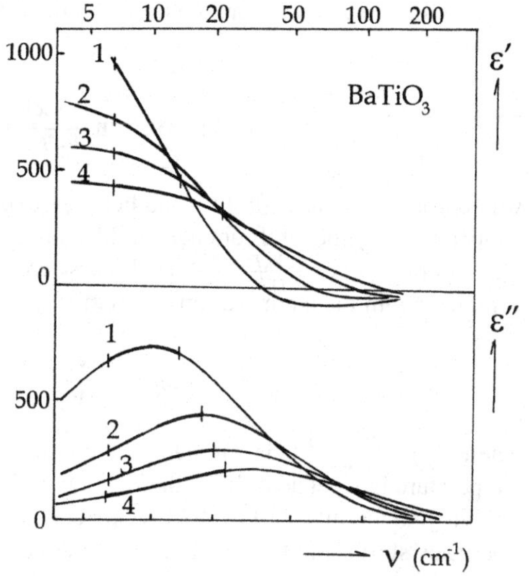

Fig. 9.3 Dielectric soft-mode spectra $\varepsilon(\omega)$ observed from $BaTiO_3$ crystals by backward-wave techniques. Curves for 1, 2, 3, and 4 were obtained at temperatures 474, 535, 582, and 667 K, respectively. From J. Petzelt, G.V. Kozlow, and A.A. Volkov, Ferroelectrics **73**, 101 (987).

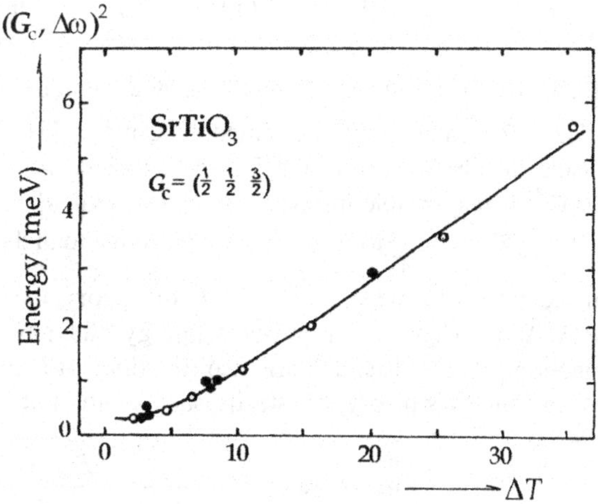

Fig. 9.4 A plot of squared soft-mode frequencies versus $T - T_c$ obtained by neutron scatterings from $SrTiO_3$. From R. Currat, K.A. Müller, W. Berlinger, and F. Desnoyer, Phys. Rev. **B17**, 2937 (1978).

ω_0 are temperature dependent, which are attributed to phonon scatterings in crystals.

Under a constant pressure, the crystal structure is strained by the energy ΔU, resulting in a small volume change. In this case, we need to confirm if such an adiabatic potential can cause a frequency change in the phonon spectra. Assuming that a strained lattice is an isotropic and continuous medium, the Gibbs potential of the lattice can be written as

$$G(\boldsymbol{k}) = \mathrm{k_B}T \ln \frac{\hbar\omega(\boldsymbol{k})}{\mathrm{k_B}T} + \frac{(\Delta V)^2}{2\kappa V},$$

where κ is the compressibility, and here $\frac{\Delta V}{V}$ expresses *volume strains* that may be temperature dependent. From the equilibrium condition, we set $\left(\frac{\partial G}{\partial V}\right)_{\mathrm{T}} = 0$ to obtain the relation $\frac{\Delta V}{V} = -\kappa \frac{\mathrm{k_B}T}{\omega(\boldsymbol{k})} \frac{\partial\omega(\boldsymbol{k})}{\partial V}$ in the first order. On the other hand, from (2.36) combined with the equipartition theorem $U_{\mathrm{vib}} = \mathrm{k_B}T$, we can write

$$\frac{\Delta V}{V} = -\kappa\Delta p = -\frac{\kappa\gamma}{V} U_{\mathrm{vib}} = -\frac{\kappa\gamma \mathrm{k_B}T}{V},$$

where $\gamma = -\frac{\mathrm{d}\ln\omega(\boldsymbol{k})}{\mathrm{d}\ln V}$ is Grüneisen's constant, which is normally assumed as temperature independent. If so, the frequency $\omega(\boldsymbol{k})$ may change only adiabatically, implying no thermal shift of $\omega(\boldsymbol{k})$ is expected in isotropic crystals; hence, a thermal frequency shift should be attributed to time-dependent ΔU_{t}.

Disregarding the volume change, we consider ΔU_{t} that can be written as

$$\Delta U_{\mathrm{t}} = V^{(3)} u_{\mathrm{L}}^2 (u_y \pm iu_z) + V^{(4)} u_{\mathrm{L}}^2 \left(u_y^2 + u_z^2\right),$$

for which $\sum_{\mathrm{T}} u_{\mathrm{T}}$ in (9.6) is replaced by a complex displacement $u_y \pm iu_z$ in a y–z plane, for calculating convenience. Cowley [4] showed that inelastic phonon scatterings by the term of $V^{(3)}$ in ΔU_{t} include $u_y \pm iu_z$, whereas the second term of $V^{(4)}$ is responsible for elastic scatterings via $u_y^2 + u_z^2$. In (9.6), the damping term $\left(V^{(1)} + V^{(3)} u_{\mathrm{L}}^2\right) \sum_{\mathrm{T}} u_{\mathrm{T}}$ is also complex, so the total damping factor can be expressed in a complex form as $\gamma = \Gamma - i\Phi$. In his theory, the secular frequency is written as $\omega(\boldsymbol{k}, \Delta\omega)$, where $\Delta\omega$ represents energy changes $\pm\hbar\Delta\omega$ during the inelastic phonon process; Γ and Φ are also functions of $\boldsymbol{k}$ and $\Delta\omega$.

In Cowley's theory, the steady-state solution of (9.9) can be written as

$$u_{\mathrm{o}\pm}(\boldsymbol{k}, \Delta\omega) = \frac{F_{\mathrm{s}\pm}/m}{-\omega(\boldsymbol{k})^2 + \omega_{\mathrm{o}}^2 - i\,\omega(\boldsymbol{k})(\Gamma - i\Phi)},$$

for which the squared characteristic frequency can be expressed effectively as

$$\omega(\boldsymbol{k}, \Delta\omega)^2 = \omega_{\mathrm{o}}^2 + \omega(\boldsymbol{k})\Phi(\boldsymbol{k}, \Delta\omega). \tag{9.14}$$

Here, the factor $\Phi(\boldsymbol{k}, \Delta\omega)$ is contributed by three parts, namely,

$$\Phi(\boldsymbol{k}, \Delta\omega) = \Phi_{\mathrm{o}} + \Phi_1(\boldsymbol{k}) + \Phi_2(\boldsymbol{k}, \Delta\omega),$$

where

$$\Phi_o = \frac{\partial \omega(\boldsymbol{k})}{\partial V}\Delta V$$

is negligible for $\Delta V = 0$ as discussed earlier in this section. The second term is expressed as

$$\Phi_1(\boldsymbol{k}) = \frac{\hbar}{N\omega(\boldsymbol{k})}\frac{\left(n+\frac{1}{2}\right)V^{(4)}u_L^2}{\omega'}\langle -\boldsymbol{K}, \boldsymbol{K}|u_T^2|\boldsymbol{K}', -\boldsymbol{K}'\rangle,$$

where n is the number of scattered phonons $(\boldsymbol{K}', \omega')$, giving rise to a temperature-dependent frequency shift with $\Phi_1(\boldsymbol{k}) \propto T$, if combining with the equipartition theorem $\hbar\omega'\left(n+\frac{1}{2}\right) = \mathrm{k_B}T$.

The expressions for $\Phi_2(\boldsymbol{k}, \Delta\omega)$ and for $\Gamma(\boldsymbol{k}, \Delta\omega)$ are similar, thereby the effective damping parameter becomes

$$\begin{aligned}2\Gamma(\boldsymbol{k}, \Delta\omega) &= \frac{\pi\hbar}{16N\omega(\boldsymbol{k})}\sum_{1,2}\frac{\left|V^{(3)}u_L^2\langle \boldsymbol{k}|u_T|\boldsymbol{K}_1, \boldsymbol{K}_2\rangle\right|^2}{\omega_1\omega_2} \\ &\times \left[(n_1+n_2+1)\{-\delta(\omega+\omega_1+\omega_2)+\delta(\omega-\omega_1-\omega_2)\}\right. \\ &\left. -(n_1-n_2)\{-\delta(\omega-\omega_1+\omega_2)+\delta(\omega+\omega_1-\omega_2)\}\right].\end{aligned}$$

Considering two temperatures $T_1 = \frac{\hbar\omega(\boldsymbol{k})}{\mathrm{k_B}}(n_1+n_2+1)$ and $T_2 = \frac{\hbar\omega(\boldsymbol{k})}{\mathrm{k_B}}(n_1-n_2)$, this damping term indicates that the phonon energy $\hbar\omega(2n_2+1)$ is transferred to the heat reservoir, which is equal to $\mathrm{k_B}(T_1-T_2)$.

It was controversial, if the soft-mode frequency converges to zero. Experimentally it is a matter of timescale of observation; however, no answer can be given to this question, because the transition threshold is always obscured by unavoidable fluctuations. Considering that $\omega(\boldsymbol{k})$ is determined at $\boldsymbol{k} = 0$ and signified by $\Delta\omega$ for a soft mode, that is, $\omega(0) = \omega_o \pm \Delta\omega$, we can write (9.14) for the effective characteristic frequency $\omega(0, \Delta\omega)$ as

$$\omega(0, \Delta\omega)^2 = \omega_o^2 + \Phi_1(0)(\omega_o \pm \Delta\omega).$$

This can be made equal to zero if we choose $\Delta\omega = \pm\frac{\Phi_1(0)}{4}$, in which case we obtain $\omega_o = \pm\frac{\Delta\omega}{2}$. Assuming that $\Delta\omega$ emerges at T_o, we may write $\omega_o^2 = A'T_o$. Further writing the second term in the above as $\Phi_1(0)(\omega_o \pm \Delta\omega) = \pm A'T$, we obtain $\omega(0, \Delta\omega)^2 = A'(T_o \pm T)$, which agree with Landau's hypothesis (9.5).

For a zone-boundary transition, a soft mode can be studied with intensities of inelastically scattered neutrons in the scattering geometry $\boldsymbol{K} + \boldsymbol{k} \to \boldsymbol{K}'$, where $\boldsymbol{K}$ and $\boldsymbol{K}'$ are wavevectors of neutrons. The timescale of neutron collision is significantly short, so that the critical anomalies can be studied by scanning scattering angles with varying temperature of the sample crystal. In this case, (9.14) can be expanded for small $|\boldsymbol{k}|$ as

$$\omega(\boldsymbol{k},\Delta\omega)^2=\omega(0,\Delta\omega)^2+\kappa \boldsymbol{k}^2+\ldots,$$

with a small constant κ, so that we can write approximate relations

$$\omega(\boldsymbol{k},\Delta\omega)^2=A'_{>}(T-T_{\rm o})+\kappa_{>}\boldsymbol{k}^2 \quad \text{for} \quad T>T_{\rm o}$$

and

$$\omega(\boldsymbol{k},\Delta\omega)^2=A'_{<}(T_{\rm o}-T)+\kappa_{<}\boldsymbol{k}^2 \quad \text{for} \quad T<T_{\rm o}, \tag{9.15}$$

which are useful formula for the intensity analysis of scattered neutrons in a fixed geometry, as shown by Fig. 9.4 [5].

9.2.3 Cochran's Model

Temperature-dependent lattice modes were observed in early spectroscopic studies on structural transitions, but it was after Cochran' theory (1690) [6] that such modes are called soft modes. Based on a simplified model of an ionic crystal, he showed that the frequency softening can occur as a consequence of counter-acting short- and long-range interactions. In this section, we review Cochran's theory but modifying that ionic displacements for a condensate in a crystal, assuming displaceable charges with effective masses that are considerably lighter than ions.

If a particle of mass m and charge $+e$ is displaced by u from its site, a *hole* of mass m' and charge $-e$ is left behind, creating a dipole moment $\sigma=eu$ that is regarded as the order variable.

Dynamically, such a dipole σ is in motion with respect to the center of mass, characterized by the reduced mass $\mu=\frac{mM}{m+M}$, where m and M are masses of the charge and the ion with a hole. Nevertheless, as $\mu\approx m$ if $m<M$, we can disregard the lattice structure in Cochran's model.

In his model, one-dimensional chain of mass particles m and $m'$is considered in an applied field$E=E_{\rm o}\exp i\omega t$. We assume that these dipoles $\sigma_{\rm i}=eu_{\rm i}$ are correlated between neighboring sites, so that the crystal as a whole is polarized as expressed by the polarization P along the direction of the chain. Accordingly, we consider a depolarizing field as expressed by $-\frac{P}{\varepsilon_{\rm o}}$ along the direction of the chain. Further, taking dipolar interactions from all other chains into account, Cochran included the Lorentz field $-\frac{P}{3\varepsilon_{\rm o}}$, assuming a uniform crystal. Acceptable in the mean-field accuracy, his theory can bring a temperature dependence into dynamic equations through the temperature-dependent polarization P.

Denoting the transverse displacements of charges $+e$ and $-e$ by $u_{\rm T}^{+}$ and $u_{\rm T}^{-}$, respectively, we can write equations of motion as

$$m\frac{d^2u_T^+}{dt^2} = -C(u_T^+ - u_T^-) + e\left(E + \frac{P}{3\varepsilon_0}\right) \quad \text{and} \quad m'\frac{d^2u_T^-}{dt^2} = -C(u_T^- - u_T^+) - e\left(E + \frac{P}{3\varepsilon_0}\right),$$

where C is a restoring constant that represents a short-range interaction. Assuming $P = \frac{e(u_T^+ - u_T^-)}{v}$, where v is the specific volume of a crystal, these equations can be combined as

$$\mu\frac{d^2P}{dt^2} = \frac{e^2}{v}\left(E + \frac{P}{3\varepsilon_0}\right) - CP. \tag{9.16a}$$

For a given field $E = E_0 \exp i\omega_T t$, we set $P = P_0 \exp i\omega_T t$ in (9.16a) and obtain the expression for the susceptibility

$$\chi(\omega_T) = \frac{P_0}{\varepsilon_0 E_0} = \frac{\frac{e^2}{\varepsilon_0 v}}{C - \frac{e^2}{3\varepsilon_0 v} - \mu\omega_T^2},$$

which shows a singularity if $\mu\omega_T^2 = C - \frac{e^2}{3\varepsilon_0 v}$. Therefore, if the condition $C = \frac{e^2}{3\varepsilon_0 v}$ is met, ω_T can be zero, for which $\chi(\omega_T)$ or $P(\omega_T)$ can be regarded to represent a soft mode.

On the other hand, if charges $e, -e$ are signified by longitudinal displacements u_L^+ and u_L^-, respectively, by defining $P = \frac{e(u_L^+ - u_L^-)}{v}$, we have the equation

$$\mu\frac{d^2P}{dt^2} = \frac{e^2}{v}\left(E - \frac{P}{\varepsilon_0} + \frac{P}{3\varepsilon_0}\right) - CP, \tag{9.16b}$$

from which the longitudinal susceptibility can be written as

$$\chi(\omega_L) = \frac{\frac{e^2}{\varepsilon_0 v}}{C + \frac{2e^2}{3\varepsilon_0 v} - \mu\omega_L^2}.$$

Since C is a positive constant, the frequency determined by $\mu\omega_L^2 = C + \frac{2e^2}{3\varepsilon_0 v}$ cannot vanish. In Cochran's theory, the soft mode should be a transverse mode, and both ω_T and ω_L can be temperature dependent.

For a binary displacive structural change, we consider that pseudospins emerged at T_c^* and then clustered in periodic lattice at T_c. In the process of condensate formation, the corresponding lattice displacement u_T of clustered pseudospins should be different from unclustered ones dynamically. Therefore, in the following, we specify these displacements by u_T and v_T, respectively, which are pinned and observed in practical experiments.

We can assume that these two modes are coupled in such a way as $v_T = cu_T$, where c is temperature dependent. For the coupled system, the equations of motion can be written as

$$\frac{d^2u_T}{dt^2} + \gamma\frac{du_T}{dt} + \gamma'\frac{dv_T}{dt} + \omega_0^2 u_T = \frac{F_{s\pm}}{m}\exp(-i\omega t)$$

and

$$\frac{dv_T}{dt} + \frac{v_T}{\tau} = F'\exp(-i\omega t),$$

for which the steady-state solutions can be expressed as

$$u_{0\pm}\left(-\omega^2 - i\omega\gamma + \omega_0^2\right) - i\omega\gamma' v_{0\pm} = \frac{F_{s\pm}}{m} \quad \text{and} \quad v_{0\pm}\left(-i\omega + \frac{1}{\tau}\right) = F'.$$

We can therefore derive the susceptibility function for the coupled oscillator,

$$\chi(\omega) = \frac{mu_{0\pm}}{F_{s\pm}} = \frac{1}{\omega_0^2 - \omega^2 - i\omega\gamma - ic\gamma' F'\dfrac{\omega\tau}{1 - i\omega\tau}}.$$

Writing $\delta^2 = c\gamma' F'$, this susceptibility formula is simplified as

$$\chi(\omega) = \frac{1}{\omega_0^2 - \omega^2 - i\omega\gamma - \dfrac{\delta^2\omega\tau}{1 - i\omega\tau}}. \tag{9.17}$$

If these two modes u and v are independent, that is, $\delta = 0$ or $c = 0$, (9.17) is characterized by a simple oscillator formula for the u-mode. If the observed susceptibility contains such an additional independent response as given by (9.13) in the vicinity of $\omega = 0$, we have

$$\chi(\omega) = \frac{1}{\omega_0^2 - \omega^2 - i\omega\gamma} + \frac{\tau}{1 - i\omega\tau}. \tag{9.18}$$

Nevertheless, at $\omega_0 \approx 0$, these two terms at a very low frequency become overlapped and indistinguishable unless $\gamma > \frac{1}{\tau}$. Experimentally however, such a response of Debye type happens to be a distinctively sharp peak and observed distinctively in zone-boundary transitions at near-zero frequency, which is called a *central peak*, as shown in Fig. 9.5 [7].

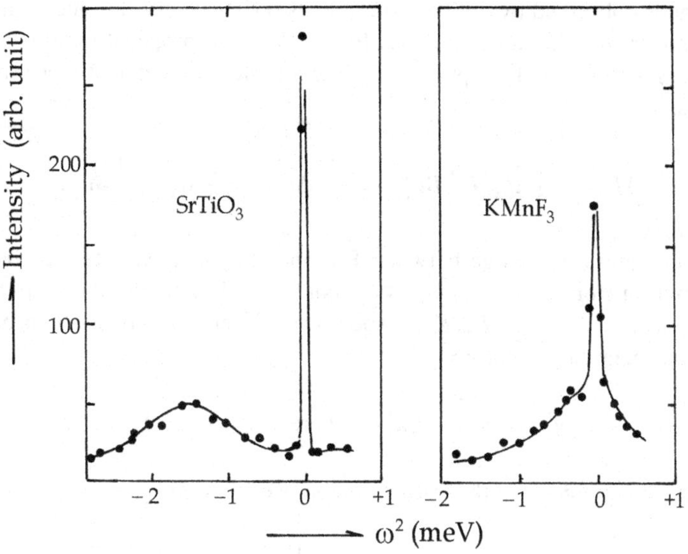

Fig. 9.5 Soft-mode spectra from $SrTiO_3$ and $KMnF_3$, consisting of a soft-mode and a central peak. From S.M. Shapiro, J.D. Axe, G. Shirane, and T. Riste, Phys. Rev. **B6**, 4332 (1972).

9.3 Symmetry Change at T_c

The presence of an additional mode ν_T near $\omega \approx 0$ is hypothetical in the foregoing one-dimensional theory. However, such a transverse mode is logical to be included in the response function at $\omega = 0$. In fact, the mode ν_T is consistent with the observed soft-mode spectra in TSCC crystals, hence indicating symmetry change of the lattice at T_c.

A structural transformation is characterized macroscopically by a symmetry change in a crystal. We consider that a symmetry change occurs at T_c as induced by emerging adiabatic potentials. Considering such a potential as given by (9.7a) and (9.7b) becomes significant at the outset of pseudospin ordering along the x-axis, we write the potential expressed as

$$\Delta U_{tr} = V^{(3)}\left(u_y^2 + u_z^2\right) + V^{(4)} u_x^2 \left(u_y^2 + u_z^2\right) \tag{9.19}$$

for lowering symmetry via phonon scatterings by transverse displacements u_y and u_z. Noted that (9.7a), (9.7b), and (9.19) belong to the same class of a cubic symmetry group, we consider that a transition may take place for the lattice structure to deform at minimal energy. Figure 9.6a, b [8, 9] illustrate symmetry changes at transition temperatures T_c, as interpreted by observed soft-mode spectra.

Soft modes observed by optical experiments on TSCC crystals and identified as B_{2u} modes for $T > T_c$ and A_1 modes for $T < T_c$, and propagating along x and z directions, respectively. Except for the critical region, the adiabatic potentials are written as

$$\Delta U_{B_u} = \frac{1}{2}Au_x^2 + \frac{1}{4}Bu_x^4 \quad \text{and} \quad \Delta U_{A_1} = \frac{1}{2}Au_z'^2 + \frac{1}{4}Bu_z'^4.$$

For the symmetry change between B_{2u} and A_1, we postulate the coordinate transformation from $\pm u_x$ to $\pm u_y' \mp u_z'$, corresponding to phonon scatterings $K_x \leftrightarrow -K_x$ and $K_y, -K_z \leftrightarrow -K_y, K_z$, respectively. Then, the process can be determined by a linear combination

$$\boldsymbol{u}' = c_x \boldsymbol{u}'_x + c_y \boldsymbol{u}'_y \quad \text{where} \quad c_x^2 + c_y^2 = 1,$$

with which the potential in the critical region is expressed by

$$\Delta U_{\text{cr}} = \frac{1}{2}A\boldsymbol{u}'^2 + \frac{1}{4}B\boldsymbol{u}'^2 u_x^2, \tag{9.20}$$

where the second quartic term represents a coupling between $\Delta U_{B_{2u}}$ and ΔU_{A_1}. Therefore, we reexpress (9.20) as

$$\Delta U_{\text{cr}} = c_x^2\left(\frac{A}{2}{u'_x}^2 + \frac{B}{4}{u'_x}^4\right) + c_y^2\left(\frac{A}{2}{u'_y}^2 + \frac{B}{4}{u'_x}^2{u'_y}^2\right).$$

We assume that the critical state can change by $c_x^2 \to 0$ and $c_y^2 \to 1$ to establish the normal state, where we should have $u_x'^2 = -\frac{A}{B}$ in thermal equilibrium. Therefore,

$$\lim_{c_y \to 1} \Delta U_{\text{cr}} \to +\frac{A}{4}u_y^2. \tag{9.21}$$

This indicates that the A_1 mode below T_c is characterized by $\omega(0)^2 = 2A(T_o - T)$, which appears to be substantiated by the curves in Fig. 9.6. It is notable at this stage that the potential (9.20) is theoretically predictable by the soliton theory as calculated by the method of reverse scattering [10].

For phonon-energy curves for $SrTiO_3$ below 105 K in Fig. 9.6b, the quadratic potential should also be related effectively to anisotropic quartic potentials, depending on $\boldsymbol{k}$. These curves are determined by $\omega_<(0,\boldsymbol{k})^2 = A(\boldsymbol{k})(T_o - T) + \kappa \boldsymbol{k}^2$ where the coefficient $A(\boldsymbol{k})$depends on the direction of $\boldsymbol{k}$. Experimentally, near-linear relations between $\omega_<(0,0)^2$ and $T_o - T$ as shown in Fig. 9.4 were obtained for many structural transitions. Consistent with the mean-field theory however, we have no significant reason to support this linear relation. Hence, we usually write $\omega_<(0,0) \propto (T_o - T)^\beta$, where β is called a *critical exponent* that can be determined from the curves, but close to $\beta = \frac{1}{2}$ in mean-field approximation.

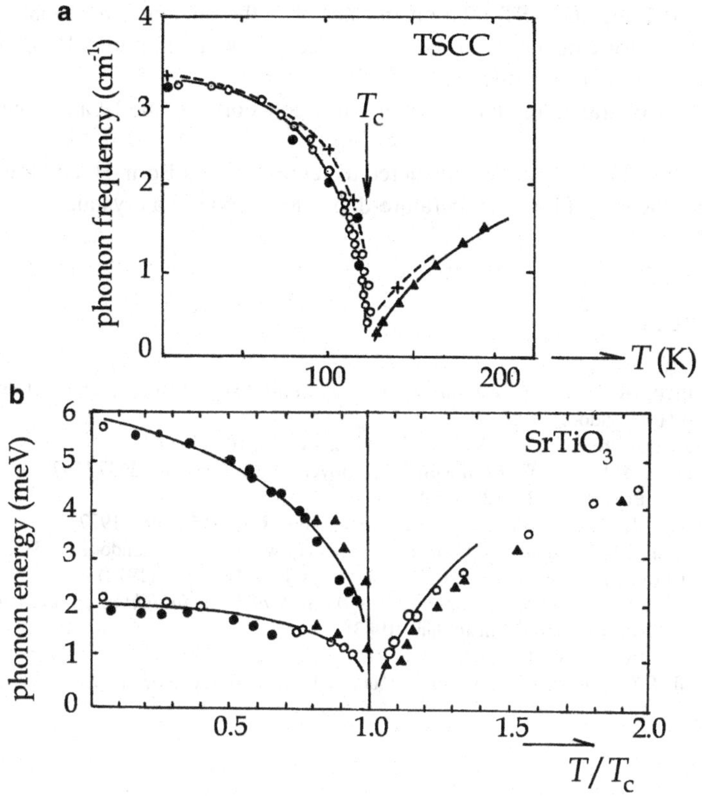

Fig. 9.6 Soft-mode spectra as a function of temperature: (**a**) from TSCC, (**b**) $SrTiO_3$. From J. F. Scott, Raman Spectroscopy of Structural Phase Transitions in *Light Scattering Near Phase Transitions*, ed. H.Z. Cummins and A.P. Levanyuk, North Holland, Amsterdam (1983). J. Feder and E. Pytte, Phys. Rev. **B1**, 4803 (1970).

Exercise 9

1. In Cochran's theory, the polarization $\boldsymbol{P}$ is a macroscopic variable. How his theory can be modified for a mesoscopic polarization?
2. In Landau's theory, he postulated that $A = A'(T - T_o)$ where A' is a positive constant; hence, A is positive for $T > T_o$ but negative for $T < T_o$. In practical crystals, A' is not necessarily a constant, depending on the wavevector $\boldsymbol{k}$ of the order parameter. Why? Discuss the mechanism for variable A'. Representing primarily a harmonic potential, but often including a quartic potential effectively, the soft-mode frequency exhibits therefore a different temperature dependence, as shown in Fig. 9.6a, b. Is higher-order potentials the acceptable explanation for different A'?

3. When deriving the LST relation in Sect. 9.1, the adiabatic potential was not considered for calculating $\varepsilon(\omega)$. Is it an acceptable assumption? If not, discuss why the potential was disregarded.
4. Review Cochran's model to justify his concepts of short- and long-range interactions.
5. Why the soft mode can be considered as representing a Born–Huang transition? Discuss the origin for a temperature-dependent mode in a crystal.

References

1. R.J. Elliott, A.F. Gibson, *An Introduction to Solid State Physics and Its Applications* (MacMillan, London, 1974)
2. J. Petzelt, G.V. Kozlov, V.V. Volkov, Ferroelectrics **73**, 101 (1987)
3. R. Currat, K.A. Müller, W. Berlinger, F. Desnoyer, Phys. Rev. **B17**, 2937 (1978)
4. A. Cowley, Prog. Phys. **31**, 123 (1968)
5. M. Iizumi, J.D. Axe, G. Shirane, K. Shimaoka, Phys. Rev. **B15**, 4392 (1977)
6. W. Cochran, *The Dynamics of Atoms in Crystals* (Edward Arnold, London, 1973)
7. S.M. Shapiro, J.D. Axe, G. Shirane, T. Riste, Phys. Rev. **B6**, 4332 (1972)
8. J.F. Scott, *Raman spectroscopy of structural phase transitions in light scattering near phase transition* (North-Holland, Amsterdam, 1983)
9. J. Feder, E. Pytte, Phys. Rev. **B1**, 4803 (1070)
10. Lamb GL Jr (1980) *Elements of soliton theory* Chap. 3, (Wiley, New York)

Chapter 10
Experimental Studies on Critical Fluctuations

In mesoscopic states of a crystal, properties of the condensates can be studied experimentally with renormalized variables, which are observable if they are pinned by intrinsic and extrinsic adiabatic potentials. Condensates in crystals are primarily longitudinal, propagating along the direction of wavevector, where the pseudospin arrangement and frequency dispersion can be sampled with varying temperature for $T < T_c$. With appropriate probes, we can obtain useful information about critical anomalies; *photons* are sensitive to the temporal variation, *neutrons* are for studying displacements in responsible constituents, and *magnetic resonance probes* yield information about structural changes. Such sampling experiments as X-ray diffraction, dielectric measurements, light and neutron inelastic scatterings, and magnetic resonances are outlined in this chapter, with relevant results to substantiate theoretical arguments.

10.1 Diffuse X-Ray Diffraction

Normally, the crystal structure is determined by X-ray diffraction studies. A collimated X-ray beam is scattered from a crystal, exhibiting a diffraction pattern on a photographic plate placed perpendicular to the beam direction. The diffraction pattern arises from elastic scatterings of X-ray photons by distributed electron densities in the crystal, keeping nuclear masses intact. In the critical region, on the other hand, the structure is modulated along symmetry directions, resulting in diffraction spots broadened in *anomalous* shape, which are called *diffuse diffraction.*

In early crystallographic studies, Bragg established the concept of *crystal planes* of constituents, like optical reflection planes, signifying periodic lattice structure in normal crystals. A collimated X-ray beam reflects from a large number of parallel crystal planes, showing an interference pattern on a photographic plate placed nearby. Figure 10.1 illustrates a practical arrangement commonly used for analyzing diffraction patterns.

M. Fujimoto, *Thermodynamics of Crystalline States*,

DOI 10.1007/978-1-4614-5085-6_10,

Fig. 10.1 X-ray diffraction setup. Incident X-ray beam $\boldsymbol{K}_o$, diffracted beam $\boldsymbol{K}$ on a conical surface. A modulated axis is indicated by $G_o \| a$.

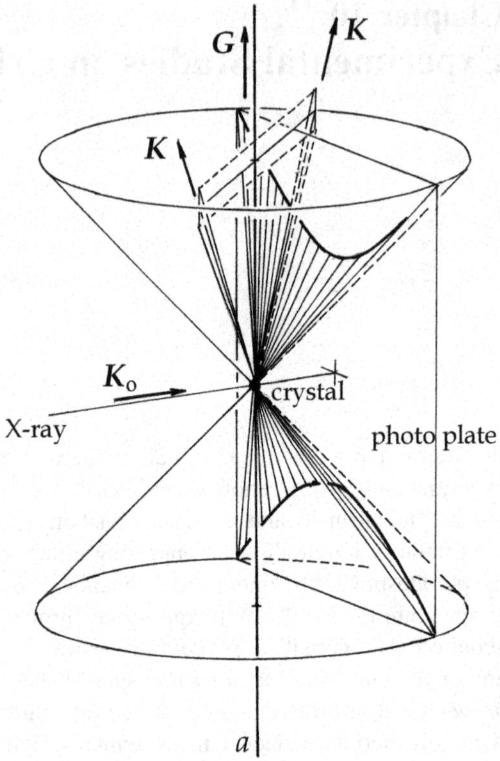

Considering that a collimated beam of monochromatic X-ray of a wavevector $\boldsymbol{K}_o$ and frequency ω_o is incident on crystal planes characterized by the normal vector $\boldsymbol{G}$, scattered wavevectors $\boldsymbol{K}$ and frequency ω are determined by Bragg's law

$$\boldsymbol{K}_o - \boldsymbol{K} = \pm \boldsymbol{G}, \tag{10.1a}$$

for which

$$|\boldsymbol{K}_o| = |\boldsymbol{K}| \quad \text{and} \quad \hbar\omega_o = \hbar\omega \tag{10.1b}$$

express the conservation laws of momentum and energy for elastic scatterings. Here, $\boldsymbol{G} = n\boldsymbol{a}$ is the lattice translation vector along the normal direction denoted by $\boldsymbol{a}$, where n is integer, representing a group of parallel crystal planes perpendicular to $\boldsymbol{a}$. We notice from (10.1b) that only the magnitude $|\boldsymbol{G}|$ is significant for the *elastically* scattered beam $\boldsymbol{K}$; diffractions occur in all directions satisfying the relation (10.1a). All of these diffracted vectors $\boldsymbol{K}$ are on *conical* surfaces with respect to the $\boldsymbol{a}$-axis, as shown in Fig. 10.1. Such conical diffractions constitute *Laue's law*, providing a useful method of constructing diffraction patterns. In this figure how the lattice modulation can occur along the $\boldsymbol{a}$-axis for a given wavevector $\boldsymbol{k}$ is illustrated.

Considering a charge element $\rho(\boldsymbol{r}_o)d^3\boldsymbol{r}$ that scatters an incident X-ray photon field $E_o \exp i(\boldsymbol{K}_o.\boldsymbol{r}_o - \omega t_o)$ at a point $\boldsymbol{r}_o$ and time t_o, the scattered field is *spherical* and expressed as

$$\boldsymbol{A}(\boldsymbol{r},t) \propto \int_{V(\boldsymbol{r}_o)} d^3\boldsymbol{r}_o \rho(\boldsymbol{r}_o) E_o \exp i(\boldsymbol{K}_o.\boldsymbol{r}_o - \omega t) \frac{\exp i\{\boldsymbol{K}.(\boldsymbol{r}-\boldsymbol{r}_o) - \omega(t-t_o)\}}{|\boldsymbol{r}-\boldsymbol{r}_o|}.$$

Hence, the scattered amplitude at a distance $r \gg r_o$ is determined by an approximate form $\boldsymbol{A}(\boldsymbol{r},t) \propto \frac{\exp i(\boldsymbol{K}.\boldsymbol{r}-\omega t)}{r} = A_o$, which can be reexpressed by

$$\frac{A_o}{E_o} \propto \int_{V(\boldsymbol{r}_o)} d^3\boldsymbol{r}_o\{\rho(\boldsymbol{r}_o)\exp i(\boldsymbol{K}_o - \boldsymbol{K}).\boldsymbol{r}_o\}. \tag{10.2}$$

Elastic scatterings from a rigid crystal are specified by the relation $\exp i(\pm \boldsymbol{G}.\boldsymbol{r}_o) = 1$; hence the scattered intensity ratio (10.2) is determined by maximum scattered amplitude, which is proportional to the net charge at the lattice point $\boldsymbol{r}_o$.

In practice, however, a collimated X-ray beam strikes a finite area on crystal planes, so that the reflected beam arises from all scatterers in the impact area. Letting positions of scatterers as $\boldsymbol{r}_{om}$ where $m = 1, 2, \ldots$ in the impact area, (10.2) can be reexpressed as

$$\frac{A_o}{E_o} = \sum_m f_m(\boldsymbol{G}) \exp(-i\boldsymbol{G}.\boldsymbol{r}_{om}), \tag{10.3}$$

where

$$f_m(\boldsymbol{G}) = \int_{V(\boldsymbol{r}_{om})} \rho(\boldsymbol{r}_{om}) d^3\boldsymbol{r}_{om}.$$

This is called the *atomic form factor*. However, in crystals electronic densities overlap significantly between neighboring atoms, so (10.3) overestimates individual contributions. In this context, we redefine the factors that are independent of overlaps of constituents.

Replacing $\boldsymbol{r}_{om}$ by a continuous variable $\boldsymbol{s} - \boldsymbol{r}_{om}$, we can write

$$\begin{aligned}\int_{V(\boldsymbol{s})} \rho(\boldsymbol{s}) \exp(-i\boldsymbol{G}.\boldsymbol{s}) d^3\boldsymbol{s} &= \int_{V(\boldsymbol{s})} \rho(\boldsymbol{s}-\boldsymbol{r}_{om}) \exp\{-i\boldsymbol{G}.(\boldsymbol{s}-\boldsymbol{r}_{om})\} d^3(\boldsymbol{s}-\boldsymbol{r}_{om}) \times \exp(-i\boldsymbol{G}.\boldsymbol{r}_{om}) \\ &= f_m(\boldsymbol{G}) \exp(-i\boldsymbol{G}.\boldsymbol{r}_{om}),\end{aligned}$$

where

$$f_{\mathrm{m}}(\boldsymbol{G}) = \int_{V(s)} \rho(\boldsymbol{s} - \boldsymbol{r}_{\mathrm{om}}) \exp\{-i(\boldsymbol{s} - \boldsymbol{r}_{\mathrm{om}})\} \mathrm{d}^3(\boldsymbol{s} - \boldsymbol{r}_{\mathrm{om}}). \tag{10.4}$$

Using (10.4), the formula (10.3) can be expressed as

$$S(\boldsymbol{G}) = \sum_{\mathrm{m}} f_{\mathrm{m}}(\boldsymbol{G}) \exp(-i\boldsymbol{G}.\boldsymbol{r}_{\mathrm{om}}) = \sum_{\mathrm{m}} f_{\mathrm{m}}(\boldsymbol{G}) \exp\left\{\frac{2\pi i}{\Omega}(x_{\mathrm{m}} h + y_{\mathrm{m}} k + z_{\mathrm{m}} l)\right\},$$

which is called the *structural form factor*; thereby the field of a scattered beam at a distant position $\boldsymbol{r}$ can be expressed as

$$\frac{E_G(\boldsymbol{r})}{E_{\mathrm{o}}} = \frac{S(\boldsymbol{G})}{r} \exp i(\boldsymbol{K}.\boldsymbol{r} - \omega t).$$

Using the structural factor, the intensity of the scattered beam is given by

$$\frac{I(\boldsymbol{G})}{I_{\mathrm{o}}} = \frac{1}{r^2} S^*(\boldsymbol{G}) S(\boldsymbol{G}). \tag{10.5}$$

In a crystal is modulated in a direction of the a-axis, the periodicity of pseudospins is modified by the wavevectors $\boldsymbol{G} \pm \mathbf{k}$ and the energies $\mp \Delta\varepsilon = \hbar(\omega - \omega_{\mathrm{o}}) = \mp \hbar \Delta\omega$, so that the scatterings should basically be inelastic. We therefore consider conservation laws

$$\boldsymbol{k}_{\mathrm{o}} - \boldsymbol{k} = \boldsymbol{k} \pm \boldsymbol{k} \quad \text{and} \quad \varepsilon(\boldsymbol{k}_{\mathrm{o}}) - \varepsilon(\boldsymbol{k}) = \mp \Delta\varepsilon(\boldsymbol{k}), \tag{10.6}$$

by which (10.2) can be revised for inelastic scatterings as

$$\begin{aligned} \frac{A_{\mathrm{o}}}{E_{\mathrm{o}}} &\propto \int \mathrm{d}^3\boldsymbol{r}_{\mathrm{o}} \rho(\boldsymbol{r}_{\mathrm{o}}) \exp i(\boldsymbol{K}_{\mathrm{o}}.\boldsymbol{r}_{\mathrm{o}} - \omega_{\mathrm{o}} t_{\mathrm{o}}) \frac{\exp i\{\boldsymbol{K}.(\boldsymbol{r} - \boldsymbol{r}_{\mathrm{o}}) - \omega(t - t_{\mathrm{o}})\}}{|\boldsymbol{r} - \boldsymbol{r}_{\mathrm{o}}|} \\ &\approx \int \mathrm{d}^3\boldsymbol{r}_{\mathrm{o}} \frac{\rho(\boldsymbol{r}_{\mathrm{o}}) \exp i(\boldsymbol{K}.\boldsymbol{r} - \omega t)}{r} \exp i\{(\boldsymbol{K}_{\mathrm{o}} - \boldsymbol{K}).\boldsymbol{r} - (\omega_{\mathrm{o}} - \omega) t_{\mathrm{o}}\}. \end{aligned}$$

Using (10.6) in this expression, we obtain

$$\begin{aligned} \frac{E_G(\boldsymbol{r})}{E_{\mathrm{o}}} \propto \frac{\exp i(\boldsymbol{K}.\boldsymbol{r} - \omega t)}{r} \sum_{\mathrm{m}} &\{f(\boldsymbol{G} + \boldsymbol{k}) \exp i(\boldsymbol{k}.\boldsymbol{r}_{\mathrm{m}} - \Delta\omega\, t_{\mathrm{o}}) \\ &+ f(\boldsymbol{G} - \boldsymbol{k}) \exp i(-\boldsymbol{k}.\boldsymbol{r}_{\mathrm{m}} + \Delta\omega\, t_{\mathrm{o}})\}, \end{aligned}$$

where

$$f(\boldsymbol{G}\pm\boldsymbol{k}) = \int\limits_{V(\boldsymbol{r}_o)} d^3\boldsymbol{r}_o\rho(\boldsymbol{r}_o)\exp i(\boldsymbol{G}\pm\boldsymbol{k}).\boldsymbol{r}_o.$$

We consider $|\boldsymbol{k}| < |\boldsymbol{G}|$ for a weak modulation, in which case $f(\boldsymbol{G}\pm\boldsymbol{k}) \approx f(\boldsymbol{G})$ and phase fluctuations $\exp i(\boldsymbol{G}\pm\boldsymbol{k}).\boldsymbol{r}_o$ can be separated to amplitude and phase modes, similar to the critical condition discussed in Chapter 6. Taking these symmetric and antisymmetric modes into account, we can derive the corresponding intensity at r, i.e.,

$$\frac{I(\boldsymbol{G}\pm\boldsymbol{k})}{I_o} = \frac{I(\boldsymbol{G})}{I_o} + \frac{2|f(\boldsymbol{G})|^2}{r^2}\sum[\cos\{\boldsymbol{k}.(\boldsymbol{r}_m - \boldsymbol{r}_n) - \Delta\omega(t_m - t_n)\} + \sin\{\boldsymbol{k}.(\boldsymbol{r}_m - \boldsymbol{r}_n) - \Delta\omega(t_m - t_n)\}]. \tag{10.7a}$$

Here, the first term on the right represents elastic scatterings, while phase fluctuations in the impact area S are explicit in the second inelastic term. Considering timescale of observation t_o, the latter represents the intensity anomaly due to diffuse diffraction, i.e.,

$$\frac{\langle I(\boldsymbol{G}\pm\boldsymbol{k})\rangle_t - \langle I(\boldsymbol{G})\rangle_t}{I_o} = \frac{2|f(\boldsymbol{G})|^2}{r^2}\frac{1}{S}\int\limits_S \{\langle\cos\phi\rangle_t + \langle\sin\phi\rangle_t\}dS, \tag{10.7b}$$

where $\phi = \boldsymbol{k}.(\boldsymbol{r}_m - \boldsymbol{r}_n) - \Delta\omega(t_m - t_n)$ represents fluctuating phases. Here, it is convenient to consider that $\boldsymbol{r}_m - \boldsymbol{r}_n = \boldsymbol{r}$ and $t_m - t_n = \tau$ are continuous space–time, i.e., $\phi = \boldsymbol{k}.\boldsymbol{r} - \Delta\omega.\tau$; (10.7b) expresses that time averages $\langle\cos\phi\rangle_t$ and $\langle\sin\phi\rangle_t$ are integrated over the area S.

Assuming a rectangular area $S = L_x L_y$ and continuous $\phi = kx - \Delta\omega.\tau$, the symmetric spatial fluctuations can be calculated as

$$\frac{1}{S}\int\limits_S \langle\cos\phi\rangle_t dS = \left(\frac{1}{L_x}\int\limits_0^{L_x}\langle\cos\phi\rangle_t dx\right)\left(\frac{1}{L_y}\int\limits_0^{L_y}dy\right). = \left\langle\frac{1}{kL_x}\int\limits_{\phi_1}^{\phi_2}\cos\phi d\phi\right\rangle_t$$
$$= \left\langle\frac{2}{\phi_2-\phi_1}\sin\frac{\phi_2-\phi_1}{2}\cos\frac{\phi_2+\phi_1}{2}\right\rangle_t.$$

Here, ϕ_1, ϕ_2 indicate limits of the space integration at x_1, x_2; hence $\phi_2 - \phi_1 = \Delta\phi = kL_x$, and $\frac{\phi_2+\phi_1}{2} = kx - \Delta\omega.\tau$. After a similar calculation for $\frac{1}{S}\int\limits_S\langle\sin\phi\rangle_t dS$, we obtain the formula for the net intensity anomaly:

$$\frac{\langle \Delta I(\boldsymbol{G})\rangle_t}{I_o} = \frac{\sin kL_x}{kL_x}\frac{\sin(t_o\Delta\omega)}{t_o\Delta\omega}\left\{\langle \cos kx\rangle_t + \langle \sin kx\rangle_t\right\}. \tag{10.7c}$$

This formula explains diffused intensities of X-ray diffraction; however such broadened spots and lines are hardly analyzed in the timescale t_o of photochemical processes in X-ray experiments.

For neutron diffraction, we can use (10.7c) for intensity anomalies of neutron scatterings. The intensity of a neutron beam is determined by counting technique characterized by a much shorter timescale than in X-ray case, but still anomalous intensities (10.7c) for neutrons cannot be analyzed.

10.2 Neutron Inelastic Scatterings

In contrast to X-ray experiments, neutron scatterers are normally associated with heavy nuclei, so symmetric and antisymmetric contributions to elastic scatterings $\Delta\boldsymbol{K} = \boldsymbol{G}$ cannot be resolved, but can be resolved for inelastic scatterings for $\Delta\boldsymbol{K} \perp \boldsymbol{G}$ within timescale of neutron counters, if scattered neutrons are specified by a small wavevector $\boldsymbol{k} = \Delta\boldsymbol{K}$ and energy $\Delta\varepsilon = \hbar\Delta\omega$.

For one-dimensional modulation, if collimated X-ray beam is parallel to the modulated axis, we obtain $\boldsymbol{K} = 0$ from (10.1a) if $\boldsymbol{K}_o = \boldsymbol{G}$, meaning no elastic scatterings. For neutrons, in contrast, inelastic scatterings can occur at the zone boundaries $\pm\frac{\boldsymbol{G}}{2}$, if $\boldsymbol{K} - \boldsymbol{K}_o = \pm\frac{1}{2}\boldsymbol{G}$, without elastic diffraction in all directions. Figure 10.2 illustrates such an arrangement that inelastic scatterings can be detected efficiently, provided that the incident beam $\boldsymbol{K}_o$ is in high-enough intensity. Otherwise, as there are some unavoidable elastic scatterings in practice, feeble inelastic intensities are undetected. Nevertheless, the latter can be detected by minimizing elastic scatterings, if the wavevector is scanned in the vicinity of $\Delta\boldsymbol{K} = 0$. Using thermal neutrons from a nuclear reactor, as characterized by their de Broglie's wavelengths comparable with lattice spacing, inelastic scatterings can reveal anomalies resolved as expressed like (10.7c).

An intense thermal neutron source was available for such experiments performed at leading nuclear reactor facilities. In this section, we only discuss the principle of the scattering experiment with representative results. In Fig. 10.3 a *triple axis spectrometer* is sketched, consisting of monochrometer, goniometer, and analyzer, which are all rotatable individually around their axes installed in parallel. The monochrometer and analyzer are diffraction devices using crystals of known lattice spacing; thereby the neutron wavelength can be selected and measured by adjusting angles θ and θ', respectively, as indicated in the figure. A sample crystal is mounted on the goniometer, which can be rotated around a symmetry axis to scan $|\boldsymbol{K}|$ in the vicinity of $\boldsymbol{K}_o || \boldsymbol{G}$, as shown in Fig. 10.2.

The wavevector and energy conservation laws can be applied to the impact between a neutron and a scatterer, i.e.,

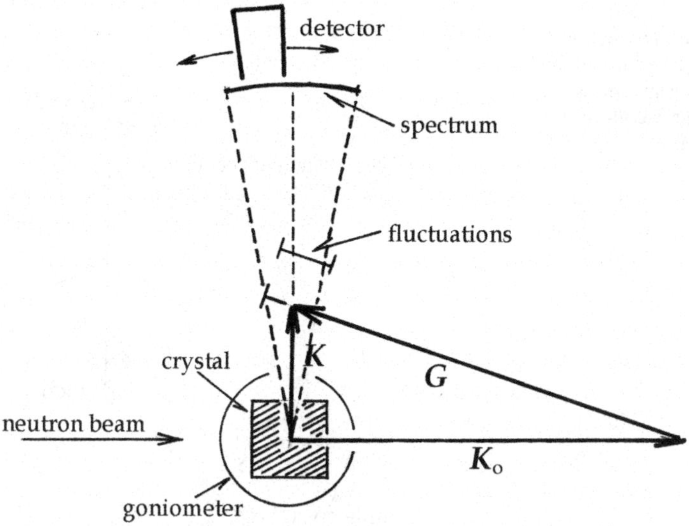

Fig. 10.2 Experimental arrangement for neutron inelastic scatterings. Incident beam $\boldsymbol{K}_o$. In perpendicular directions $\boldsymbol{K}$, only inelastic scatterings can be observed with no diffraction. The fluctuation spectrum can be recorded by scanning the scattering angle.

$$\boldsymbol{K}_o - \boldsymbol{K} = \boldsymbol{G}_i \pm \boldsymbol{k} \quad \text{and} \quad \varepsilon(\boldsymbol{K}_o) - \varepsilon(\boldsymbol{K}) = \mp\Delta\varepsilon(\boldsymbol{k}) = \mp\hbar\Delta\omega, \tag{10.8}$$

where $\Delta\varepsilon(\boldsymbol{k})$ represents an amount of energy transfer. Here, we have written the lattice vector as $\boldsymbol{G}_i$ due to distributed collisions, where the phase transition occurs near the Brillouin zone boundaries at $\pm\frac{G}{2}$. The zone-boundary transition is signified by the scatterer of two unit cells, and hence called a cell-doubling transition. Approaching the phase translation, the vector $\boldsymbol{G}_i$ is not exactly equal to $\pm\frac{G}{2}$, but shifting from the boundary by a perturbation attributed to Born–Huang's principle.

Expressing incident neutrons by the wave function $\Psi \approx \exp i(\boldsymbol{K}_o.\boldsymbol{r}_o - \omega_o t_o)$, the scattered neutron from a scatterer at a lattice point $\boldsymbol{r}_o$ can be described by the wavefunction

$$\Psi(\boldsymbol{r},t) \approx \int\limits_{V(\boldsymbol{r}_o)} d^3\boldsymbol{r}_o\rho(\boldsymbol{r}_o)\frac{\exp i\{\boldsymbol{K}.(\boldsymbol{r}-\boldsymbol{r}_o) - \omega(t-t_o)\}}{|\boldsymbol{r}-\boldsymbol{r}_o|}$$
$$\times \exp i\{(\boldsymbol{K}-\boldsymbol{K}_o).\boldsymbol{r}_o - (\omega-\omega_o)t_o\},$$

where $\boldsymbol{r}_o$ and t_o represent the space–time of the impact. In the direction perpendicular to $\boldsymbol{K}_o$, we require the condition for no elastic scatterings, i.e., $\boldsymbol{K} = \mp\boldsymbol{k}$ and $\boldsymbol{K}_o = -\boldsymbol{G}_i$, for which case $\Psi(\boldsymbol{r},t)$ can be modified as

Fig. 10.3 Triple-axis spectrometer for neutron scattering experiments. The monochrometer, goniometer, and analyzer are all rotatable around their axes in parallel.

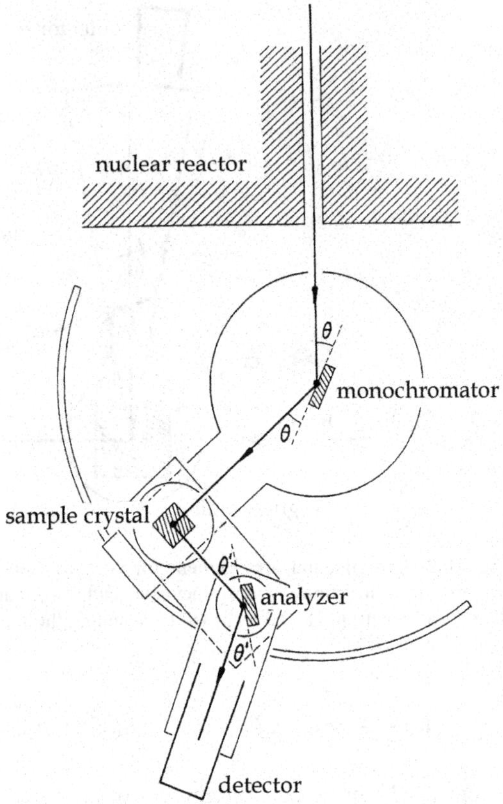

$$\Psi(\boldsymbol{r},t) \approx \int_{V(\boldsymbol{r}_o)} \mathrm{d}^3\boldsymbol{r}_o \rho(\boldsymbol{r}_o) \frac{\exp i\{\mp \boldsymbol{k}.(\boldsymbol{r}-\boldsymbol{r}_o) - \omega(t-t_o)\}}{|\boldsymbol{r}-\boldsymbol{r}_o|} \times \exp i\{(\boldsymbol{G}_i \pm \boldsymbol{k}).\boldsymbol{r}_o \mp \Delta\omega.t_o\}.$$

The structural factor can be defined for a finite area that includes many scatterers at $\boldsymbol{r}_{om}$ as

$$f(\boldsymbol{G}_i \pm \boldsymbol{k}) = \int_{V(\boldsymbol{r}_o)} \mathrm{d}^3\boldsymbol{r}_o \rho(\boldsymbol{r}_o) \exp i(\boldsymbol{G}_i \pm k).\boldsymbol{r}_o,$$

and express the wavefunction, by assuming $|\boldsymbol{r}-\boldsymbol{r}_{om}| \approx r$ and $t_{om} \approx t_o$, as

$$\Psi(\boldsymbol{G}_i \pm \boldsymbol{k}) \propto \frac{\exp i\{\mp \boldsymbol{k}.\boldsymbol{r} \pm \omega(t-t_o)\}}{r} \times \sum_{\mathrm{m}} \{f(\boldsymbol{G}_i + \boldsymbol{k}) \exp i(\boldsymbol{k}.\boldsymbol{r}_{om} - \Delta\omega.t_{om}) + f(\boldsymbol{G}_i - \boldsymbol{k}) \exp i(-\boldsymbol{k}.\boldsymbol{r}_{om} + \Delta\omega.t_{om})\}.$$

Writing $n_m(\boldsymbol{G}_i \pm \boldsymbol{k}) = f(\boldsymbol{G}_i \pm \boldsymbol{k}) \exp i(\pm \boldsymbol{k}.\boldsymbol{r}_{om} \mp \Delta\omega.t_{om})$, the intensity of scattered neutrons can be given by

$$I(\boldsymbol{G}_i \pm \boldsymbol{k}) = \left\langle |\Psi(\boldsymbol{G}_i \pm \boldsymbol{k})|^2 \right\rangle_t \propto \left\langle \sum_{m} {}_{,n} \{ n_m^*(\boldsymbol{G}_i + \mathbf{k}) n_n(\boldsymbol{G}_i + \boldsymbol{k}) + n_m^*(\boldsymbol{G}_i - \boldsymbol{k}) n_n(\boldsymbol{G}_i - \boldsymbol{k}) \} \right\rangle_t$$

$$= \left\langle \sum_{m} |n_m(\boldsymbol{G}_i \pm \boldsymbol{k})|^2 \right\rangle_t$$

$$+ \left\langle \sum_{m} \neq n \{ n_m^*(\boldsymbol{G}_i + \boldsymbol{k}) n_n(\boldsymbol{G}_i + \boldsymbol{k}) + n_m^*(\boldsymbol{G}_i - \boldsymbol{k}) n_n(\boldsymbol{G}_i - \boldsymbol{k}) \} \right\rangle_t .$$

The first term represents elastic scatterings at $\pm \mathbf{k}$, whereas the second one indicates inelastic contributions. The latter can be expressed as

$$\langle \Delta I(\boldsymbol{G}_i \pm \boldsymbol{k}) \rangle_t \propto \Big\langle \sum_{m \neq n} \exp i\boldsymbol{G}_i.(\boldsymbol{r}_{om} - \boldsymbol{r}_{on})$$
$$\times \Big[|f(\boldsymbol{G}_i + \boldsymbol{k})|^2 \exp i\{\boldsymbol{k}.(\boldsymbol{r}_{om} - \boldsymbol{r}_{on}) - \Delta\omega(t_{om} - t_{on})\}$$
$$+ |f(\boldsymbol{G}_i - \boldsymbol{k})|^2 \exp i{-\boldsymbol{k}}.(\boldsymbol{r}_{om} - \boldsymbol{r}_{on}) + \Delta\omega(t_{om} - t_{on})\} \Big] \Big\rangle_t .$$

Considering that $\boldsymbol{r}_{om} - \boldsymbol{r}_{on} = \boldsymbol{r}$ and $t_{om} - t_{on} = t$ are continuous space–time, the intensity anomaly $\langle \Delta I(\boldsymbol{G}_i \pm \boldsymbol{k}) \rangle_t$ is signified by phase fluctuations $\phi = \boldsymbol{k}.\boldsymbol{r} - \Delta\omega.t$ and $-\phi = -\boldsymbol{k}.\boldsymbol{r} + \Delta\omega.t$.

If these scattering factors $f(\boldsymbol{G}_i \pm \boldsymbol{k})$ can be approximated for a small $\boldsymbol{k}$ as

$$|f(\boldsymbol{G}_i \pm \boldsymbol{k})|^2 = |f(\boldsymbol{G}_i)|^2 \pm 2i\boldsymbol{k}.\nabla_{\boldsymbol{k}} |f(\boldsymbol{G}_i)|^2,$$

we can express the intensity anomaly as

$$(\Delta I(\boldsymbol{G}_i \pm \boldsymbol{k}))_t = A \langle \cos\phi \rangle_t + P \langle \sin\phi \rangle_t, \tag{10.9}$$

where

$$A \propto \int_{V(\boldsymbol{r})} |f(\boldsymbol{G}_i)|^2 \cos(\boldsymbol{G}_i.\boldsymbol{r}) \mathrm{d}^3\boldsymbol{r} \quad \text{and} \quad \mathrm{P} \propto \int_{\mathrm{V}(\boldsymbol{r})} 2\boldsymbol{k}.\nabla_{\boldsymbol{k}} |\mathrm{f}(\boldsymbol{G}_\mathrm{i})|^2 \cos(\boldsymbol{G}_\mathrm{i}.\boldsymbol{r}) \mathrm{d}^3\boldsymbol{r}.$$

We have considered that the fluctuations are pinned at the zone boundary. It is noted that the anomalies (10.9) are identical to (10.7c) for X-ray diffraction.

Expression (10.9) indicates that the intensity anomaly is contributed by two modes of phase fluctuations $\langle \cos\phi \rangle_t$ and $\langle \sin\phi \rangle_t$ at intensities A and P. In Chapter 6, we have discussed such distributed intensities with respect to the phase $\phi = \boldsymbol{k}.\boldsymbol{r} + \phi_o(t_o)$ where $\phi_o(t_o)$ is an arbitrary phase depending on t_o, as shown in Fig. 6.3. Figure 10.4 [1] shows examples of anomalous intensities of scattered

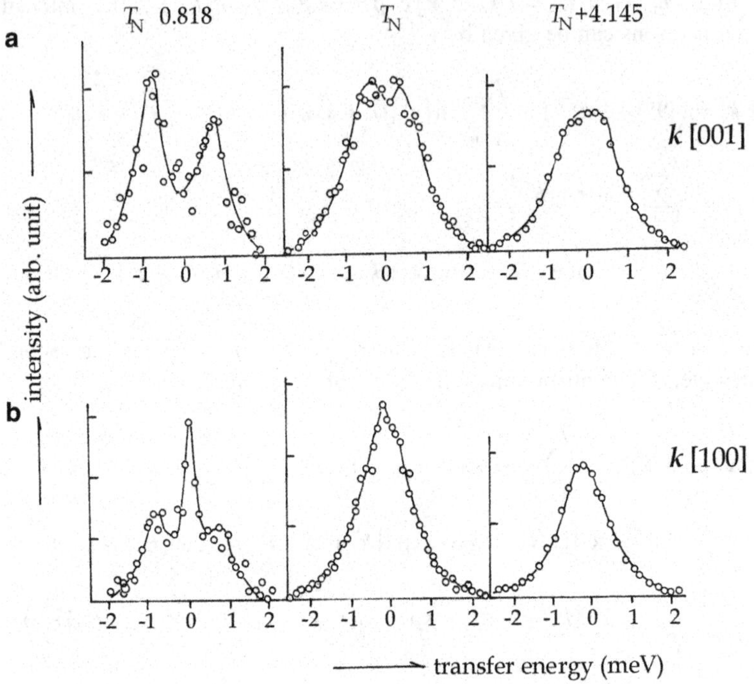

Fig. 10.4 Typical intensity distributions observed from a magnetic MnF_2 crystal at the Neél temperature T_N, showing A and P absorption peaks. Here, the experimental result is quoted as an example of neutron inelastic scattering experiments, assuming that neutron–nucleus collisions constitute the scattering mechanism. For magnetic crystals such as MnF_2, we normally consider neutron–magnetic spin collisions, which are nonetheless unidentifiable in neutron experiments; the difference is however insignificant. From M.P. Schulhof, P. Heller, R. Nathans and A. Linz, Phys. Rev. **B1**, 2403 (1970).

neutrons, which were observed from magnetic MnF_2 crystals near the critical Neél temperature T_N. These anomalies observed in different scattering geometries show clearly anisotropic intensities composed of *A* and *P* modes. For observing such fluctuations, it is noted that the timescale t_o in the scattering process should be comparable to $\Delta\omega^{-1}$ for the energy transfer; otherwise the scattering anomalies vanish for $t_o > \Delta\omega^{-1}$.

10.3 Light Scattering Experiments

Using intense coherent light waves from a *laser* oscillator, we can normally detect feeble inelastic scatterings in enhanced intensity from a crystal, permitting detailed studies of scattered photon spectra. Incident light induces dielectric fluctuations due to random impacts with constituents, producing the *Rayleigh radiation* that is attributed to elastic collisions, while inelastic scatterings also occur in detectable intensity, yielding information about structural changes in dielectric crystals. Capable to study inelastic scatterings, laser spectroscopy provides useful results for studying modulated structures.

10.3.1 Brillouin Scatterings

When sound waves are *standing* in one dimension inside liquid, a monochromatic light beam incident to it in perpendicular direction generates a diffraction pattern that is characterized by modulated liquid densities. In dielectric crystals, induced dielectric fluctuations are responsible for similar phenomena, known as the *Brillouin scatterings.* A dielectric crystal exhibits such scatterings in the presence of sound waves, for which a *photo-elastic* mechanism is responsible. In this case, inelastic collisions between photons and constituent ions generate such Brillouin scatterings, yielding useful information about a modulated structure analyzable with Born–Huang' principle.

In Fig. 10.5a, we consider that an intense monochromatic light of wavevector $\boldsymbol{K}_0$ polarized along the y-direction is incident upon the (101) plane of a sample crystal, and that the scattered light of $\boldsymbol{K}$ is detected in the perpendicular direction $[\bar{1}01]$. In this case, phonons of wavevectors $\pm\boldsymbol{Q}$ should participate in the scattering process as shown in Fig. 10.5b, illustrating a scattering geometry for $\boldsymbol{K}$, $\boldsymbol{K}_0$, and $\pm\boldsymbol{Q}$. It is noted that owing to the photoelastic properties determined by the lattice symmetry, the phonon waves are not necessarily longitudinal but transversal as well; consequently the scattered light $\boldsymbol{K}$ can be polarized in two directions v–v and h–h, as indicated in Fig. 10.5a. The symmetry relation between incident and scattered lights should be indicated by group operations; however we simply consider that the scattered components fluctuate in vertical and horizontal directions. Namely, we can write $\Delta p_v = \alpha_v E$ and $\Delta p_h = \alpha_h E$, where α_v and α_h are polarizabilities, where the amplitude of light is E.

Denoting the polarized laser light by the electric field $\boldsymbol{E}_0 \exp i(\boldsymbol{K}_0.\boldsymbol{r}_0 - \omega_0 t_0)$ at the impact position and time $(\boldsymbol{r}_0, t_0)$, the induced polarization is $\Delta\boldsymbol{p}(\boldsymbol{r}_0, t_0) = (\alpha)$ $\boldsymbol{E}_0 \exp i(\boldsymbol{K}_0.\boldsymbol{r}_0 - \omega_0 t_0)$, where (α) represents a tensor with components α_v, α_h. As given in Sect. 10.1, the scattered field can be expressed for a distant point $|\boldsymbol{r} - \boldsymbol{r}_0| \approx \boldsymbol{r}$ as

$$\begin{aligned} \boldsymbol{E}(\boldsymbol{r}, t) \approx & \frac{\exp i(\boldsymbol{K}.\boldsymbol{r} - \omega t)}{r} \iint \Delta\boldsymbol{p}(\boldsymbol{r}_0, t_0).\boldsymbol{E}_0 \\ & \times \exp i\{(\boldsymbol{K}_0 - \boldsymbol{K}).\boldsymbol{r}_0 - (\omega_0 - \omega)t_0\} \mathrm{d}^3\boldsymbol{r}_o \mathrm{d}t_o. \end{aligned} \tag{10.10}$$

If $\omega = \omega_0$, (10.9) is maximum at all wavevectors for $|\boldsymbol{K}| = |\boldsymbol{K}_0|$, representing elastic collisions known as the Rayleigh scatterings. Defining the scattering angle φ by

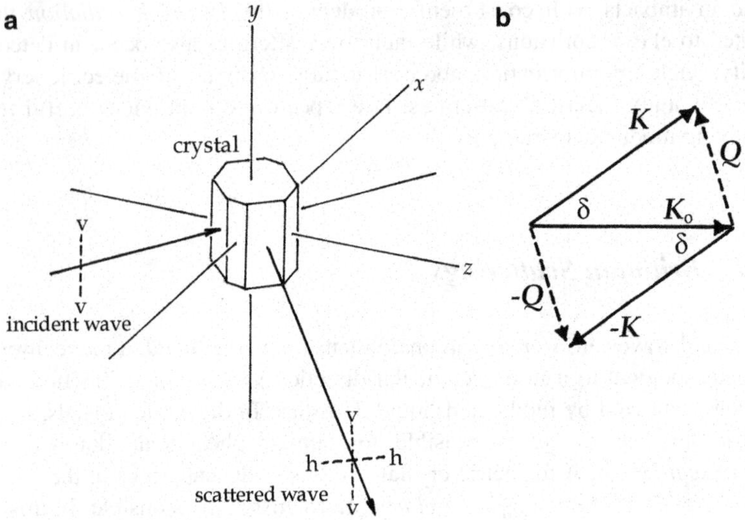

Fig. 10.5 (**a**) A light scattering arrangement for Brillouin scatterings. Incident light is polarized in the y-direction as marked v. . .v, while scattered light is depolarized to v. . .v and h. . .h directions. (**b**) A general Brillouin scattering geometry with phonon excitations $\pm \boldsymbol{Q}$.

$\boldsymbol{K}_0.\boldsymbol{r}_0 = \frac{2\pi r_0}{\lambda}\cos\varphi$, where λ is the wavelength, the Rayleigh intensity $I_R(\boldsymbol{K})$ can be expressed for $r_0 \ll \lambda$ as

$$\frac{I_R(\boldsymbol{K})}{I(\boldsymbol{K}_0)} \propto \frac{2\pi^2 v^2}{r^2\lambda^4}\left(1+\cos^2\varphi\right),$$

where the two terms on the right represent intensities for $\boldsymbol{E}_v$ and $\boldsymbol{E}_h$, respectively.

For inelastic Brillouin scatterings, the conservation rules are

$$\boldsymbol{K}_0 - \boldsymbol{K} = \pm\boldsymbol{Q} \quad \text{and} \quad \omega_0 - \omega = \mp\Delta\omega,$$

where $\pm$ **Q** and $\mp\Delta\omega$ are the wavevectors and frequencies of phonons in acoustic mode, as illustrated in Fig. 10.5b. The angle δ between $\boldsymbol{K}_0$ and $\boldsymbol{K}$ is called the scattering angle, with which the magnitude of $\boldsymbol{Q}$ is expressed as

$$|\boldsymbol{Q}| = 2|\boldsymbol{K}_0|\sin\frac{\delta}{2}. \tag{10.11a}$$

Experimentally, to keep incident photons away from the detecting photon counter, we normally select $\delta = 90°$, but the Rayleigh scatterings cannot be avoided in practice, obscuring Brillouin spectra. In isotropic media characterized by the optical index n, for laser light of wavelengths λ we have $|\boldsymbol{K}_0| = \frac{2\pi n}{\lambda}$, and Brillouin scatterings show frequency shifts given by

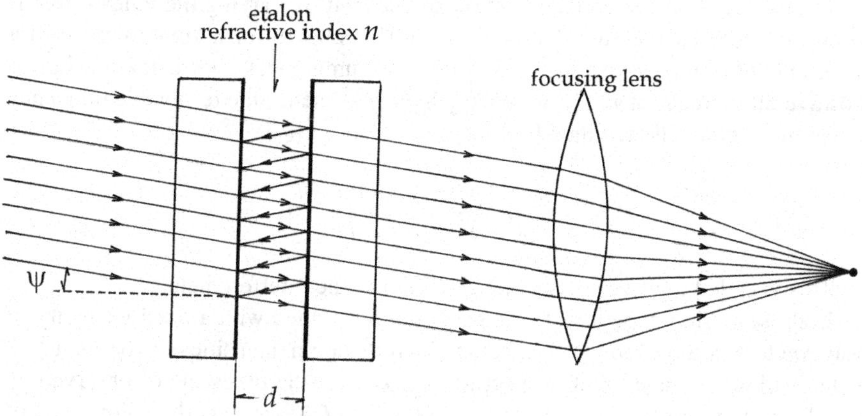

Fig. 10.6 Fabry–Pérot interferometer.

$$\Delta\nu_{\mathrm{B}} = \pm\frac{2\pi v}{\lambda}\sin\frac{\delta}{2}. \tag{10.11b}$$

The speed v of phonon propagation can be evaluated from the Brillouin shift $\Delta\nu_{\mathrm{B}}$ using (10.11b).

Brillouin intensities can be expressed as

$$I_{\mathrm{B}}(\boldsymbol{K},\pm\boldsymbol{Q}) \propto \frac{1}{r^2}\int\!\!\int \left(\Delta p_{\mathrm{i}}\Delta p_{\mathrm{j}}\right)\exp i\{\pm\boldsymbol{Q}.(\boldsymbol{r}'_{\mathrm{o}}-\boldsymbol{r}_{\mathrm{o}}) \mp \Delta\omega(t'_{\mathrm{o}}-t_{\mathrm{o}})\}\mathrm{d}^3(\boldsymbol{r}'_{\mathrm{o}}-\boldsymbol{r}_{\mathrm{o}})\mathrm{d}(t'_{\mathrm{o}}-t_{\mathrm{o}}), \tag{10.11c}$$

where $\left(\Delta p_{\mathrm{i}}\Delta p_{\mathrm{j}}\right)$ is a *dyadic tensor* with elements, e.g., $\Delta p_{\mathrm{v}}\Delta p_{\mathrm{h}}$, for which the time average over fluctuations should be calculated with respect to the timescale of measurement.

While Brillouin scatterings can always be observed in crystals, we are interested in soft acoustic modes signified by temperature-dependent frequencies. Therefore, the experimental task is to find the direction of a sound wave $\boldsymbol{Q}$ where $\Delta\nu_{\mathrm{B}}$ can be temperature dependent. On the other hand, the critical fluctuations exhibited by Brillouin intensities $I_{\mathrm{B}}(\boldsymbol{K},\pm\boldsymbol{Q})$ are difficult to isolate from the Rayleigh line, if $\Delta\nu_{\mathrm{B}}$ is too close to zero.

For a typical dielectric crystal, if assuming $n \approx 1.5$, $v \approx 2\times 10^3\mathrm{m}\cdot\mathrm{s}^{-1}$, and $\lambda = 514.5$ nm, the Brillouin shift can be estimated to be of the order of $\Delta\nu_{\mathrm{B}} \simeq 6 \times 10^9\mathrm{Hz} = 0.2\ \mathrm{cm}^{-1}$. A Fabry–Pérot interferometer, as sketched in Fig. 10.6, is commonly used for Brillouin's spectra, where a device consisting of parallel semitransparent planes with a narrow gap d can enhance interference after multiple reflections. Denoting the index of refraction of air by n, we can write the interference condition for an incident angle ψ as

$$2nd\cos\psi = m\lambda \quad (m\text{: integers}).$$

In this device, the wavelength λ can be scanned by varying the index n that is controllable by air pressure. Normally $\psi = 0^\circ$ is chosen for normal incidence for such an interferometer, in which case $\frac{1}{\lambda} = \frac{m}{2d}$. Assuming that a wave of $\lambda - \Delta\lambda$ gives positive interference at $m + 1$, we have $\frac{1}{\lambda - \Delta\lambda} = \frac{m+1}{2d}$; hence the resolvable frequency range of spectra is determined by

$$\frac{1}{\lambda - \Delta\lambda} - \frac{1}{\lambda} = \frac{1}{2d}.$$

This is called the free spectral range (FSR) of the interferometer.

Light scattering experiments are generally performed with a fixed geometry of wavevectors as shown in Fig. 10.5b, for which Brillouin intensities can be evaluated with (10.11c) averaged with respect to $t'_o - t_o$ over the timescale of observation. For the scattering geometry where $\boldsymbol{K}_o = \boldsymbol{G}$, where $\boldsymbol{G}$ represents the unique axis of modulation, (10.11c) gives rise to intensity anomalies similar to (10.8) for neutron inelastic scatterings. For such light scatterings, the equation of motion can be written for induced dipole displacements as

$$\mathrm{m}(\ddot{u}_\mathrm{i} + \gamma\dot{u}_\mathrm{i} + \omega_o^2 u_\mathrm{i}) = eE_\mathrm{v}\exp(-it\Delta\omega),$$

where e and m are the charge and mass of the dipole moment $\Delta p_\mathrm{i} = eu_\mathrm{i}$. The steady-state solution is derived as

$$\Delta p_\mathrm{i} = \frac{e^2}{\mathrm{m}}\frac{E_\mathrm{v}}{-\Delta\omega^2 + \omega^2 \mp i\gamma\Delta\omega} = (\alpha'_\mathrm{i} \mp i\alpha''_\mathrm{i})E_\mathrm{v},$$

where

$$\alpha'_\mathrm{i} \mp i\alpha''_\mathrm{i} = \frac{\dfrac{e^2\varepsilon_o}{\mathrm{m}}}{-\Delta\omega^2 + \omega^2 \mp i\gamma\Delta\omega}$$

are complex polarizabilities due to E_v. Therefore, (10.11c) can be expressed as

$$\Delta I_\mathrm{i}(\boldsymbol{K} \pm \boldsymbol{k}) \propto \frac{E_\mathrm{v}^2}{r^2}\{(\alpha''_\mathrm{i}\alpha_\mathrm{v})\langle\cos\phi\rangle_\mathrm{t} + (\alpha'_\mathrm{i}\alpha_\mathrm{v})\langle\sin\phi\rangle_\mathrm{t}\},$$

where $\phi = \boldsymbol{k}.\boldsymbol{r} - \Delta\omega.\tau$ is the fluctuating phase in the range $0 \le \phi \le 2\pi$. We already discussed such anomalies in Chapter 6. Experimentally these terms above give rise to a single peak A at $\phi = 0$ and separated peaks P at $\phi = \pm\frac{\pi}{2}$, if the wavevector $\boldsymbol{k}$ represents a temperature-dependent soft mode. Accordingly, anomalies in Brillouin spectra depend on the anisotropic factors $\alpha''_\mathrm{i}\alpha_\mathrm{v}$ and $\alpha'_\mathrm{i}\alpha_\mathrm{v}$ involved in a soft mode. Figure 10.7a,b [2, 3] shows a typical result of Brillouin spectra that were obtained from KDP crystals. The frequency shifts of Brillouin lines in Fig. 10.7a are

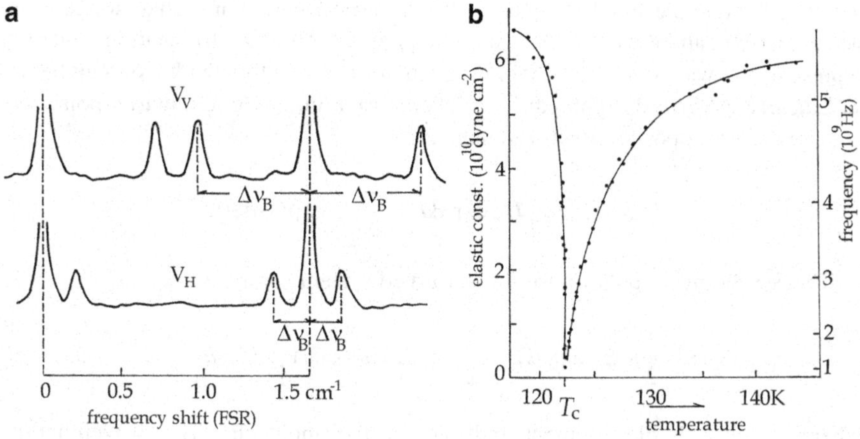

Fig. 10.7 (**a**) Examples of Brillouin spectra from KDP at room temperature. Observed in v. . .v and h. . .h directions. (**b**) Soft-mode frequency vs. temperature. From E.M. Brody and Cummins, Phys. Rev. **9**, 179 (1974).

different, depending on polarized components of scattered light while exhibiting clearly temperature dependences as shown in Fig. 10.7b. The temperature-dependent frequency shift $\Delta\nu_B$ represents clearly soft modes of $\pm\boldsymbol{k}$ and $\mp\Delta\omega$, while sinusoidal fluctuations turn theoretically to elliptical ones with increasing $E_v{}^2$. However, small Brillouin shifts near the critical temperature are masked by the intense Rayleigh line near the center, making the detail unknown. Line shapes of Brillouin lines are theoretically asymmetrical, appearing however as practically symmetrical in observed spectra.

10.3.2 Raman Scatterings

Light scatterings can also be observed from ions that are independent from acoustic phonon modes. Although primarily independent from the modulated lattice, the scatterings show a frequency shift due to perturbation from the lattice in optic vibrations. Such a frequency shift called the Raman shift can be analyzed as related with optic phonons, so the results may be useful to study lattice contributions, if exhibiting a temperature dependence.

Following Placzek [4], assuming that the lattice vibrations are characterized by a single frequency ω, we discuss in this section Raman scatterings of a polarized monochromatic light. For a constituent ion in the lattice, we write wave equations

$$H\varphi_0 = \varepsilon_0\varphi_0 \quad \text{and} \quad H\varphi_1 = \varepsilon_1\varphi_1$$

for the ground and excited energies ε_o and ε_1, respectively. For such an ion, the net wavefunction can be written as $\varphi_o\chi$ and $\varphi_1\chi$ in adiabatic approximation, where χ represents the wavefunction of vibrating lattice specified by the phonon number n.

Linearly polarized light can be decomposed into two circularly polarized components in opposite directions, namely,

$$\boldsymbol{E}\cos\Omega t = \frac{1}{2}\boldsymbol{E}_+\exp i\Omega t + \frac{1}{2}\boldsymbol{E}_-\exp(-i\Omega t).$$

The Schrödinger equation for the perturbed system can be written as

$$\left\{\mathsf{H} - \frac{1}{2}\boldsymbol{p}.\boldsymbol{E}_+\exp i\Omega t - \frac{1}{2}\boldsymbol{p}.\boldsymbol{E}_-\exp(-i\Omega t)\right\}\Psi = i\hbar\frac{\partial\Psi}{\partial t},$$

where $\boldsymbol{p}$ is a dipole moment induced in the molecule. The wavefunction representing the ground state perturbed by $\boldsymbol{E}$ can be written as

$$\Psi_o = \{\psi_o + \psi_{o+}\exp i\Omega t + \psi_{o-}\exp(-i\Omega t)\}\exp\left\{-it\frac{\varepsilon_o + n\hbar\omega}{\hbar}\right\}.$$

Accordingly, in the first-order approximation we have

$$\{\mathsf{H} - (\varepsilon_o + n\hbar\omega \pm \hbar\Omega)\}\psi_{o\pm} = \boldsymbol{p}.\boldsymbol{E}_\pm\Psi_o,$$

relating the unperturbed ground state ψ_o with polarized ground states $\psi_{o\pm}$. We can therefore write

$$\psi_o = \varphi_o\chi_o \quad \text{and} \quad \psi_{o\pm} = \varphi_o(c_+\chi_+ + c_-\chi_-) + c_o\psi_o$$

where $$c_\pm = \frac{\langle\pm|\boldsymbol{p}|0\rangle.\boldsymbol{E}_\pm}{\hbar(\pm\omega\pm\Omega)}.$$

The dipole moment in the ground state can then be expressed with components

$$p_i(t) = \langle 0|p_i|0\rangle + \sum_j\{\alpha_{ij}(\Omega)E_{+j}\exp i\Omega t + \alpha_{ij}(-\Omega)E_{-j}\exp(-i\Omega t)\},$$

where

$$\alpha_{ij}(\Omega) = \frac{\langle 0|p_i|+\rangle\langle +|p_j|0\rangle}{\hbar(\omega+\Omega)} + \frac{\langle 0|p_j|+\rangle\langle +|p_i|0\rangle}{\hbar(\omega-\Omega)} + \frac{\langle 0|p_i|-\rangle\langle -|p_j|0\rangle}{\hbar(-\omega+\Omega)} + \frac{\langle 0|p_j|-\rangle\langle -|p_i|0\rangle}{\hbar(-\omega-\Omega)}$$

constitutes the *polarizability tensor* with respect to perpendicular directions $i, j = x, y$ in the plane of $\boldsymbol{E}_\pm$. In this tensor, the relations $\alpha_{ij} = \alpha_{ij}{}^*$ and $\alpha_{ij} = \alpha_{ji}{}^*$ can be verified, so the tensor (α_{ij}) is *real* and *hermitic*.

We can write similar expressions for the excited molecular state, i.e.,

$$\Psi_1 = \{\psi_{1+} \exp i\Omega t + \psi_{1-} \exp(-i\Omega t)\} \exp\left(-it\frac{\varepsilon_1 + n\hbar\omega}{\hbar}\right)$$

where $quad\psi_{1\pm} = \dfrac{\langle\pm|\boldsymbol{p}|0\rangle.\boldsymbol{E}_\pm}{\hbar(\pm\omega \pm \Omega)}$.

Using this Ψ_1, the induced transition probability for emission processes can be calculated with the matrix elements

$$\begin{aligned}\langle\Psi_o^*|\boldsymbol{p}|\psi_{1\pm}\rangle &= \exp\frac{i\varepsilon_{1o}t}{\hbar}\int\{c_o^*\langle\psi_o^*|\boldsymbol{p}|\psi_1\rangle + c_+^*\langle\psi_+^*|\boldsymbol{p}|\psi_1\rangle + c_-^*\langle\psi_-^*|\boldsymbol{p}|\psi_1\rangle\}dv\\ &= c_o^*\langle\psi_o^*|\boldsymbol{p}|\psi_1\rangle \exp\frac{i\varepsilon_{1o}t}{\hbar} + \frac{\langle 0|\boldsymbol{p}^*|\pm\rangle\langle\pm|\boldsymbol{p}|0\rangle}{\hbar(\pm\omega+\Omega)} \exp it\left(\Omega \pm \frac{\varepsilon_{1o}}{\hbar}\right)\\ &+ \frac{\langle 0|\boldsymbol{p}^*|\pm\rangle\langle\pm|\boldsymbol{p}|0\rangle}{\hbar(\pm\omega-\Omega)} \exp it\left(-\Omega \pm \frac{\varepsilon_{1o}}{\hbar}\right),\end{aligned}$$

where $\varepsilon_{1o} = \varepsilon_1 - \varepsilon_o$. The first term on the right side is independent of the phonon energy $\hbar\omega$, whereas the second and third terms correspond to Raman emission processes at $\Omega = \omega_{1o} \pm \omega$ that are signified by transitions $\Delta n = \pm 1$. Therefore, two satellite lines correspond to the emission and absorption of $\hbar\omega$, as illustrated by Fig. 10.8a, which are traditionally called Stokes and anti-Stokes lines for phonons $\pm\hbar\omega$, respectively. Figure 10.8b [5] shows an example of Raman results, showing soft modes on both sides of transition temperature T_c. Placzek's theory can be evaluated with the hermitic tensor α_{ij} for a given crystal, thereby determining anisotropic Raman shifts. However, such a tensor is generally contributed by an internal field of correlations, which is temperature dependent in a modulated crystal. Therefore, Raman lines can be temperature dependent, if α_{ij} is involved in a soft mode. Figure 10.9a shows a typical example of Raman studies from ferroelectric $PbTiO_3$, showing ferroelectric soft modes characterized empirically as $\Delta\omega_R \propto \sqrt{T_c - T}$ below the critical temperatures $T_c = 500\,K$, which was in fact found consistent with neutron scattering results. Supported by the neutron inelastic scattering result for $T > T_c$ in Fig. 10.9b, the observed soft mode frequency is characterized by $\omega_o^2 \propto |T - T_c|$, which is compatible with the mean-field theory.

Further notable is that Raman spectroscopy was successfully used for perovskites $SrTiO_3$ and $KTaO_3$, whose critical regions were studied with an applied electric field. With this technique, Fleury et al. [6] identified the soft mode in these perovskites.

10.4 Magnetic Resonance

The magnetic resonance technique utilizes either nuclear moments or paramagnetic impurity ions, to probe local magnetic fields in a crystal undergoing a structural transition. In the former case, naturally abundant isotopes can be used as spin probes; in contrast, in the latter, magnetic ions with unpaired electrons are embedded as

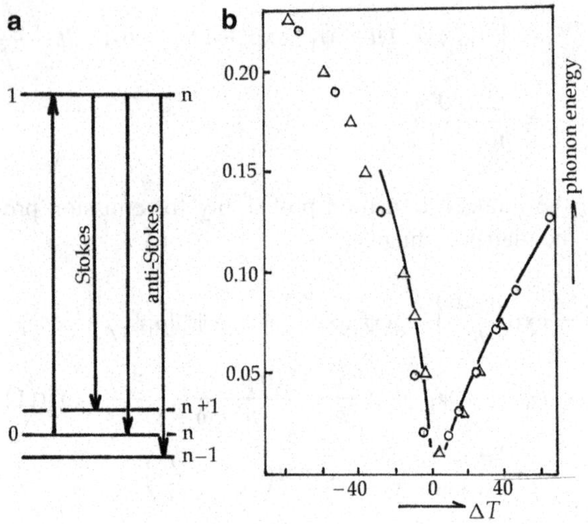

Fig. 10.8 (**a**) Raman transitions with Stokes and Anti-Stokes lines. (**b**) An example of Raman spectra. Soft-modes from ferroelastic LaP_5O_{14} and $La_{0.5}Nd_{0.5}P_5O_{14}$. From J.C. Toledano, E. Errandonea and I.P. Jaguin, Solid State Comm. **20**, 905 (1976).

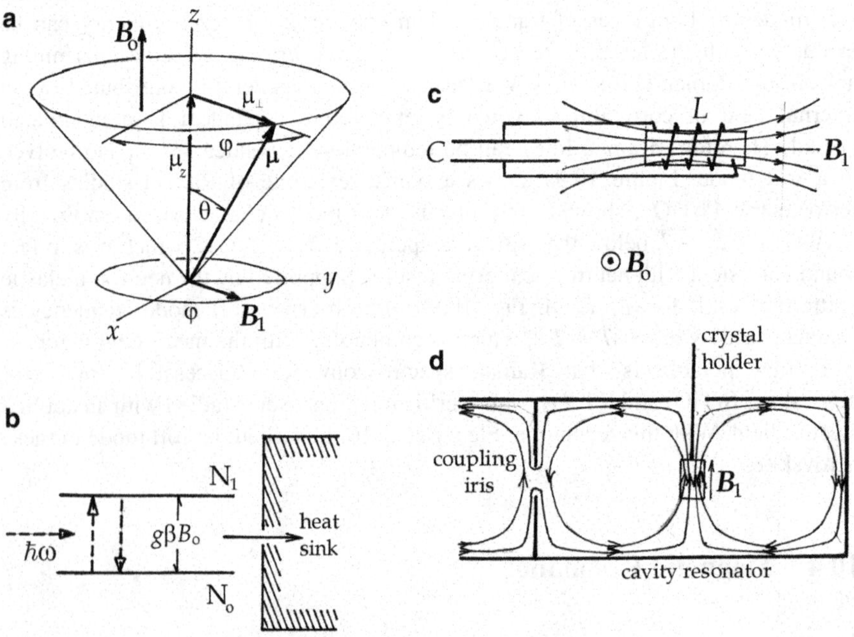

Fig. 10.9 (**a**) Larmor's precession. (**b**) Magnetic resonance in thermal environment. (**c**) An *LC* resonator for magnetic resonance at radio frequencies. (**d**) A microwave resonator for a paramagnetic resonance experiment.

impurities in normal crystals. For such chemically *doped* crystals, we have to be cautious to determine if doping causes only insignificant effect on sample crystals. Nevertheless, if so determined, the method provides very useful information about the structural change. On the other hand, nuclear probes can be used without doubt about modified crystals, although usable isotopes are limited to their natural abundance.

10.4.1 Principles of Nuclear Magnetic Resonance and Relaxation

Although well known in molecular beam spectroscopy, in condensed matter the equation of motion of magnetization was established by Bloch [7] for nuclear spins, which is applicable to paramagnetic ions in crystals as well. Nuclear magnetic moments are considered as primarily independent from the hosting lattice; however their time-dependent energy can be transferred to the lattice by a relaxation process.

Microscopically, the elemental magnetic moment $\boldsymbol{\mu}$ is in Larmor precession in an applied uniform magnet field $\boldsymbol{B}_o \| z$ at a constant frequency $\omega_L = \gamma \boldsymbol{B}_o$, where γ is the gyromagnetic ratio, as shown in Fig. 10.9a. Its z-component μ_z is kept constant while in precession around the z-direction; the perpendicular component $\mu_\perp$ is driven by an oscillating magnetic field $\boldsymbol{B}_1$ that is applied perpendicular to $\boldsymbol{B}_o$. The linearly oscillating field $\boldsymbol{B}_1$ is composed of two circularly rotating components as

$$B_1 \propto \cos \omega t = \frac{1}{2}(\exp i\omega t + \exp(-i\omega t)),$$

hence exerting a torque $-\mu_\perp B_1$ on $\boldsymbol{\mu}$ by the circular field in the first term, if $\omega = \omega_L$, to increase the angle θ of precession. This phenomenon is called the magnetic resonance. Considering such $\boldsymbol{\mu}$ in a crystal, the motion of macroscopic nuclear magnetization $\boldsymbol{M} = \sum \boldsymbol{\mu}$ should be discussed, for which Bloch wrote the equations for M_z and $M_\perp$ separately as

$$\frac{\mathrm{d}M_z}{\mathrm{d}t} = -\frac{M_z - M_o}{T_1} \quad \text{and} \quad \frac{\mathrm{d}M_\perp}{\mathrm{d}t} = -\frac{M_\perp}{T_2}. \tag{10.12}$$

The first equation signifies the relaxation process to thermal equilibrium with the lattice, but T_1 is called the *spin–lattice relaxation time*. Since M_z represents the macroscopic magnetization energy of nuclear magnetic moments $\boldsymbol{\mu}$, the time T_1 can be taken for the thermal relaxation in thermodynamic environment. The second one describes a synchronization of $\mu_\perp$ until the net precession of $M_\perp = \sum \mu_\perp$ becomes in phase; the phasing time is indicated by T_2. Bloch considered that spin–spin interactions of magnitude ΔB among $\mu_\perp$ are responsible for diverse precessions of $\boldsymbol{\mu}$, for which T_2 is called the *spin–spin relaxation time*. For random precessions of $\boldsymbol{\mu}$ in the presence of ΔB, it is necessary to have the condition

$$\frac{1}{T_2} = \gamma\Delta B \ll \gamma B_1. \tag{10.13}$$

This criterion (10.12) is referred to as the condition for *slow passage*.

In these two magnetic fields $\boldsymbol{B}_0$ and $\boldsymbol{B}_1$, the Bloch equations are written as

$$\frac{dM_\pm}{dt} \pm \gamma B_0 M_\pm + \frac{M_\pm}{T_2} = -i\gamma B_1 M_z \exp(\mp i\omega t)$$

and

$$\frac{dM_z}{dt} + \frac{M_z - M_0}{T_1} = \frac{i}{2}\gamma B_0\{M_+ \exp(-i\omega t) + M_- \exp(i\omega t)\}. \tag{10.14}$$

In equations (10.3) known also as the Bloch equations, $M_\pm$ are the transverse components of $\boldsymbol{M}$ synchronized with the rotating fields $B_1 \exp(\mp i\omega t)$.

A steady-state solution under the slow passage condition is given by $M_z = \text{const.}$; hence from the first equation we obtain

$$M_\pm = \frac{\gamma B_1 M_z \exp(\mp i\omega t)}{\mp\omega + \gamma B_0 + \frac{i}{T_2}}.$$

Using this result into (10.11), we can derive

$$\frac{M_z}{M_0} = \frac{1 + (\omega - \omega_L)^2 T_2^2}{1 + (\omega - \omega_L)^2 + \gamma^2 T_1 T_2} \quad \text{and} \quad \frac{M_\pm}{M_0} = \frac{\{(\omega - \omega_L)T_2 + i\}\gamma B_1 \exp(\mp i\omega t)}{1 + (\omega - \omega_L)^2 T_2^2 + \gamma^2 B_1^2 T_1 T_2}.$$

If $(\gamma B_1 T_1 T_2)^2 \ll 1$, from the first relation we have $M_z \approx M_0$, for which the angle θ can be expressed as

$$\tan\theta = \frac{M_\pm}{M_0} \approx \frac{\gamma B_1 T_2}{1 + (\omega - \omega_L)^2 T_2^2}.$$

Defining the high-frequency susceptibility by writing as $M_\pm = \chi(\omega) B_1 \exp(\mp i\omega t)$, and writing $M_0 = \chi_0 B_0$, we obtain the susceptibility in complex form $\chi(\omega) = \chi'(\omega) - i\chi''(\omega)$, where the real and imaginary parts are

$$\frac{\chi'(\omega)}{\chi_0} = \frac{\omega_L(\omega - \omega_L)}{(\omega - \omega_L)^2 + \delta\omega^2 + \gamma^2 B_1^2 T_1 \delta\omega} \quad \text{and}$$
$$\frac{\chi''(\omega)}{\chi_0} = \frac{\omega_L \delta\omega}{(\omega - \omega_L)^2 + \delta\omega^2 + \gamma^2 B_1^2 T_1 \delta\omega}. \tag{10.15}$$

Here, we have set $\delta\omega = \frac{1}{T_2}$ for convenience. Magnetic resonance specified by $\omega = \omega_L$ can be detected either with $\chi'(\omega_L)$ or $\chi''(\omega_L)$; the former shows frequency dispersion in the vicinity of ω_L, whereas the latter is related to the loss of high-frequency energy, as illustrated by the curves in Fig. 9.2. These expressions in

(10.15) are normally called *dispersion* and *absorption* of magnetic resonance, which can be measured as changes in resonant circuits when scanning the oscillator frequency ω or applied magnetic field B_o across the resonance condition $\omega_L = \gamma B_o$. Nevertheless, in practice, most experiments are using the field-scan method at a fixed oscillator frequency ω. At this point, remarkable is that these relaxation times T_1 and T_2 should be functions of temperature T at which the experiment is performed, although implicit in the foregoing theory.

10.4.2 Paramagnetic Resonance of Impurity Probes

The paramagnetic moment $\boldsymbol{\mu}$ is associated with the angular momentum of electrons, i.e., $\boldsymbol{\mu} = \gamma \boldsymbol{J} = \gamma\hbar(\boldsymbol{L} + \boldsymbol{S})$, where $\boldsymbol{L}$ and $\boldsymbol{S}$ are total orbital and spin angular momenta. The energy levels in a field $\boldsymbol{B}_o$ are given by $\varepsilon_m = -\gamma\hbar J_m \boldsymbol{B}_o$ where $J_m = J, J-1, \ldots, -J$, for which the selection rule is $\Delta J_{m,m-1} = \pm 1$. Hence, the magnetic resonance takes place when the radiation quantum $\hbar\omega$ is equal to the energy gap between adjacent levels, i.e.,

$$\hbar\omega = \varepsilon_m - \varepsilon_{m-1} = \hbar\gamma B_o \Delta J_{m,m-1} = \hbar\gamma B_o \quad \text{and} \quad \omega = \gamma B_o.$$

The latter is exactly the same as the Larmor frequency ω_L. Figure 10.9b shows the quantum view of magnetic resonance. For an electron, the Larmor frequency is expressed as $\omega_L = g\beta B_o$, where $\beta = \frac{\gamma}{2\hbar}$ and g are Bohr's *magneton* and Landé's factor, respectively; g = 2 for an electron.

Quantum mechanically, the magnetic resonance can be described as absorption and emission of a photon $\hbar\omega$ with probabilities

$$w_{m-1,m} = \frac{\pi B_1^2}{2\hbar^2} |\langle m|\mu_\perp|m-1\rangle|^2 f(\omega),$$

where $f(\omega) = \frac{1}{T_2}$ is a shape function normalized as $\int f(\omega)d\omega = 1$. Under practical experimental conditions where ω is a radio- or microwave frequency, *induced transitions* are predominant, and the *spontaneous emission* is negligible. Further, for these magnetic moments are in a crystal at a normal temperature *T*, the population N_m at an energy level ε_m is determined by the Boltzmann statistics, i.e., $N_m \propto \exp\left(-\frac{\varepsilon_m}{k_B T}\right)$. Therefore, at resonance $\omega = \omega_L$ the energy transfer rate to the magnetic system can be written as

$$w_{m,m-1}(\hbar\omega_L)(N_m - N_{m-1}) = w_{m,m-1}(\hbar\omega_L)N_m\left(1 - \exp\left(-\frac{\hbar\omega_L}{k_B T}\right)\right)$$
$$\approx w_{m.m-1}(\hbar\omega_L)N_m\frac{\hbar\omega_L}{k_B T} = N_m\frac{\pi\omega_L^2 B_1^2}{k^B T}|\langle m|\mu_\perp|m-1\rangle|^2 f(\omega_L).$$

This should be equivalent to the macroscopic energy loss $\frac{1}{2}\omega_L\chi''(\omega_L)B_1^2$ in a resonator, and therefore we can write

$$\chi''(\omega_L) = N_m \frac{\pi\omega_L}{k_B T}|\langle m|\mu_\perp|m-1\rangle|^2 f(\omega_L),$$

indicating that magnetic resonance intensity is appreciable at low temperatures.

The magnetic resonance signified by a complex susceptibility can be measured from a sample crystal placed in an inductor $L = \chi(\omega)L_o$, where L_o is the inductance of the empty coil, producing a field B_1 by an oscillating current on L_o. With a conventional laboratory magnet of $B_o \sim 10^4$ gauss, the Larmor frequency ω_L is of the order of 1–10 MHz for nuclear resonance, whereas ω_o is a microwave frequency of 9–35 GHz for paramagnetic resonance. Accordingly, we use an *LC* resonator or a cavity resonator in these measurements, respectively, as illustrated by Fig. 10.9c, d, where sample crystals are placed at a location of B_1 in maximum strength. Such a loaded inductor combined with a capacitor *C* for a radio-wave measurement, and a resonant cavity for microwaves, can be expressed by an *impedance* *Z*, i.e.,

$$Z = R + i\omega L + \frac{1}{i\omega C} = R + \omega\chi'' L_o + i\left(\omega\chi' L_o - \frac{1}{\omega C}\right),$$

where *R* is the resistance in the resonator. Figure 10.10 shows a block diagram of a microwave bridge, which is an impedance bridge commonly used for paramagnetic resonance experiments. In balancing such a bridge, the real and imaginary parts of *Z* can be measured independently. Although these parts are related mathematically, it is convenient in practice for $\chi'(\omega)$and $\chi''(\omega)$ to be measured separately.

10.4.3 The Spin Hamiltonian

Although a magnetic impurity ion in a nonmagnetic crystal is primarily independent of the lattice, its electronic density is deformed by the spin–orbit coupling, so that the local symmetry is modified. The impurity ion is then modified by a static potential at its site, called *crystalline potential*, signified by modified local symmetry. We postulate that such a potential can be orthorhombic with respect to impurity center, and expressed as

$$V(x,y,z) = Ax^2 + By^2 + Cz^2, \tag{10.16}$$

where x, y, and z are coordinates along symmetry axes of $V(x,y,z)$. Here, for these coefficients we have the relation $A + B + C = 0$, because such a static potential should satisfy the Laplace equation $\nabla^2 V = 0$. Further, such a local symmetry axes may not necessarily be the same as the symmetry axes of the lattice, since the unit

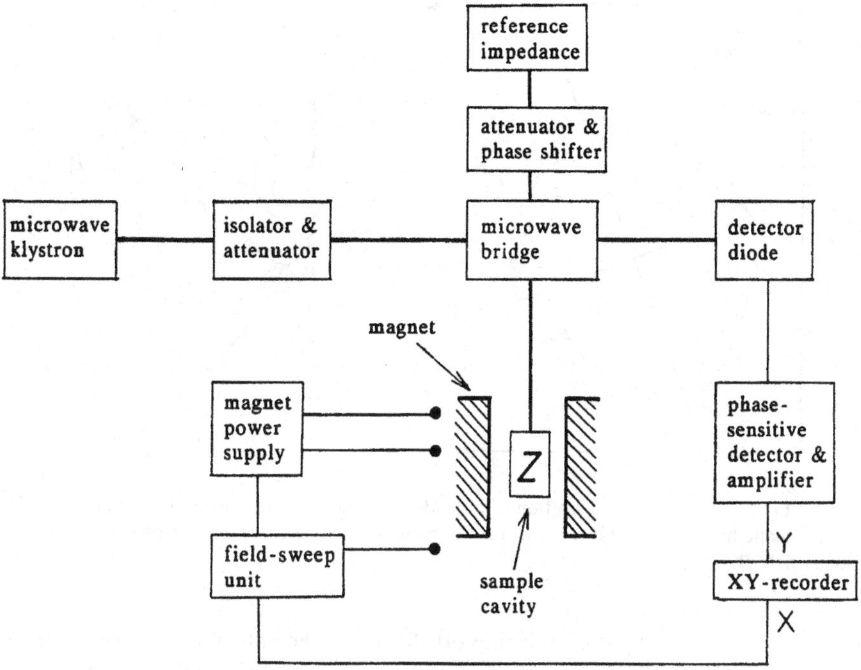

Fig. 10.10 A microwave bridge spectrometer for a paramagnetic resonance.

cell contains usually even numbers of equivalent constituent ions. In this case, the coordinates x, y, and z and the symmetry axes a, b, and c of the crystal are related by a coordinate transformation.

Most paramagnetic impurities for typical probing are ions in the groups of *transition elements*. Considering ions of the iron groups, electrons are characterized by their net orbital and spin angular momenta, $\boldsymbol{L}$ and $\boldsymbol{S}$, which are coupled as $\lambda\boldsymbol{L}.\boldsymbol{S}$, where the spin–orbit coupling constant λ is of the order of 300 cm^{-1}. On the other hand, the crystalline field energy is typically of the order of thousand cm^{-1}. Therefore, in the first approximation, we can consider that $\boldsymbol{L}$ is quantized in precession around the crystalline field, while the spin $\boldsymbol{S}$ is left as free from the lattice, which is usually described as *orbital quenching*. In this approximation, the spin–orbit coupling energy can be expressed as the first-order perturbation

$$E_{\text{LS}}^{(1)} = \lambda\left(L_x S_x + L_y S_y + L_z S_z\right), \tag{10.17}$$

assuming λ as constant. The second-order perturbation energy can be calculated as

$$E_{\text{LS}}^{(2)} = \frac{\lambda^2}{\Delta\varepsilon}\sum_{\text{i,j}} S_\text{i} S_\text{j}\left(\int_v \psi_\text{o}^* L_\text{i}\psi_\varepsilon \text{d}v\right)\left(\int_v \psi_\varepsilon{}^* L_\text{j}\psi_\text{o}\text{d}v\right),$$

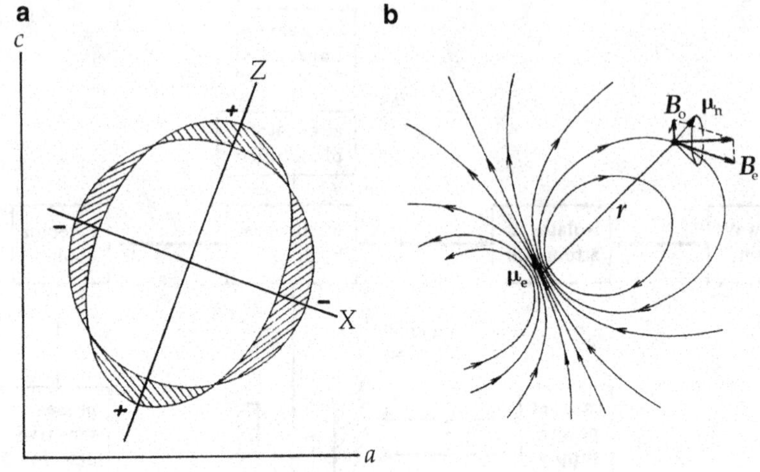

Fig. 10.11 (**a**) A symmetric distortion of a quadratic tensor of zero trace in two dimension. (**b**) The magnetic hyperfine field $\boldsymbol{B}_e$ is shown at the position of a nuclear spin $\boldsymbol{\mu}_n$ in precession around the net field $\boldsymbol{B}_e + \boldsymbol{B}_o$.

which is determined by nonvanishing off-diagonal elements of $L_i L_j$ between states with an energy gap $\Delta\varepsilon$. Hence, this can be expressed in a convenient form as

$$E_{LS}^{(2)} = \sum_{i,j} S_i D_{ij} S_j = \langle \boldsymbol{S}|\boldsymbol{D}|\boldsymbol{S}\rangle, \tag{10.18}$$

where $\boldsymbol{D}$ is a tensor $(D_{ij}) = \frac{\lambda^2}{\Delta\varepsilon}\left(\int_v \psi_o^* L_i \psi_\varepsilon \mathrm{d}v\right)\left(\int_v \psi_\varepsilon^* L_j \psi_o \mathrm{d}v\right)$ that is called the fine-structure tensor, for which we can verify the relation trace:

$$\text{Trace}\,\boldsymbol{D} = \sum_i D_{ii} = 0 \quad \text{or} \quad D_{xx} + D_{yy} + D_{zz} = 0. \tag{10.19}$$

The tensor $\boldsymbol{D}$ represents the ionic charge cloud exhibiting an ellipsoidal deformation in the crystalline field, as illustrated in Fig. 10.11a. In this case, the spin $\boldsymbol{S}$ modified by the spin–orbit coupling interacts with the crystalline potential as described by (10.17).

In an applied magnetic field $\boldsymbol{B}_o$, the spin can interact with $\boldsymbol{B}_o$ as expressed by the *Zeeman energy* $\mathsf{H}_Z = -\beta(\boldsymbol{L} + 2\boldsymbol{S}).\boldsymbol{B}_o$, where the orbital angular momentum $\boldsymbol{L}$ is quenched by the crystalline field; therefore H_Z must be expressed to the second-order accuracy. In the first-order approximation H_Z is determined by the effective magnetic moment of $\boldsymbol{S}$ only; $E_Z^{(1)} = -2\beta\boldsymbol{S}.\boldsymbol{B}_o$ where $g_o = 2$. Assuming that $B_o \| z$, the second-order energy is expressed as $E_Z^{(2)} = \frac{2\lambda}{\Delta\varepsilon} S_z B_o$, so the Zeeman energy determined by $E_Z^{(1)} + E_Z^{(2)}$ is given by

$$E_z = -g_z \beta S_z B_0 \quad \text{where} \quad g_x = g_y = 2\left(1 - \frac{\lambda}{\Delta\varepsilon}\right).$$

If assuming $\boldsymbol{B}_0 \| x$, or y, by a similar calculation we can derive different $g_{x,y}$ expressed by

$$E_{x,y} = -g_{x,y} \beta S_{x,y} B_0 \quad \text{where} \quad g_x = g_y = 2\left(1 - \frac{\lambda}{\Delta\varepsilon}\right).$$

Here, we considered for simplicity that the crystalline potential is uniaxial along the z-direction. These g-factors depend on the direction of applied field, so the Zeeman energy can generally be expressed as a tensor product by defining a tensor quantity $g = \left(g_{ij}\right)$, i.e.,

$$\mathsf{H}_Z = -\beta\langle \boldsymbol{S}|\boldsymbol{g}|\boldsymbol{B}_0\rangle. \tag{10.20a}$$

For experimental convenience, we usually specify the applied field as $\boldsymbol{B}_0 = B_0 \boldsymbol{n}$ with the unit vector $\boldsymbol{u}$ along the direction. It is therefore convenient to express (10.20a) as

$$\mathsf{H}_Z = -g_n \beta \boldsymbol{S}.\boldsymbol{B}_0 \quad \text{where} \quad g_n^2 = \langle \boldsymbol{n}|\boldsymbol{g}^2|\boldsymbol{n}\rangle. \tag{10.20b}$$

In paramagnetic resonance, we consider the Hamiltonian consisting of (10.17) and (10.19b), i.e.,

$$\mathsf{H} = -g_n \beta \boldsymbol{S}.\boldsymbol{B}_0 + \langle \boldsymbol{S}|\boldsymbol{D}|\boldsymbol{S}\rangle, \tag{10.21}$$

which is called the *spin Hamiltonian*.

For experimental convenience, the spin is considered as quantized with respect to the applied field $\boldsymbol{B}_0$, and so we express $S_n = M\hbar$, where the quantum number M are $\pm$ half integers. From (10.19a) and (10.19b), we have $\boldsymbol{g}_n\boldsymbol{S} = \langle \mathbf{S}|\boldsymbol{g}\rangle$; hence the second term in (10.20) can be modified as $\langle S|\boldsymbol{gDg}^\dagger|S\rangle/g_n^2$. However, since $\boldsymbol{g}$ and $\boldsymbol{D}$ are coaxial tensors, the tensor $\boldsymbol{D}' = \boldsymbol{gDg}^\dagger/g_n^2$ is usually written as $\boldsymbol{D}$, replacing what was originally defined as the fine structure tensor. Accordingly, (10.21) is written for $\boldsymbol{S}$ quantized with $\boldsymbol{B}_0$ as

$$\mathsf{H} = -g_n \beta S_n B_0 + \langle \boldsymbol{n}|\boldsymbol{D}|\boldsymbol{n}\rangle S_n^2.$$

The quantities $g_n^2 = \langle \boldsymbol{n}|\boldsymbol{g}^2|\boldsymbol{n}\rangle$ and $\langle \boldsymbol{n}|\boldsymbol{D}|\boldsymbol{n}\rangle S_n^2$ can be determined from magnetic resonance spectra recorded for all directions of $\boldsymbol{B}_0$, which can be numerically diagonalized for the *principal axes x, y*, and *z*.

10.4.4 Hyperfine Interactions

The interaction between the magnetic moment $\boldsymbol{\mu}_e$ and nuclear magnetic moment $\boldsymbol{\mu}_n$ located within the electron orbital can be determined from paramagnetic resonance spectra, yielding significant structural information of the impurity ion. As illustrated in Fig. 10.11b, such an interaction called the *hyperfine interaction* can be expressed classically as a dipole–dipole interaction, but quantum theoretically contributed also by the charge density at the position of $\boldsymbol{\mu}_n$ that is called a *contact* interaction. The hyperfine interaction is thus characterized by their orientation and expressed as $\langle \boldsymbol{\mu}_e | \boldsymbol{A} | \boldsymbol{\mu}_n \rangle$, but trace $\boldsymbol{A} \neq 0$ because of the contact contribution. Writing $\langle \boldsymbol{\mu}_e | = \beta \langle \boldsymbol{S} | \boldsymbol{g}$ and $\langle \mu_n | = \gamma \langle \boldsymbol{I} |$ where $\boldsymbol{I}$ is the nuclear spin, the hyperfine interaction can be expressed as

$$\mathsf{H}_{\mathrm{HF}} = \beta\gamma \langle \boldsymbol{S} | \boldsymbol{g}\boldsymbol{A} | \boldsymbol{I} \rangle. \tag{10.22}$$

As the spin $\boldsymbol{S}$ is quantized along $\boldsymbol{B}_o$, (10.21) can be expressed as

$$E_{\mathrm{HF}}^{(1)} = -\gamma \boldsymbol{B}_e.\boldsymbol{I},$$

where $\boldsymbol{B}_e = \beta M \sqrt{\left\langle \boldsymbol{n} \middle| \boldsymbol{g}\boldsymbol{A}\boldsymbol{A}^\dagger \boldsymbol{g}^\dagger \middle| \boldsymbol{n} \right\rangle}$ is the effective magnetic field due to $\boldsymbol{\mu}_e$, as shown in Fig. 10.11b; the nuclear moment $\gamma \boldsymbol{I}$ can be considered to be in precession around $\boldsymbol{B}_e$, if $B_e \gg B_o$. In this case, the first-order hyperfine energy is given by

$$E_{\mathrm{HF}}^{(1)} = -g_n \beta K_n M m' \quad \text{where} \quad g_n^2 K_n^2 = \left\langle \boldsymbol{n} \middle| \boldsymbol{g}\boldsymbol{A}\boldsymbol{A}^\dagger \boldsymbol{g}^\dagger \middle| \boldsymbol{n} \right\rangle. \tag{10.23}$$

In fact, the quantity K_n defined here is in energy unit, while in magnetic resonance practice it is more convenient to express it in magnetic field unit. In the field unit, we note that the hyperfine energy can simply be written as $E_{\mathrm{HF}}^{(1)} = -K_n M m'$. In any case, for practical convenience we may define the hyperfine tensor $K = \frac{gA}{g_n}$, so (10.21) is expressed as $\mathsf{H}_{\mathrm{HF}} = \langle S | K | S \rangle$. Further noted is that the quadratic form of K_n^2 can be diagonalized, but the principal axes do not coincide with those for the crystalline potential.

10.5 Magnetic Resonance in Modulated Crystals

Paramagnetic resonance spectra of transition ions are generally complicated due to the fact that the applied magnetic field may not be strong enough to avoid forbidden transitions. Nevertheless, the transition anomalies can be detected in anomalous g-factors, fine- and hyperfine-structures in selected directions of $\boldsymbol{B}_o$, although complete analysis is not always possible.

We assume that a probe spin $\boldsymbol{S}$ can be modulated in a mesoscopic phase by the order variable $\boldsymbol{\sigma}(\phi)$ in such a way that

$$\boldsymbol{S}' = \boldsymbol{a}.\boldsymbol{S} \quad \text{where} \quad \boldsymbol{a} = \boldsymbol{1} + \langle \boldsymbol{\sigma} | \boldsymbol{e}. \tag{10.24}$$

Here $\boldsymbol{1}$ and $\boldsymbol{e}$ are the tensors for identity and strains in the crystal; the latter deforms local symmetry, restricted by trace $\boldsymbol{e} = 0$. The spin Hamiltonian can therefore be modified as

$$\mathsf{H}' = -\beta \langle \boldsymbol{S}' | \boldsymbol{g} | \boldsymbol{B}_o \rangle + \langle \boldsymbol{S}' | \boldsymbol{D} | \boldsymbol{S}' \rangle + \langle \boldsymbol{S}' | \boldsymbol{K} | \boldsymbol{I} \rangle.$$

Using (10.23), this can be expressed as

$$\mathsf{H}' = \mathsf{H} + \mathsf{H}_1,$$

where

$$\mathsf{H} = -\beta \langle \boldsymbol{S} | \boldsymbol{g} | \boldsymbol{B}_o \rangle + \langle \boldsymbol{S} | \boldsymbol{D} | \boldsymbol{S} \rangle + \langle \boldsymbol{S} | \boldsymbol{K} | \boldsymbol{I} \rangle$$

and

$$\mathsf{H}_1 = -\sigma\beta \left\langle \boldsymbol{S} \middle| \boldsymbol{e}^\dagger \boldsymbol{g} \middle| \boldsymbol{B}_o \right\rangle + \sigma \left\langle \boldsymbol{S} \middle| \boldsymbol{e}^\dagger \boldsymbol{D} + \boldsymbol{D}\boldsymbol{e} \middle| \boldsymbol{S} \right\rangle + \sigma^2 \left\langle \boldsymbol{S} \middle| \boldsymbol{e}^\dagger \boldsymbol{D}\boldsymbol{e} \middle| \boldsymbol{S} \right\rangle + \sigma \left\langle \boldsymbol{S} \middle| \boldsymbol{e}^\dagger \boldsymbol{K} \middle| \boldsymbol{I} \right\rangle. \tag{10.25}$$

These terms in (10.24) are basic formula for anomalies in paramagnetic spectra in modulated phases.

In practice, we analyze such anomalies in terms of quantum numbers M and m' for $\boldsymbol{S}$ and $\boldsymbol{I}$. The Zeeman and hyperfine anomalies expressed by the first and last terms are given by $-g_n\beta M B_o$ and $K_n M m'$, respectively, where the spin is linearly modulated by σ at a binary transition, hence exhibiting symmetrical and antisymmetrical fluctuations. On the other hand, the fine-structure term of $\boldsymbol{D}$ is quadratic with respect to σM, being characterized by σ and σ^2.

For the hyperfine term in (10.24), we obtain the expression

$$K_n'^2 = \left\langle \mathrm{n} \middle| \boldsymbol{a}^\dagger \boldsymbol{K}^\dagger \boldsymbol{K} \boldsymbol{a} \middle| \mathrm{n} \right\rangle = \langle \mathrm{n} | \boldsymbol{K}^2 | \mathrm{n} \rangle + \sigma \left\langle \mathrm{n} \middle| \boldsymbol{e}^\dagger \boldsymbol{K}^2 + \boldsymbol{K}^2 \boldsymbol{e} \middle| \mathrm{n} \right\rangle + \sigma^2 \left\langle \mathrm{n} \middle| \boldsymbol{e}^\dagger \boldsymbol{K} \boldsymbol{e} \middle| \mathrm{n} \right\rangle;$$

therefore for a binary splitting characterized by $\pm\sigma$ we can write

$$K'_n(+)^2 - K'_n(-)^2 = 2\sigma \left\langle \mathrm{n} \middle| \boldsymbol{e}^\dagger \boldsymbol{K}^2 + \boldsymbol{K}^2 \boldsymbol{e} \middle| \mathrm{n} \right\rangle.$$

The hyperfine anomaly can therefore be expressed by

$$\Delta K'_{\mathrm{n}} = K'_{\mathrm{n}}(+) - K'_{\mathrm{n}}(-) = \frac{2\sigma}{\bar{K}'_{\mathrm{n}}}\left\langle \mathrm{n}\middle| \boldsymbol{e}^{\dagger}\boldsymbol{K}^{2} + \boldsymbol{K}^{2}\boldsymbol{e}\middle|\mathrm{n}\right\rangle, \tag{10.26}$$

where $\bar{K}'_{\mathrm{n}} = \frac{1}{2}(K'_{\mathrm{n}}(+) + K'_{\mathrm{n}}(-))$, signifying the anomaly as proportional to σ, similar to the g_{n} anomaly.

On the other hand, the fine-structure anomaly derived from (10.24) is

$$\mathsf{H}_{1\mathrm{F}} = \sigma\left\langle S_{\mathrm{n}}\middle|\boldsymbol{e}^{\dagger}\boldsymbol{D} + \boldsymbol{D}\boldsymbol{e}\middle|S_{\mathrm{n}}\right\rangle + \sigma^{2}\left\langle S_{\mathrm{n}}\middle|\boldsymbol{e}^{\dagger}\boldsymbol{D}\boldsymbol{e}\middle|S_{\mathrm{n}}\right\rangle,$$

yielding a modulated energy

$$\Delta E_{\mathrm{F}}^{(1)} = (a_{\mathrm{n}}\sigma + b_{\mathrm{n}}\sigma^{2})M^{2},$$

where

$$a_{\mathrm{n}} = \left\langle \mathrm{n}\middle|\boldsymbol{e}^{\dagger}\boldsymbol{D} + \boldsymbol{D}\boldsymbol{e}\middle|\mathrm{n}\right\rangle \quad \text{and} \quad \mathrm{b}_{\mathrm{n}} = \left\langle \mathrm{n}\middle|\boldsymbol{e}^{\dagger}\boldsymbol{D}\boldsymbol{e}\middle|\mathrm{n}\right\rangle.$$

Hence, the magnetic resonance condition for the selection rules $\Delta M = \pm 1$ is given by

$$\hbar\omega = g_{\mathrm{n}}\beta B_{\mathrm{n}} + (D_{\mathrm{n}} + \Delta D_{\mathrm{n}})(2M + 1),$$

where

$$\Delta D_{\mathrm{n}} = a_{\mathrm{n}}\sigma + b_{\mathrm{n}}\sigma^{2}. \tag{10.27a}$$

It is noted that a binary structural change is signified by a *mirror plane* characterized by $\sigma \to \pm\sigma$. Therefore, the critical region is signified by symmetric and antisymmetric fluctuations between $+a_{\mathrm{n}}\sigma + b_{\mathrm{n}}\sigma^{2}$ and $-a_{\mathrm{n}}\sigma + b_{\mathrm{n}}\sigma^{2}$, so the binary fluctuations are given specially by

$$\Delta D_{\mathrm{n}}(A) = 2a_{\mathrm{n}}\sigma_{\mathrm{A}} \quad \text{and} \quad \Delta D_{\mathrm{n}}(\mathrm{P}) = 2a_{\mathrm{n}}\sigma_{\mathrm{P}} \tag{10.27b}$$

in all directions of $\boldsymbol{B}_{\mathrm{o}}$. For fluctuations in amplitude and phase modes, we have $\sigma_{\mathrm{A}} = \sigma_{\mathrm{o}}\cos\phi$ and $\sigma_{\mathrm{P}} = \sigma_{\mathrm{o}}\sin\phi$. In Fig. 10.12, $\Delta D_{\mathrm{n}}(P)$ in (10.26b) and ΔD_{n} in (10.26a) are sketched according to magnetic resonance practice. These drawings are made on derivatives of distributed intensities with respect to the microwave frequency ν, representing $\Delta\nu = \nu_{1}\cos\phi$ and $\Delta\nu = \nu_{1}\cos\phi + \nu_{2}\cos^{2}\phi$. It is only a matter of technical convenience that magnetic resonance spectra are displayed as the first derivative.

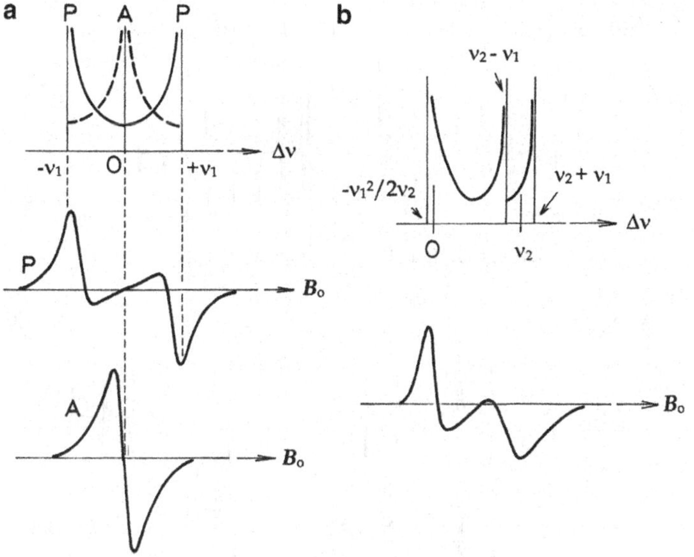

Fig. 10.12 Magnetic resonance anomalies displayed by first derivatives of the absorption $\chi''(B_o)$. (**a**) Line shape given by (10.26b). (**b**) Line shape given by (10.26a).

10.6 Examples of Transition Anomalies

Example 1 Mn^{2+} Spectra in TSCC

The ferroelectric phase transition in TSCC crystals at $T_c = 130\,\mathrm{K}$ was thoroughly investigated with paramagnetic impurities, Mn^{2+}, Cr^{3+}, Fe^{3+}, VO^{2+}, etc. Although trivalent impurities show complex spectra as associated with unavoidable charge defects, there should be no problem with divalent ions. Mn^{2+} ions are particularly useful, among others, because of the simple spectral analysis in TSCC. In Fig. 10.13a [8] a Mn^{2+} spectrum when B_o applied in the *bc* plane is shown, where two lattice sites in a unit cell give identical spectra at temperatures above T_c. The *g*-tensor and the hyperfine tensor with a ^{55}Mn nucleus $(I = 5/2)$ were found as isotropic, while the spectra in all directions of B_o were dominated by the fine structure as shown by Fig. 10.13b [8]. In Fig. 10.14a [9] the observed *D* plotted in the solid curves are for $T > T_c$, where these lines split into two broken lines at temperatures $T < T_c$, showing anomalous line shape with decreasing temperature. Figure 10.14b [9] shows changing line shapes for the transition $\Delta M = 1$ and $\Delta m' = 0$ at 9.2 and 35 GHz with decreasing temperature. The shape and splitting explained by (10.26b) with different timescales in these resonance experiments are noted. Also noticeable from Fig. 10.14b is that the central line A is more temperature dependent than the stable domain lines P.

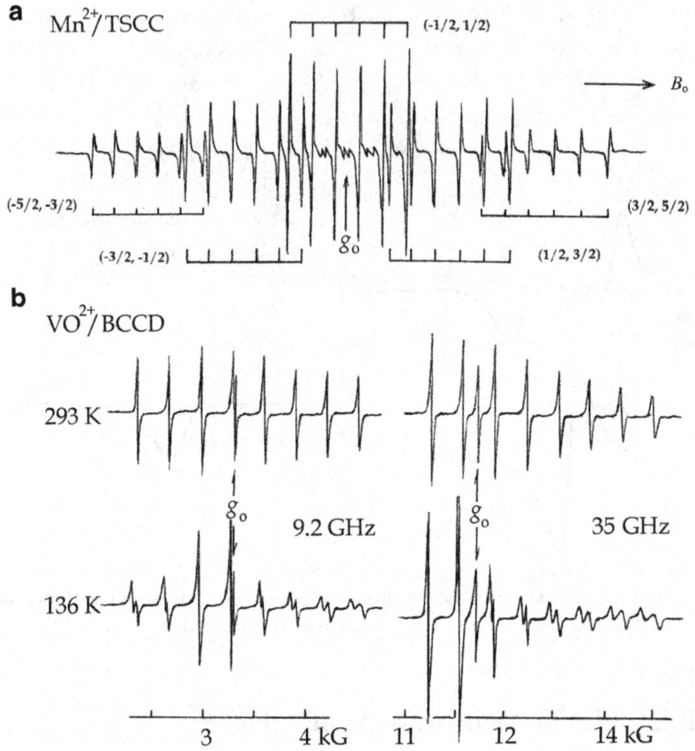

Fig. 10.13 (**a**) A representative resonance spectrum of Mn^{2+} ions in TSCC. From M. Fujimoto, S. Jerzak and W. Windsch, Phys. Rev. **B34**, 1668 (1986). (**b**) Representative VO^{2+} spectra in BCCD. Spectra observed at different microwave frequencies are compared at temperatures above and below $T_c = 164$ K. From M. Fujimoto and Y. Kotake, J. Chem. Phys. **90**, 532 (1989).

Example 2 Mn^{2+} Spectra in BCCD

Betaine Calcium Chloride Dihydrate (BCCD) crystals exhibit sequential phase changes, as in Fig. 10.15c, showing commensurate and incommensurate phases that are indicated by C and I, where soft modes were found at thresholds in some of these transitions. Mn^{2+} impurities substituted for Ca^{2+} ions exhibited spectra distinguishing these phases, however too complex for complete analysis. Nevertheless, Mn^{2+} spectra are characterized by larger fine-structure splitting than in TSCC. Due to the large crystalline potential, the spectra exhibit both allowed and forbidden lines in comparable intensities, making analysis difficult, except for the direction of $\boldsymbol{B}_o$ parallel to the principal axes x, y, and z, where only allowed transitions are observed. In some other directions $\boldsymbol{n}$, transition anomalies were revealed in resonance lines for $M = \pm\frac{5}{2}$, as shown in Fig. 10.15b. Here six components of ^{55}Mn hyperfine lines are in anomalous shape as illustrated by Fig. 10.12b, which is characterized by fluctuations of $a_n\sigma + b_n\sigma^2$, indicating that such resonance lines do not represent domain splitting.

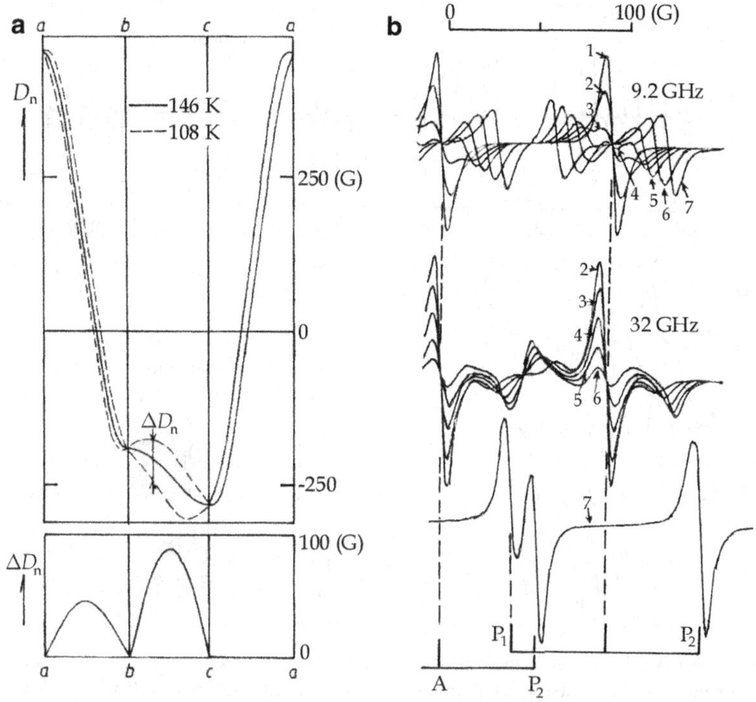

Fig. 10.14 (**a**) Angular dependence of the fine structure splitting D_n and anomaly D_n in Mn^{2+} spectra in TSCC. (**b**) Representative change of a Mn^{2+} hyperfine line with temperature. The temperature was lowered from T_c downward as $1 \rightarrow 2 \rightarrow 3 \rightarrow$ ....7. No anomaly was observed 7. From M. Fujimoto, S. Jerzak and W. Windsch, Phys. Rev. **B34**, 1668 (1986).

Example 3 VO^{2+} Spectra in BCCD

VO^{2+} ions substituted for Ca^{2+} in a BCCD crystal show simple spectra as in Fig. 10.13b, which are dominated by the hyperfine structure of ^{51}V nucleus (I = 7/2). Eight hyperfine component lines observed at 9.2 and 35 GHz are in different shapes, which are similar but with different separations, as can be explained by the formula (10.25). The g_n-factor and hyperfine splitting K_n are both anisotropic, as shown in Fig. 10.16a [10], and the anomalies ΔK_n are of type (10.25) in all directions of $\boldsymbol{B}_o$ being observed in the transition line for m' in the best resolution. Observed temperature dependence of a hyperfine line is shown in Fig. 10.16b [10].

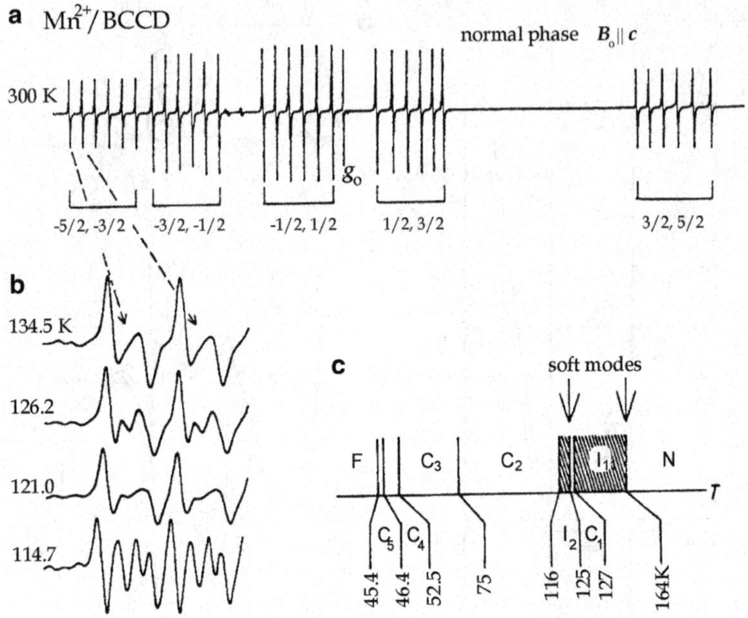

Fig. 10.15 (**a**) A Mn^{2+} spectrum from the normal phase of BCCD for $B_0 \| c$. (**b**) A change in anomalous lines with decreasing temperature. (**c**) Sequential transitions in BCCD. N, I, C, and F indicate normal, incommensurate, commensurate, and ferroelectric phases, respectively. From M. Fujimoto and Y. Kotake, J. Chem. Phys. **91**, 6671 (1989).

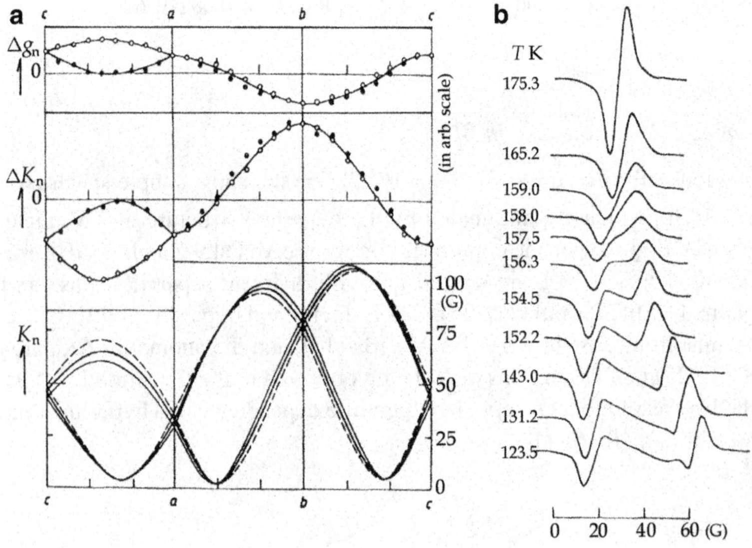

Fig. 10.16 (**a**) Angular dependences of Δg_n, ΔK_n, and K_n obtained from VO^{2+} spectra in BCCD at 130 K. (**b**) Change in line shape of a VO^{2+} hyperfine line with decreasing temperature. From M. Fujimoto and Y. Kotake, J. Chem. Phys. **91**, 6671 (1989).

References

1. M.P. Schulhof, P. Heller, R. Nathans, A. Linz, Phys. Rev. **B1**, 2403 (1970)
2. E.M. Brody, H.Z. Cummins, Phys. Rev. Lett. **21**, 1263 (1968)
3. E.M. Brody, H.Z. Cummins, Phys. Rev. **9**, 179 (1974)
4. M. Born, K. Huang, *Dynamical Theory of Crystal Lattices.* (Oxford Univ. Press, 1968), p. 204
5. J.C. Toledano, E. Errandonea, J.P. Jaguin, Solid State Comm. **20**, 905 (1976)
6. P.A. Fleury, J.F. Scott, J.M. Worlock, Phys. Rev. Lett. **21** (1968)
7. F. Bloch, Phys. Rev. **70**, 460 (1946)
8. M. Fujimoto, S. Jerzak, W. Windsch, Phys. Rev. **B34**, 1668 (1986)
9. M. Fujimoto, Y. Kotake, J. Chem. Phys. **90**, 532 (1989)
10. M. Fujimoto, Y. Kotake, J. Chem. Phys. **91**, 6671 (1989)

Chapter 11
Magnetic Crystals

Magnetic crystals are composed of magnetic ions at all lattice or sublattice sites, possessing microscopic magnetic moments $\boldsymbol{\mu}_n$. On lowering temperature, these crystals are magnetized by ordered $\boldsymbol{\mu}_n$. In this case, magnetic interactions due to quantum-mechanical electron exchanges between ions are responsible for ordered $\boldsymbol{\mu}_n$. The magnetization of a crystal comprises differently oriented domains, where their volume ratio can be modified by an applied magnetic field.

Unlike electric dipolar order with inversion symmetry, magnetic moment $\boldsymbol{\mu}_n$ is in Larmor precession in the magnetic field, where inversion symmetry is held by reversing applied magnetic field. Although primarily independent of the lattice, motion of $\boldsymbol{\mu}_n$ carried by constituent ions suffers strains in the crystal structure; consequently, magnetic ordering is not quite analogous to dipolar ordering. In this chapter, reviewing the origin of magnetic moments $\boldsymbol{\mu}_n$, magnetic correlations in crystals are discussed, for which both crystalline and magnetic symmetries play basic roles.

In this book, we discuss only basic physics of magnetism, as related to structural transitions, leaving many of the details to specific articles. Therefore, our arguments in this chapter remain sketchy at the level of introductory textbooks [1, 2].

11.1 Microscopic Magnetic Moments in Crystals

Magnetic ions of transition- and rare-earth groups are characterized by the presence of unpaired electrons. In *incompletely filled* 3*d* and 4*f* inner *shells* of a magnetic ion, odd number of electrons is responsible for magnetism of various types in these crystals.

Assuming that electrons are charged mass particles orbiting independently around the nucleus, their energies and angular momentum, ε_i and $\boldsymbol{l}_i$, are conserved quantities. In this case, the total $\sum_i \varepsilon_i$ and $\sum_i \boldsymbol{l}_i$ are constant, ignoring their mutual interactions. In quantum mechanics, on the other hand, each electron has an intrinsic *spin angular*

M. Fujimoto, *Thermodynamics of Crystalline States*,

DOI 10.1007/978-1-4614-5085-6_11,

momentum $\boldsymbol{s}_i$ so that we need to consider their total spin $\boldsymbol{S} = \sum_i \boldsymbol{s}_i$ and total orbital angular momentum $\boldsymbol{L} = \sum_i \boldsymbol{l}_i$. Primarily, this assumption, called Russell–Saunders coupling, is significant to specify ionic states. In this scheme, $\boldsymbol{L}$ and $\boldsymbol{S}$ are quantized along an arbitrary direction z in space so that their steady z-components, M_L and M_S, are specified by $+L, L-1, \ldots, -L$ and $+S, S-1, \ldots -S$, respectively. Further, the total angular momentum $\boldsymbol{J} = \boldsymbol{L} + \boldsymbol{S}$ is conserved in the absence of an applied magnetic field, where the component J_z along the z direction can take discrete values $M_J = L+S, L+S-1, \ldots, |L-S|$. Specified by J, the ionic energies are determined by the principal quantum number n and J; the total energy $\sum_i \varepsilon_i$ is M_J-fold degenerate, which is referred to as the *J-multiplet*.

Primarily independent of $\boldsymbol{S}$ though, the orbital angular momentum $\boldsymbol{L}$ interacts with the spin momentum $\boldsymbol{S}$ *magnetically*. Considering classical definitions for $\boldsymbol{L} = m(\boldsymbol{r} \times \boldsymbol{v})$ and $\boldsymbol{j} = e\boldsymbol{v}$, where e and m are charge and mass of an electron, Biot–Savart law states that the orbital current $\boldsymbol{j}$ generates a magnetic field proportional to $\left\langle \frac{1}{r^3} \right\rangle (\boldsymbol{r} \times \boldsymbol{j}) = \frac{e}{m} \left\langle \frac{1}{r^3} \right\rangle \boldsymbol{L}$ at the center of the closed orbit. Accordingly, the magnetic interaction energy with the spin momentum $\boldsymbol{S}$ can be expressed as

$$\mathsf{H}_{LS} = \lambda \boldsymbol{L}.\boldsymbol{S}, \tag{11.1}$$

which is called the *spin–orbit coupling* energy. In the above model, the constant λ is difficult to calculate accurately though its experimental values obtained from atomic spectra can be used in (11.1) as adequate for most applications.

In crystals, however, electronic states of an ion are perturbed by the anisotropic crystal field, in which electronic orbits cannot be considered as spherical around the nucleus. In other words, an ion is deformed by the lattice potential in such a way that the ground state is signified by the expectation value $\langle \boldsymbol{L} \rangle = 0$. Using the basic definition $L = i\hbar(\boldsymbol{r}) \times \nabla$, we can verify that

$$\langle \boldsymbol{L} \rangle_o = \int_v \psi_o \boldsymbol{L} \psi_o \mathrm{d}v = \int_v \psi_o^* \boldsymbol{L} \psi_o^* \mathrm{d}v = -\left(\int_v \psi_o \boldsymbol{L} \psi_o \mathrm{d}v \right)^* = -\langle \boldsymbol{L} \rangle_o^*, \tag{11.2}$$

for the ground state ψ_o so that $\boldsymbol{L}$ can be considered as insignificant in crystals as signified by $\langle \boldsymbol{L} \rangle_o = 0$, which is known as *orbital quenching*.

For multiplet states signified by $\boldsymbol{J} = \boldsymbol{L} + \boldsymbol{S}$, the spin–orbit coupling H_{LS} is considered as a perturbation. Since $\boldsymbol{J}$ is a constant in this case, we can write the equation $\frac{\mathrm{d}}{\mathrm{d}t}(\boldsymbol{L} + \boldsymbol{S}) = 0$, and hence

$$\frac{\mathrm{d}\boldsymbol{L}}{\mathrm{d}t} = -\frac{\mathrm{d}\boldsymbol{S}}{\mathrm{d}t}. \tag{i}$$

Considering H_{LS} as responsible for the motion described by (i), we can write

$$\frac{\mathrm{d}\boldsymbol{L}}{\mathrm{d}t} = (\lambda \boldsymbol{S} \times \boldsymbol{L}) \quad \text{and} \quad \frac{\mathrm{d}\boldsymbol{S}}{\mathrm{d}t} = (\lambda \boldsymbol{L} \times \boldsymbol{S}),$$

Fig. 11.1 Larmor precession of L and S around J.

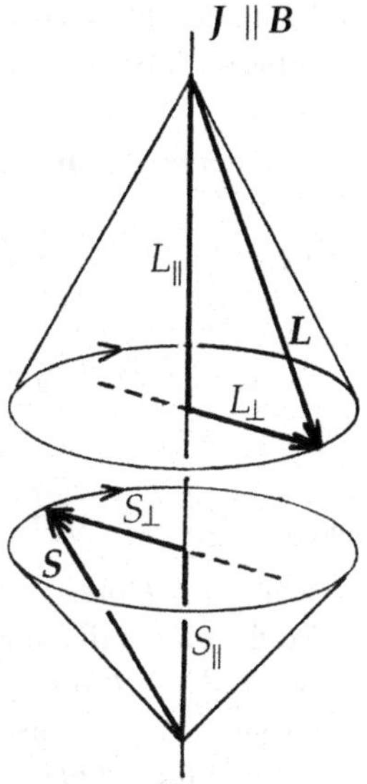

where the right-hand sides represent *torques* on $\boldsymbol{L}$ and $\boldsymbol{S}$ exerted by the local magnetic fields $\lambda\boldsymbol{S}$ and $\lambda\boldsymbol{L}$, respectively. Consequently, equations of motion for $\boldsymbol{L}$ and $\boldsymbol{S}$ can be expressed as

$$\frac{\mathrm{d}\boldsymbol{L}}{\mathrm{d}t} = \lambda(\boldsymbol{J} \times \boldsymbol{L}) \quad \text{and} \quad \frac{\mathrm{d}\boldsymbol{S}}{\mathrm{d}t} = \lambda(\boldsymbol{J} \times \boldsymbol{S}), \tag{ii}$$

indicating that $\boldsymbol{L}$ and $\boldsymbol{S}$ are in *precession* around $\boldsymbol{J}$, as illustrated in Fig. 11.1.

The ground state of the ion signified by the smallest component $M_J = |L - S|$ is perturbed by $\mathrm{H_{LS}}$ for $\lambda>0$ to a lower value, where the perturbed state can be visualized by such a parameter a as given by the relations

$$\boldsymbol{L} = a\boldsymbol{J} \quad \text{and} \quad \boldsymbol{S} = (1 - a)\boldsymbol{J}. \tag{11.3a}$$

Regarding to constant $\boldsymbol{L} + \boldsymbol{S} = \boldsymbol{J}$, the factor a indicates a degree of orbital quenching modified by $\mathrm{H_{LS}}$. Using quantum relations $\boldsymbol{J}^2 = \hbar^2 J(J+1)$ and $\boldsymbol{L}^2 = \hbar^2 L(L+1)$with the relation $\mathbf{J}.\boldsymbol{L} = a\boldsymbol{J}^2 = \boldsymbol{L}^2 + \boldsymbol{S}.\boldsymbol{L}$, we obtain

$$a = \frac{J(J+1) + L(L+1) - S(S+1)}{2J(J+1)}. \tag{11.3b}$$

It is noted that vectors J, L, and $\boldsymbol{S}$ can usually be defined either as angular momenta or *quantum numbers*, but (11.3b) remains the same, as the factor $\hbar^2$cancels out in any case.

Using the latter definition, the magnetic moment of an ion in the ground state can be expressed as

$$\boldsymbol{\mu} = -\gamma(\boldsymbol{L} + 2\boldsymbol{S}) = -\gamma(2 - a)\boldsymbol{J} = -g_J\beta\boldsymbol{J}, \tag{11.4a}$$

where

$$g_J = \frac{3J(J+1) - L(L+1) + S(S+1)}{2J(J+1)}, \tag{11.4b}$$

and $\beta = \dfrac{e\hbar}{2mc} = 0.927 \times 10^{-20}$emu is *Bohr magneton*. The expression (11.4b) is known as the *Landé factor*.

To determine the total spin $\boldsymbol{S}$ in incomplete shell specified by L, electrons must obey *Pauli exclusion principle*; only one electron can occupy each of single electronic states. Consider $3d$ shell signified by the principal quantum number $n = 3$ and the orbital $L = 2$, for example, we can accommodate up to $2(2L + 1) = 10$ electrons. Accordingly, the number of electrons in the incomplete $3d$ shell of transition elements can be either one of the number n in the range $1 \le n \le 9$. Such an n should be determined by empirical *Hund's rule* in atomic spectroscopy for arranging electrons at maximize $M_S = \sum_{\text{shell}} m_s$. Using this rule, we have $M_S = \pm\frac{1}{2}$ for odd n and $M_S = 0$ for even n. Therefore, for $\lambda > 0, n = 1, 2, 3, 4$ correspond to the lowest spin–orbit coupling, whereas for $\lambda < 0, n = 6, 7, 8, 9$. The latter case can be rephrased by *holes* $n' = 10 - n = 4, 3, 2, 1$ to be consistent with $\lambda > 0$. If $n = 5$, in particular, the lowest energy state is ^6S; hence $\mathsf{H}_{LS} = 0$, signifying $L = 0$ and $S = \frac{1}{2}$. Table 11.1 lists multiplet states of transition group ions and experimental values of λ. It is noted that $3d$ electrons of transition ions orbit closely to the outer shell so that their orbital momentum is strongly quenched in a crystal field. On the other hand, the $4f$ shell of rare-earth ions is relatively deep inside ions so that the quenching of $\boldsymbol{L}$ is weak and hence insignificant. Table 11.2 shows multiplet states for rare-earth elements.

Table 11.1 Incomplete $3d$ shells in transition group ions

Ion	n	State	λcm^{-1}
Ti^{3+}	1	^2D	154
V^{3+}	2	^3F	104
Cr^{3+}	3	^4F	87
Mn^{2+}	4	^5D	85
V^{2+}	3	^4F	55
Cr^{2+}	4	^5D	57
Fe^{3+}, Mn^{2+}	5	^6S	0
Fe^{2+}	6	^5D	−100
Co^{2+}	7	^4F	−180
Ni^{2+}	8	^3F	−335
Cu^{2+}	9	^2D	−828

Table 11.2 Spin–orbit coupling constants

Ion	Ti^{3+}	V^{3+}	Cr^{3+}	Mn^{3+}	V^{2+}	Cr^{2+}	Fe^{2+}	Co^{2+}	Ni^{2+}	Cu^{2+}
d-shell	d^1	d^2	d^3	d^4	d^3	d^4	d^6	d^7	d^8	d^9
J-state	^2D	^3F	^4F	^5D	^4F	^5D	^5D	^4F	^3F	^2D
$\lambda(\text{cm}^{-1})$	154	104	87	85	55	57	−100	−180	−335	−828

11.2 Paramagnetism

In a magnetic crystal, magnetic ions at lattice sites cannot be independent from each other, as weak magnetic dipole–dipole interactions should also be considered in principle. Nevertheless, assuming the latter as negligible at elevated temperatures, magnetic ions are considered in paramagnetic states, where their electronic energy splits into multiplet levels by an applied magnetic field $\boldsymbol{B}$, namely,

$$\varepsilon(J, M_J) = -g_J \beta B M_J, \;\; M_J = J, J-1, \ldots, -J.$$

These levels are populated thermodynamically as proportional to the Boltzmann factors $\exp\{-\varepsilon(J, M_J)/\text{k}_\text{B}T\}$.

For the paramagnetic state of a crystal of N ions, the macroscopic magnetization can be calculated as

$$M(B,T) = \text{N}\frac{\sum_{-J}^{+J}(-g_J\beta M_J)\exp\frac{g_J\beta M_J B}{\text{k}_\text{B}T}}{\sum_{-J}^{+J}\exp\left(-\frac{g_J\beta M_J B}{\text{k}_\text{B}T}\right)}.$$

Here, $Z = \sum_{-J}^{+J}\exp\frac{g_J\beta B M_J}{\text{k}_\text{B}T}$ is the partition function of magnetic ions. Writing $x = g_J\beta B/\text{k}_\text{B}T$ for convenience,

$$Z = \sum_{-J}^{+J}\exp x M_J = \exp(-xJ) + \cdots + \exp xJ = \frac{\exp(-xJ) - \exp x(J+1)}{1-\exp x}$$
$$= \frac{\sinh(J+\frac{1}{2})x}{\sinh\frac{1}{2}x}.$$

Therefore,

$$M(B,T) = \text{N}\text{k}_\text{B}T\frac{\partial \ln Z}{\partial B} = \text{N}\text{k}_\text{B}T\frac{\partial \ln Z}{\partial x}\frac{\partial x}{\partial B}$$
$$= g_J\beta\left\{\frac{(J+\frac{1}{2})\cosh(J+\frac{1}{2})x}{\sinh(J+\frac{1}{2})x} - \frac{\cosh\frac{1}{2}x}{2\sinh\frac{1}{2}x}\right\},$$

or

$$M(B,T) = g_J \beta J \mathrm{B}_J(x), \tag{11.5a}$$

where

$$\mathrm{B}_J(x) = \frac{2J+1}{2J}\coth\frac{(2J+1)x}{2J} - \frac{1}{2J}\coth\frac{x}{2J}. \tag{11.5b}$$

The function $\mathrm{B}_J(x)$is known as the *Brillouin function.*

Experimentally, (11.5ab) were found to explain measured results of M/M_o versus B/T from some rare-earth crystals, in particular, as illustrated in Fig. 11.2; here, M_o is the value of M at $B = 0$. In the limit of $x \to 0$, we can show that $\mathrm{B}_J(x) \cong \frac{J+1}{3J}x$ approximately, with which we can derive the *Curie law*. The response function of a *paramagnetic* crystal, called *susceptibility*, is defined as

$$\chi_m = \frac{M(B,T)}{B} = \lim_{B\to 0} \mathrm{N} g_J \beta J \mathrm{B}_J\left(\frac{g_J \beta J B}{\mathrm{k_B} T}\right) = \frac{\mathrm{N} g_J^2 \beta^2 J(J+1)}{3\mathrm{k_B} T}.$$

This susceptibility represents the Curie law expressed as

$$\chi = \frac{C}{T}, \quad \text{where } C = \frac{\mathrm{N} g_J^2 \beta^2 J(J+1)}{3\mathrm{k_B}} \tag{11.6a}$$

is called Curie's constant; the law is correctly applicable in the limit of $B \to 0$. Experimentally, however, the formula

$$\chi_m = \frac{C}{T} + \alpha_J \quad \text{and} \quad C = \frac{\mathrm{N}\beta^2 P^2}{3\mathrm{k_B} T}, \quad P^2 = g_J J(J+1) \tag{11.6b}$$

can be used for data analysis, where α_J is constant and P is the *effective Bohr magneton number*. Table 11.2 lists rare-earth ions, showing their ground states, comparing observed values of P with those calculated by (11.6b).

If $x \to \infty$, on the other hand, we obtain $\mathrm{B}_J(x) \to 1$, indicating that all magnetic moments are aligned in parallel with B, which is called *saturated magnetization.*

11.3 Spin–Spin Correlations

Magnetic moments of constituent ions are primarily independent from the lattice, where the magnetic and the crystal symmetries coexist. Inversion symmetry $\boldsymbol{\mu}_n \to -\boldsymbol{\mu}_n$ holds by reversing its direction of an applied field, that is, $\boldsymbol{B} \to -\boldsymbol{B}$. Characterized by fast motion at elevated temperatures, the average $\langle \boldsymbol{\mu}_n \rangle$ vanishes in

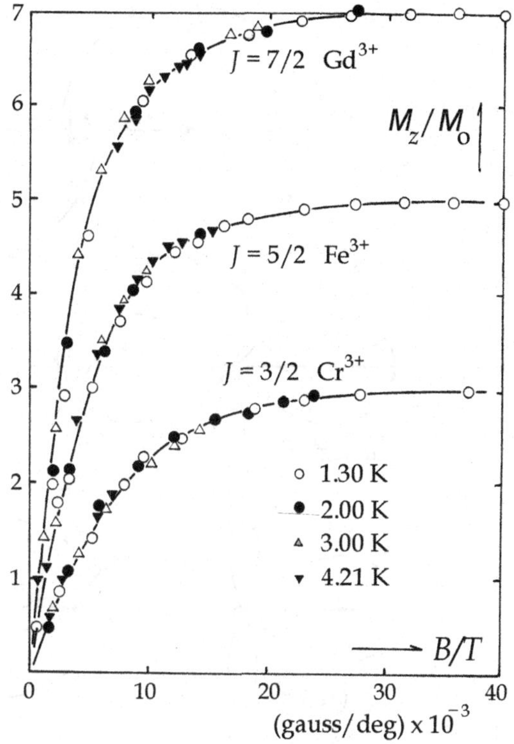

Fig. 11.2 M_z/M_o versus B/T of paramagnetic ions Cr^{3+}, F e^{3+}, and Gd^{3+}.

the paramagnetic phase. However, if the motion is slowed down on lowering temperature, these $\boldsymbol{\mu}_n$ are observed like classical vectors. We consider a potential ΔU_n at the site n that emerges in the direction of magnetization to drive $\boldsymbol{\mu}_n$ at and below the critical temperature T_c, similar to binary pseudospins. Owing to the relation $\boldsymbol{\mu}_n = g_S \beta \boldsymbol{S}_n$, inversion symmetry $\boldsymbol{S}_n \rightarrow -\boldsymbol{S}_n$ holds by reversing $\boldsymbol{B}$, where $\boldsymbol{\mu}_n$ and $\boldsymbol{S}_n$ are both in *precession* around the direction of $\boldsymbol{B}$. Signified by precession, magnetic moments should be in relative motion with respect to the ions, where the mutual interactions are attributed to *electron exchange* between $\boldsymbol{\mu}_n$.

In early physics, mutual magnetic dipole–dipole interactions were considered as too weak for magnetic ordering, but, instead, Heisenberg proposed in 1929 that electron exchange between ions in short distance can be responsible for ferromagnetic interactions, referring to its correct order of magnitude. Following his theory, we consider the exchange interaction for two Fe^{3+} ions in iron and similar cases in other magnetic crystals, in analogy of the covalently bonded hydrogen molecule H_2. Figure 11.3a, b show two $1s$ orbitals in H_2 and two adjacent orbitals of $d(3z^2 - r^2)$ in an iron crystal, respectively, showing that the magnitude of overlapped wavefunctions is essential for magnetic interactions.

Denoting interacting atoms by A and B, the wavefunctions of two electrons can be written as $\varphi_{A,B}(\boldsymbol{r}_i)\chi_{A,B}(s_{i,j})$, where $i = 1, 2$ and $j = +\frac{1}{2}, -\frac{1}{2}$. The total wavefunction can be expressed in determinant form to satisfy Pauli's principle, namely,

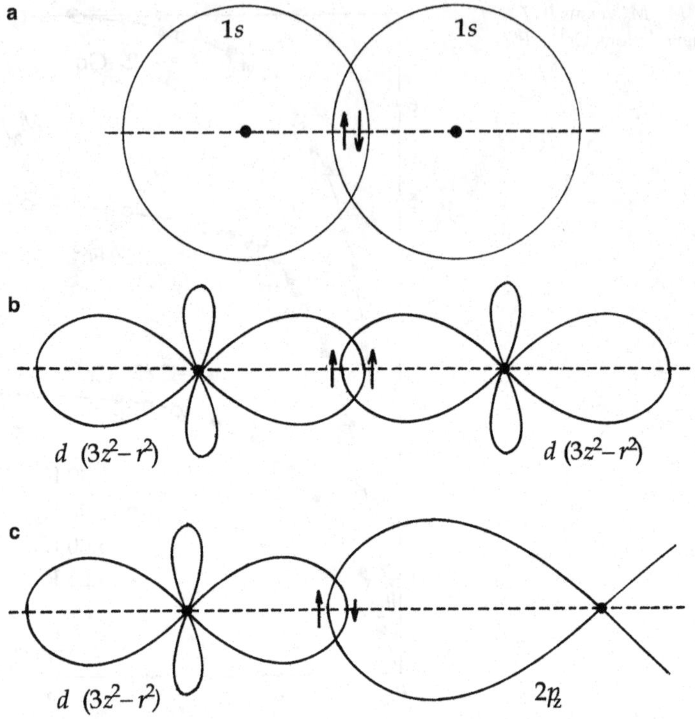

Fig. 11.3 Electron exchange between two ions (**a**) two 1 *s* orbitals in H_2, (**b**) two $3d\ (3z^2 - r^2)$ orbital of Fe^{3+}, (**c**) one $3d$ orbital of magnetic ion and p orbital of a negative ion.

$$\psi \propto \begin{vmatrix} \varphi_A(r_1)\chi_A(s_1) & \varphi_B(r_1)\chi_B(s_1) \\ \varphi_A(r_2)\chi_A(s_2) & \varphi_B(r_2)\chi_B(s_2) \end{vmatrix},$$

indicating that the sign of ψ changes by exchanging electrons between $\boldsymbol{r}_1$ and $\boldsymbol{r}_2$. Also noted is that combined spin functions of $\chi(+\frac{1}{2}) = 1, \chi(-\frac{1}{2}) = 0$ and of $\chi(+\frac{1}{2}) = 0, \chi(-\frac{1}{2}) = 1$ signify $m_s = +\frac{1}{2}$ and $m_s = -\frac{1}{2}$ states, respectively, which are denoted by α and β in the following discussion.

Considering four states $\psi_{\alpha\alpha}, \psi_{\alpha\beta}, \psi_{\beta\alpha}$ and $\psi_{\beta\beta}$ as the basis functions, we calculate the matrix of Hamiltonian H for two electrons, including their kinetic energies and all Coulomb potentials. The diagonal element $\mathsf{H}_{\alpha\alpha}$, for instance, is written as

$$\begin{aligned} \mathsf{H}_{\alpha\alpha} &= \sum_{s_1,s_2} \iint \Psi_{\alpha\alpha}{}^{*}\mathsf{H}\Psi_{\alpha\alpha} d^3\boldsymbol{r}_1 d^3\boldsymbol{r}_2 \\ &= \iint \varphi_A{}^{*}(\boldsymbol{r}_1)\varphi_B{}^{*}(\boldsymbol{r}_2)\mathsf{H}\varphi_A(\boldsymbol{r}_1)\varphi_B(\boldsymbol{r}_2) d^3\boldsymbol{r}_1 d^3\boldsymbol{r}_2 \\ &- \iint \varphi_A{}^{*}(\boldsymbol{r}_1)\varphi_B{}^{*}(\boldsymbol{r}_2)\mathsf{H}\varphi_B(\boldsymbol{r}_1)\varphi_A(\boldsymbol{r}_2) d^3\boldsymbol{r}_1 d^3\boldsymbol{r}_2 \end{aligned}$$

where φ_A and φ_B are orthogonal functions for $\mathsf{H}_{\alpha\alpha}$. Writing these terms in $\mathsf{H}_{\alpha\alpha}$ as K_{AB} and $-J_{AB}$, the interaction $\mathsf{H}(A,B)$ can be expressed by a determinant

$$\mathsf{H}(A,B) = \begin{vmatrix} K_{AB} - J_{AB} & 0 & 0 & 0 \\ 0 & K_{AB} & -J_{AB} & 0 \\ 0 & -J_{AB} & K_{AB} & 0 \\ 0 & 0 & 0 & K_{AB} - J_{AB} \end{vmatrix}. \tag{11.7a}$$

Here, J_{AB} is referred to as the *exchange integral*. Transforming this determinant to a diagonal form, we obtain two eigenvalues: $\varepsilon_3 = K_{AB} - J_{AB}$ and $\varepsilon_1 = K_{AB} + J_{AB}$, whose eigenfunctions are

$$\Psi_{\alpha\alpha},\ \frac{1}{\sqrt{2}}\left(\Psi_{\alpha\beta} + \Psi_{\beta\alpha}\right),\ \Psi_{\beta\beta} \quad \text{for} \quad s_1 + s_2 = 1$$

and

$$\frac{1}{\sqrt{2}}\left(\Psi_{\alpha\beta} - \Psi_{\beta\alpha}\right) \quad \text{for} \quad s_1 + s_2 = 0, \tag{11.7b}$$

representing parallel and antiparallel spins, respectively. Referring to the total spin $S = s_1 + s_2 = 1$ and 0, the first three functions and the second function in (11.7b) are referred to as *triplet* and *singlet* states, respectively, whose energy gap is given by $2|J_{AB}|$.

The exchange integral can be calculated as

$$J_{AB} = \iint \varphi_A{}^*(\boldsymbol{r}_1)\varphi_B(\boldsymbol{r}_2)\frac{e^2}{|\boldsymbol{r}_1 - \boldsymbol{r}_2|}\varphi_B(\boldsymbol{r}_1)\varphi_A(\boldsymbol{r}_2)\mathrm{d}^3\boldsymbol{r}_1\mathrm{d}^3\boldsymbol{r}_2, \tag{11.8}$$

which is numerically evaluated as positive for *s* orbitals in H_2 and negative for *p* orbitals in O_2. The sign of J_{AB} determines if the ground state is singlet or triplet. For the ferromagnetic state of iron crystals, Heisenberg's theory predicted that the exchange integral estimated in Fig. 11.3b is positive for parallel spins, giving rise to ferroelectric spin arrangement in crystals.

Exchanging *unpaired* electrons between neighboring magnetic ions binds them covalently together with parallel spins, whereas negative J_{AB} gives rise to antiparallel spins. Such an exchange can occur indirectly in other compounds via an ion in-between. In crystals of perovskite KMF_3, where the intervening ion M can be either one of $Mn^{2+}, Fe^{2+}, Co^{2+}, Ni^{2+}$, exhibiting *antiferromagnetism.* These crystals show antiferromagnetic phases below critical temperatures T_N, called the Neél temperature; Fig. 11.4 illustrates the arrangement of magnetic spins in these crystals. In $KMnF_3$, for example, below $T_N = 88\mathrm{K}$, unpaired electrons are exchanged through $Mn^{2+} \ldots F^- \ldots Mn^{2+}$, as shown in Fig. 11.3c. In this case, the integral J_{AB} was considered as negative for antiferromagnetic spin arrangement.

Spin–spin correlations associated with exchanging unpaired electrons can be indicated by $s_{Az}s_{Bz} = \pm\frac{1}{4}$, referring the A–B axis to as *z*, for the two spin states to be triplet and singlet. Nonetheless, this relation can be generalized to the scalar

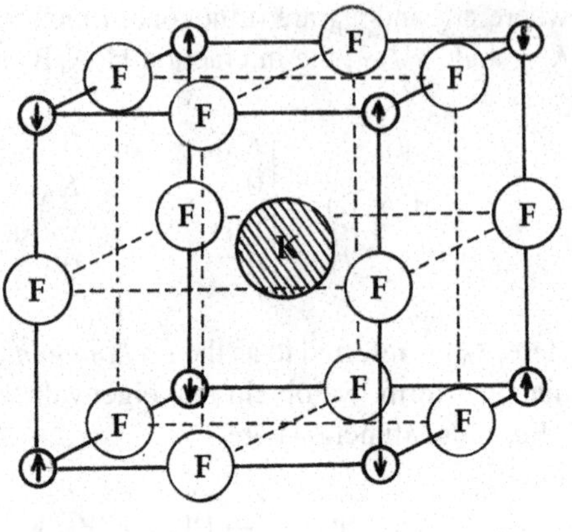

Fig. 11.4 Structure of a KM F_3 crystal. M represents Mn, Fe, Co, or Ni.

product $s_A.s_B = \pm\frac{1}{4}$of two classical vectors s_A and s_B, which is valid if both spins are in rapid precession, so $\langle s_{Ax}s_{Bx} + s_{Ay}s_{By}\rangle = 0$ in the first order. Therefore, the exchange integral J_{AB} can be replaced conveniently by $\frac{1}{2}(1 \pm 4s_A.s_B)J_{AB}$, for which $s_A.s_B = \frac{1}{4}$ and $-\frac{1}{4}$ specify for triplet and singlet states, respectively. Hence, the correlation energy can be written as

$$\mathsf{H}_{ex} = -2J_{AB}s_A.s_B \tag{11.9}$$

excluding the constant term. This formula (11.9) for spin–spin correlations is identical to previous one (5.10) for the nonmagnetic correlation, except for the factor 2. Noted from (11.8), the integral J_{AB} is a function of distance $|\boldsymbol{r}_A - \boldsymbol{r}_B|$ and highly directional in crystal space via an intervening ion, if any, between magnetic ions. The latter indirect exchange is called *superexchange*. For thermodynamics, the sign of J_{AB} in any case is significant to determine the spin–spin coupling to lower energy by H_{ex}.

11.4 Spin Clusters and Magnetic Symmetry

Evidenced by easy magnetization phenomena, no appreciable adiabatic potentials can be expected in magnetic crystals for modulating $\boldsymbol{\mu}_n$, unless an external magnetic field $\boldsymbol{B}$ is applied in parallel to the easy axis. We therefore consider that $\boldsymbol{B}$ should be applied in the direction of magnetization axis; otherwise, magnetic symmetry is independent of crystalline symmetry. In this context, such an easy direction can be identified as an axis for electron exchanges in the lattice, which were substantiated by neutron diffraction experiments.

In an axis for electron exchanges, collective motion of $\boldsymbol{\mu}_n$ occurs as described by Fourier series, showing their propagation in the periodic lattice. In the condensate model, it is a valid assumption that such a propagation is driven by an adiabatic potential that synchronize collective $\boldsymbol{\mu}_n$ in this direction. Considering the moment $\boldsymbol{\mu}_n$ as relative displacements from the site n, the collective $\boldsymbol{\mu}_n$ should be in phase with sinusoidal lattice excitations $\boldsymbol{u}_n$. Disregarding uncertainties in the transition region, such a collective $\boldsymbol{\mu}_n$ mode can be written as

$$\boldsymbol{\mu}_n = \boldsymbol{\mu}_o \exp i(\boldsymbol{q}.\boldsymbol{r}_n - \omega t_n)$$

or directly by the Fourier amplitude

$$\boldsymbol{\mu}_o = \boldsymbol{\mu}_n \exp\{-i(\boldsymbol{q}, \boldsymbol{r}_n - \omega t_n)\}.$$

Including space–time uncertainties $\Delta\boldsymbol{r}_n$ and Δt_n at sites n, it is convenient to express these modes by the phase variables $\phi_n = \boldsymbol{k}.\boldsymbol{r}_n - \omega t_n$ with uncertainties $\Delta \phi_n = \boldsymbol{k}.\Delta\boldsymbol{r}_n - \omega\Delta t_n$.

In an applied field $\boldsymbol{B}$, the magnetic moments $\boldsymbol{\mu}_n$ are in precession around its direction, where the angles of precession are considered to synchronize with lattice translation, as sketched in Fig. 11.5. Writing these moments as classical vectors $\boldsymbol{\mu}_m = \boldsymbol{\mu}\boldsymbol{e}_m$ and $\boldsymbol{\mu}_n = \boldsymbol{\mu}\boldsymbol{e}_n$, where $|\boldsymbol{e}_m| = |\boldsymbol{e}_n| = 1$ for unit vectors, the short-range interaction energy at the site n is given by

$$\mathsf{H}_n = -\sum_m 2J_{mn}\boldsymbol{\mu}_m.\boldsymbol{\mu}_n = -\mu^2 \sum_m 2J_{mn}\boldsymbol{e}_m.\boldsymbol{e}_n, \tag{11.10}$$

where J_{mn} is an exchange integral between m and n sites.

Using this for short-range correlations for a magnetic cluster, the summation over m covers the nearest-neighbor sites. Assuming μ as constant, we write for these unit vectors to be *real* ones, that is,

$$\boldsymbol{e}_m = \boldsymbol{e}_{+q} \exp i\phi_m + \boldsymbol{e}_{-q} \exp(-i\phi_m) \quad \text{and} \quad \boldsymbol{e}_n = \boldsymbol{e}_{+q} \exp i\phi_n + \boldsymbol{e}_{-q} \exp(-i\phi_n),$$

where $\phi_m = \boldsymbol{q}.\boldsymbol{r}_m - \omega t_m$ and $\phi_n = \boldsymbol{q}.\boldsymbol{r}_n - \omega t_n$ are phases at sites m and n, respectively. In Chap. 5, we showed that the average of H_n over the timescale of observation t_o is given by

$$\langle \mathsf{H}_n \rangle_t = -2\boldsymbol{\mu}^2 \Gamma_t \boldsymbol{e}_{+q}.\boldsymbol{e}_{-q} J(\boldsymbol{q}), \tag{11.11a}$$

where

$$J(\boldsymbol{q}) = \sum_m J_{mn} \exp i\boldsymbol{q}.(\boldsymbol{r}_m - \boldsymbol{r}_n) \tag{11.11b}$$

and

$$\Gamma_t = \frac{1}{2t_o} \int_{-t_o}^{+t_o} \exp\{-i(t_m - t_n)\} d(t_m - t_n) = \frac{\sin \omega t_o}{\omega t_o}. \tag{11.11c}$$

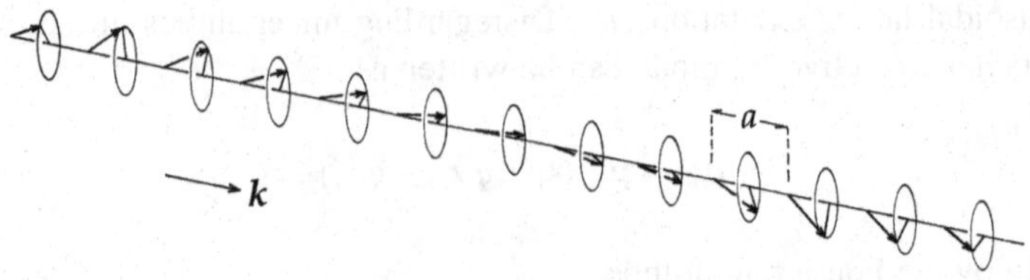

Fig. 11.5 One-dimensional array of spins in Larmor precession in applied magnetic field $\boldsymbol{B} \parallel \boldsymbol{k}$.

Equations (11.11a) and (11.11b) represent such a cluster centered at $\boldsymbol{\mu}_n$ with limited number of interacting neighbors, which can be considered as a *seed* for magnetic *condensation*. For $J(\boldsymbol{q})$, however, we consider electron exchanges only with nearest neighbors to keep local symmetry unchanged, as next-nearest-neighbor interactions J' disrupt local symmetry.

In the presence of a magnetic field $\boldsymbol{B}$, we should have collective motion of magnetic moments $\boldsymbol{\mu}_n$ in propagation mode. Minimizing $J(\boldsymbol{q})$ with respect to $\boldsymbol{q}$, we can determine the axis of propagation $\boldsymbol{k}$ for $\boldsymbol{B} = 0$, which should be the same axis for minimum strains in the lattice. If $\boldsymbol{B} \neq 0$, on the other hand, the collective motion is described in terms of phases ϕ_n; all $\boldsymbol{\mu}_n$ waves are synchronized with $\Delta\phi_n = 0$ to ensure stability of the crystal. This process can be observed as thermal relaxation for $\boldsymbol{B} \| \boldsymbol{k}$. Further, such a mode of $\boldsymbol{\mu}_n$ below T_c is commensurate with the lattice, as signified by either $\boldsymbol{k} = 0$ or $\boldsymbol{k} = \boldsymbol{G}/2$, corresponding to *ferromagnetic* or *antiferromagnetic* arrangements.

11.5 Magnetic Weiss Field

Implied by Larmor precession of $\boldsymbol{\mu}_n$ around an applied field $\boldsymbol{B}$, the magnetic moment of a cluster $\boldsymbol{m}_n = \sum_{i,\text{cluster}} \boldsymbol{\mu}_i$ can propagate as $\boldsymbol{m}_n \exp i\phi_n$ in the direction of $\boldsymbol{B}$. The space–time reference for $\boldsymbol{m}_n$ can be reset for motion in this direction by replacing ϕ_n by a continuous phase angle in the range $0 \leq \phi < 2\pi$. For such a cluster, the amplitude $|\boldsymbol{m}_n|$ is larger than individual $|\boldsymbol{\mu}_i|$. We consider that $\boldsymbol{m}_n$ can propagate collectively in the direction of $\boldsymbol{k} \| \boldsymbol{B}$, being driven by an adiabatic potential $\Delta U(\phi)$ originated from *spin correlations* between clusters. The potential $\Delta U(\phi)$ can be expressed as proportional to $\boldsymbol{m}_m.\boldsymbol{m}_n$, corresponding to $\boldsymbol{u}_m.\boldsymbol{u}_n$ of displacement correlations in the lattice.

Expressed by $\boldsymbol{m}_n = -g\beta \boldsymbol{S}_n$, where $\boldsymbol{S}_n = \sum_i \boldsymbol{s}_i$, the energy of spin correlations among $\mathbf{S}_n$ can be given as

$$\mathsf{H}_{mn} = -\sum_{m,n} 2J_{mn} \boldsymbol{S}_m.\boldsymbol{S}_n, \tag{11.12}$$

where J_{mn} is the exchange integral between two clustered spins $\boldsymbol{S}_m$ and $\boldsymbol{S}_n$. Equation (11.12) is the formula acceptable for most magnetic studies. Such clustered spin $\boldsymbol{S}_n$ is invariant of the magnetic transformation, but not necessarily governed by lattice symmetry in modulated crystals.

Figure 11.5 shows schematically a wave of $\boldsymbol{m}_n$ propagating along the direction of $\boldsymbol{B}$, which is nevertheless perturbed by lattice displacements. Assuming that ΔU_m (ϕ) is given by an Eckart potential along the direction of $\boldsymbol{B}$, we can write the equation for the parallel component of $\boldsymbol{m}_n$as (iv) in Chap. 8, that is,

$$\frac{\mathrm{d}^2 m_p(x)}{\mathrm{d}x^2} + \{-p^2 + p(p+1)\mathrm{sech}^2 x\} m_p(x) = 0,$$

where $p = 1, 2, \ldots, n$ indicate serial peaks of Eckart potentials replacing the cnoidal potential.

The magnetization in the mesoscopic states can be written as

$$M(x) = p\, m_p(x),$$

where the factor p indicates a peak density of $\mathrm{sech}^2 x_n$ potential for $M(x)$, which is in fact a function of temperature in thermodynamics.

Such magnetic correlations as (11.12) are considered for the effective internal field $\boldsymbol{B}_{\mathrm{int}}$ at site n that was defined by

$$\mathsf{H}_n = -\boldsymbol{\mu}_n . \boldsymbol{B}_{\mathrm{int}} \quad \text{where} \quad \boldsymbol{B}_{\mathrm{int}} = \sum_m 2J_{mn} \boldsymbol{\mu}_m$$

represents the Weiss field. Postulated by Weiss originally as $\boldsymbol{B}_{\mathrm{int}} \propto \boldsymbol{M}$, where $\boldsymbol{M}$ is the macroscopic magnetization, we can assume $\boldsymbol{B}_{\mathrm{int}} = -2pJ(0)\langle\boldsymbol{\mu}_m\rangle$, including the factor p and $J(0)$ for $\boldsymbol{k} = 0$ or $\boldsymbol{G}/2$. Eckart potentials may be interpreted as representing domain boundaries in multi-domain structure; hence we can write $\boldsymbol{B}_{\mathrm{int}} = -\boldsymbol{M}$, as determined by the demagnetization field due to magnetic charge density on the surface of one-dimensional domain.

In general, $\boldsymbol{M}$ is uniform over an ellipsoidal volume, for which we can write $\boldsymbol{M} = \chi_\mathrm{o}(\boldsymbol{B} + \lambda \boldsymbol{M})$, where χ_o can be determined by the paramagnetic susceptibility given by the Curie law $\chi_\mathrm{o} = C/T$. Thus, we obtain the equation $M = \frac{C}{T}(B + \lambda \mathrm{M})$; hence

$$\chi = \frac{M}{B} = \frac{C}{T - T_\mathrm{o}} \quad \text{where} \quad T_\mathrm{o} = C\lambda. \tag{11.13}$$

This relation is the Curie–Weiss law indicating a singularity at $\chi \to \infty$, that is, $T \to T_\mathrm{o}$.

Using the Brillouin theory of paramagnetism, where $\boldsymbol{B}_{\mathrm{int}}$ is considered as if applied externally, we write

$$M(B,T) = pg\beta\langle S_{\parallel}\rangle = pg\beta S\mathrm{B}_S\left(\frac{g\beta SB_{\mathrm{int}}}{\mathrm{k_B}T}\right) \quad \text{where} \quad B_{\mathrm{int}} = g\beta 2pJ(0)\langle S_{\parallel}\rangle.$$

Therefore,

$$\langle S_{\parallel}\rangle = S\mathrm{B}_S\left(\frac{g^2\beta^2 S2pJ(0)\langle S_{\parallel}\rangle}{\mathrm{k_B}T}\right). \tag{11.14}$$

For a small value of $g\beta SB_{\mathrm{int}}/\mathrm{k_B}T$ in the vicinity of critical temperature T_c, the Brillouin function can be approximated as

$$\mathrm{B}_S \cong \frac{S+1}{3S}\frac{g\beta SB_{\mathrm{int}}}{\mathrm{k_B}T} = \frac{g^2\beta^2 S(S+1)}{3\mathrm{k_B}}\frac{2pJ(0)}{T}\frac{\langle S_{\Pi}\rangle}{S}.$$

Denoting $T_c = \frac{2g^2\beta^2 S(S+1)J(0)}{3\mathrm{k_B}}$, we obtain the relation $\frac{\langle S_{\parallel}\rangle}{S} = \frac{T_c}{T}\frac{p\langle S_{\parallel}\rangle}{S}$ from (11.14), which implies that $\langle S_{\parallel}\rangle = S$ if $T = pT_c$. We can assume $p = 1$ if the spin–spin interaction in (11.12) is limited to the shortest distance, which is consistent with assuming uniform $\boldsymbol{M}$.

Equation (11.14) can be solved numerically for $\langle S_{\parallel}\rangle/S$, in similar manner discussed in Sect. 4.2. Figure 11.6a shows a graphical sketch, where the curve of $\mathrm{B}_S\left(\ldots\frac{\langle S_{\parallel}\rangle}{S}\right)$ versus $\frac{\langle S_{\parallel}\rangle}{S}$ crosses with the straight line of the 45° slope at $T = T_c$. If $T<T_c$, we have nonzero solution $\langle S_{\parallel}\rangle \neq 0$, whereas $\langle S_{\parallel}\rangle = 0$, if $T>T_c$. Therefore, the temperature $T = T_c$ in this analysis should be the critical temperature.

In Fig. 11.6b shown is another plot of $\frac{\langle S_{\parallel}\rangle}{S}$ versus $\frac{T}{T_c}$, where the numerical curve calculated for $S = \frac{1}{2}$ and $p = 1$ fits reasonably well to the experimental data for Ni. However, calculated curves for larger values of S, as $S \to \infty$, deviate substantially from $S = \frac{1}{2}$ so that the latter curve, supported experimentally, implies that the nearest-neighbor correlations are overwhelming in iron group magnet. In this context, spin–spin correlations in long range is considered as negligible at least in Ni.

Further notable from the above analysis with the Brillouin function is that for small values $\frac{\langle S_{\parallel}\rangle}{S}$, the slope of Brillouin function against $\frac{T}{T_c}$ for Ni is almost vertical in the vicinity of T_c for $S = 1/2$, as seen from Fig. 11.6b. In contrast, the calculated curve for $S \to \infty$ looks closely like a graph of an equation $y = \tanh y$ that represents a typical binary ordering. The difference between these curves suggests that an internal Weiss field exists in the latter case but absent in the former case for Ni and Fe. A large S can logically be attributed to virtually all spins as $\boldsymbol{S} = \sum \boldsymbol{s}_n$; on the other hand, there is no concern about the cluster size in the Brillouin theory. Hence, a difference may be attributed to long-range electronic spin–spin correlations among $\boldsymbol{s}_n$ constituting $\boldsymbol{S}$, analogous to pseudospin clusters in dielectric crystals. Nevertheless, such an interpretation is no real, unless verified experimentally in the timescale of observation.

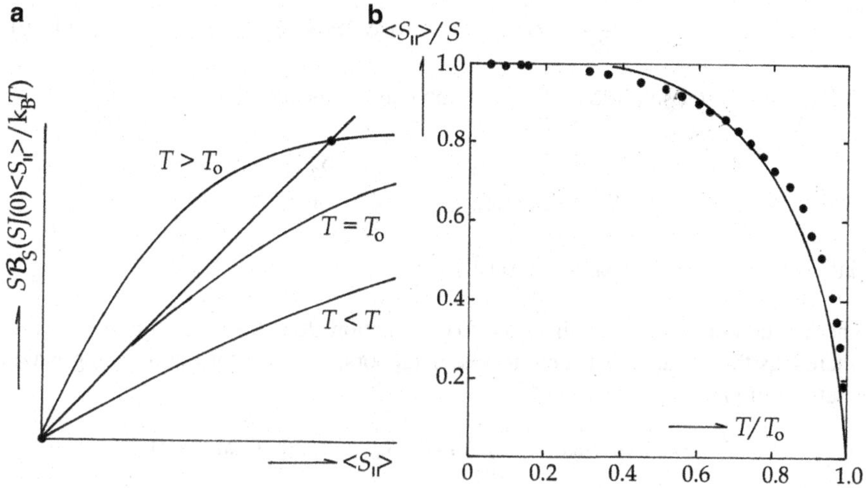

Fig. 11.6 Brillouin plots. (**a**) $S\mathrm{B}_S\left(SJ(0)\langle S_{\parallel}\rangle/\mathrm{k_B}T\right)$ versus $\langle S_{\parallel}\rangle$, (**b**) $\langle S_{\parallel}\rangle/S$ versus T/T_{o}.

In any case, in a given magnetic crystal, the Weiss field of correlated magnetic spins $\boldsymbol{S}_n$ of magnetic ions is a realistic concept for magnetic phase transitions.

11.6 Spin Waves

It is noted that the magnetic spin correlation (11.12) is invariant for transformation along the easy direction, where the Weiss field $\boldsymbol{B}_{\mathrm{int}}$ is characterized by inversion symmetry. The magnetic field drives spins $\boldsymbol{S}_n$ in Larmor precession, whose collective motion is driven by an adiabatic potential synchronized with $\boldsymbol{B}_{\mathrm{int}}$, as sketched in Fig. 11.5. Dominated by nearest-neighbor interactions, nonlinear effect is regarded as negligible for such spin waves of $\boldsymbol{S}_n$. Such a wave is commensurate at $\boldsymbol{k}=0$ and $\boldsymbol{k}=\boldsymbol{G}/2$, exhibiting ordered spin arrangements, as verified by neutron diffraction experiments.

Spin waves should dominate magnetic crystals at temperatures for $T<T_{\mathrm{o}}$, for which the short-range interactions (11.12) can be expressed as $-\,2J\boldsymbol{S}_n.(\boldsymbol{S}_{n-1}+\boldsymbol{S}_{n+1})$ with J for the nearest neighbors, and hence the Weiss field is expressed as

$$\boldsymbol{B}_{\mathrm{int}\,n}=-\frac{2J}{g\beta}(\boldsymbol{S}_{n-1}+\boldsymbol{S}_{n+1}). \tag{11.15}$$

The spin angular momentum is determined by the torque $\boldsymbol{\mu}_n\times\boldsymbol{B}_{\mathrm{int}\,n}$ so that we can write the equation of motion as

$$\frac{dS_n}{dt} = 2J(S_n \times S_{n-1} + S_n \times S_{n+1}). \qquad (11.16)$$

By considering small $|S_n|$, this equation can be linearized as,

$$S_{n,x} = 2JS(2S_{n,y} - S_{n-1,y} - S_{n+1,y}),$$
$$S_{n,y} = 2JS(2S_{n,x} - S_{n-1,x} - S_{n+1,x}),$$

and
$$S_{n,z} = S(\text{const.})$$

where x and y are transverse directions to propagation along the z-axis. Letting $z = ka$, where a is the distance between nearest neighbors, we look for the traveling wave solutions of (11.16)

$$S_{n,x} = u \exp i(nka - \omega t) \quad \text{and} \quad S_{n,y} = \upsilon \exp i(nka - \omega t),$$

where u and υ are constants. Therefore,

$$-\mathrm{i}\omega u = 4JS(1 - \cos ka)\upsilon \quad \text{and} \quad -\mathrm{i}\omega\upsilon = -4JS(1 - \cos ka)u,$$

from which we obtain the dispersion relation

$$\omega = 4JS(1 - \cos ka) \approx 2JSa^2k^2 \quad \text{for} \quad ka \ll 1. \qquad (11.17)$$

The dispersion relation (11.17) of such spin waves was actually convinced with neutron inelastic scattering experiments, showing significant dispersion relation $\omega = \omega(\boldsymbol{k})$ resulted from neutron inelastic scatterings for momentum change $\hbar\Delta\boldsymbol{K} = \hbar\boldsymbol{k}$ along specific symmetrical axes $\boldsymbol{k}$ signified by anisotropic J. Figure 11.7a shows a neutron diffraction pattern from iron crystals, measured with varying scattering angle, where crystal planes of ordered spins are identified in strong intensities. The dispersion relation (11.17) was verified by scanning scattering angle in the vicinity of $\Delta\boldsymbol{K} = \boldsymbol{k}$. Figure 11.7b is an example of experimental spin-wave dispersion, showing a reasonable agreement with (11.17) (Fig. 11.8).

11.7 Magnetic Anisotropy

Calculating with the formula (5.18) of Sect. 5.5, magnetic spins are clustered via electron exchange $J(\boldsymbol{k})$ for minimum correlations in a specific direction determined by $\nabla_k J(\boldsymbol{k}) = 0$. Determined by this equation, the vector $\boldsymbol{k}$ should be parallel to a symmetry axis for *easy magnetization*. Figure 11.9 shows magnetization curves of a single crystal of Fe, where the easy axis is the [100] direction. On the other hand, for $\boldsymbol{B} \parallel [111]$, the magnetization versus $\boldsymbol{B}$ curve is at a slower rate to reach saturation with increasing $\boldsymbol{B}$ than $\boldsymbol{B} \parallel [100]$.

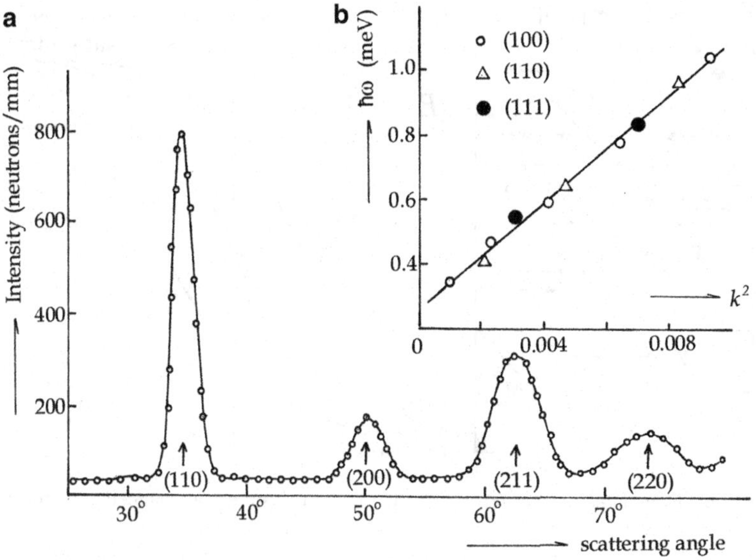

Fig. 11.7 (**a**) Neutron diffraction pattern from Fe. After C. G. Shull, E. O. Wollan and W. C. Koehler, Phys. Rev. **84**, 912 (1951) (**b**) Magnon energy $\hbar\omega$ versus k^2 for ferromagnet $MnPt_3$.

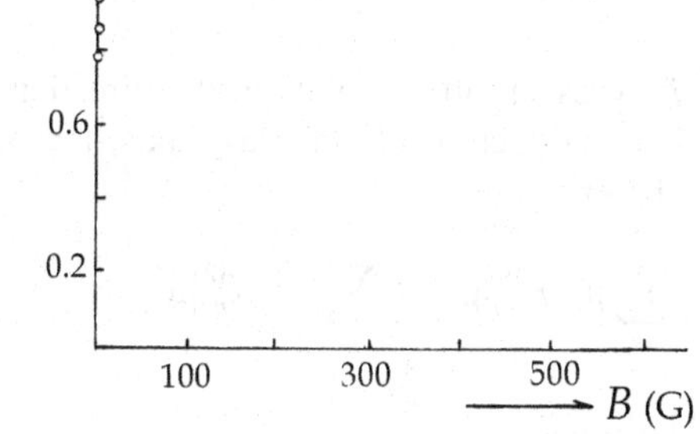

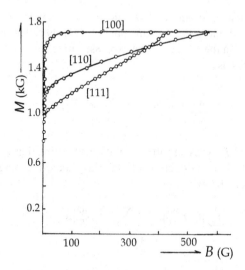

Fig. 11.8 Magnetization curves for iron. The direction [100] is for easy magnetization.

A magnetized crystal can be strained, depending on the direction of $\boldsymbol{B}$. In Fig. 11.9, the graphic area under a magnetization curve represents a free energy for magnetization, that is,

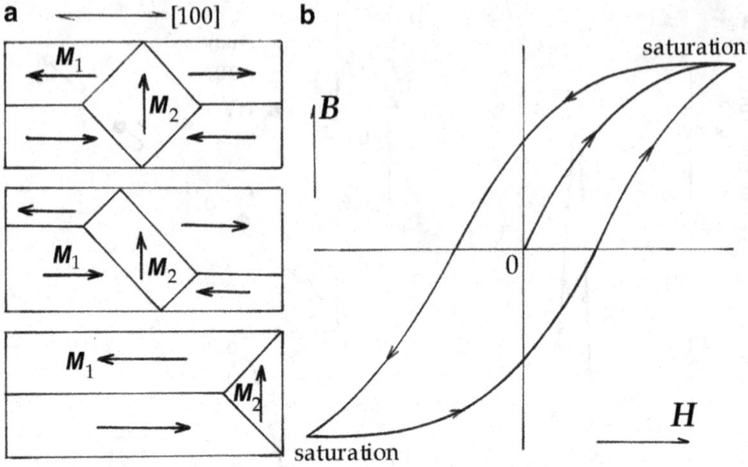

Fig. 11.9 (**a**) Domain patterns in ferromagnetic iron. (**b**) Hysteresis in magnetizing process.

$$\Delta F = -\int_0^B \boldsymbol{M}.\mathrm{d}\boldsymbol{B}. \tag{11.18}$$

The process for reaching saturated $\boldsymbol{M}$ is quasi-static in practice, during which the lattice is strained with increasing $\boldsymbol{B}$; this phenomenon is called *magnetostriction.*

The strain energy in the lattice can be expressed in terms of small displacements $\boldsymbol{u}_n$ of the coordinates as

$$\Delta U = \sum_{n,ij} u_{n,i} K_{n,ij}^{(1)} u_{n,j} + \sum_{n,ij} u_{n,i}^2 K_{n,ij}^{(2)} u_{n,j}^2 + \cdots \tag{11.19a}$$

where indexes i and j represent lattice points in the direction of $\boldsymbol{B}$. On the other hand, with $\boldsymbol{B}$ applied in a direction off the easy axis, the magnetic correlation energy among $\boldsymbol{\mu}_n$ can be expressed by

$$\Delta U_{\mathrm{mag}} = \sum_{n,ij} \mu_{n,i} K_{n,ij}^{(1)} \mu_{n,j} + \sum_{n,ij} \mu_{n,i}^2 K_{n,ij}^{(2)} \mu_{n,j}^2 + \cdots, \tag{11.19b}$$

which is related to (11.19a) as

$$\Delta U + \Delta U_{\mathrm{mag}} \rightarrow \text{minimum}, \tag{11.20}$$

owing to Born–Huang principle; the total strain energies should be minimized under the equilibrium condition with the surroundings. If $\boldsymbol{B} = 0$, the magnetic tensor $K_{n,ij}$ in (11.19b) should be symmetric and conformal with the strain tensor $K_{n,ij}$ in (11.19a), whereas for $\boldsymbol{B} \neq 0$, inversion symmetry of $\boldsymbol{\mu}_n$ should be established with regard to $\boldsymbol{B}$ so that K_{ij} and K_{ij} are generally not conformal. The tensor $K_{n,ij}$ in (11.19b) is therefore asymmetrical in each domain, that is, $K_{n,ij} \neq K_{n,ji}$. Writing

$$K_{n,ij} = \frac{1}{2}\left(K_{n,ij} + K_{n,ji}\right) + \frac{1}{2}\left(K_{n,ij} - K_{n,ji}\right),$$

the tensor $K_{n,ij}$ is obviously a symmetrical tensor if $K_{n,ij} = K_{n,ji}$, but antisymmetrical if $K_{n,ij} = -K_{n,ji}$. In the latter case, the spin–spin correlations are characterized by off-diagonal elements $J_{mn} = -J_{nm}$, which was proposed by Dzyaloshinsky.

Spin–spin correlations can be evaluated by applying an external field in an easy direction of magnetization, where the Weiss field can be defined to evaluate spin–spin correlations in the mean-field approximation. In a general direction of $\boldsymbol{B}$, however, we need to include additional interactions between ΔU and ΔU_{mag}; accordingly, we write the spin–spin interaction as

$$\mathsf{H}_{SS} = -2\sum_{m.n} J_{mn}\boldsymbol{S}_m.\boldsymbol{S}_n + \sum_{m.n}\left(\sum_{i.j} S_{m,i}K_{mn}S_{n.j}\right). \tag{11.21a}$$

If $K_{mn} = K_{nm}$, the quantity in the brackets can be expressed as proportional to $-\sum_n \boldsymbol{B}_{\text{aniso}}.\boldsymbol{S}_n$, where $\boldsymbol{B}_{\text{aniso}} = \left\langle \sum_{m,i} S_{m,i}K_{mn}\right\rangle$ is another internal field due to anisotropic spin arrangements in the mean-field approximation. Combining with the magnetic Weiss field $\boldsymbol{B}_{\text{int}} = 2\left\langle \sum_m J_{mn}\boldsymbol{S}_m\right\rangle$, H_{SS} in (11.21a) can be interpreted as

$$\mathsf{H}_{SS} = -\sum_n \left(\boldsymbol{B}_{\text{int}} + \boldsymbol{B}_{\text{aniso}}\right).\boldsymbol{S}_n \tag{11.21b}$$

When $\boldsymbol{B}$ is parallel to the easy axis, we have $\boldsymbol{B}_{\text{int}} \parallel \boldsymbol{B}_{\text{aniso}}$; otherwise, $\boldsymbol{B}_{\text{int}} \parallel$ (easy axis) and $\boldsymbol{B}_{\text{aniso}} \parallel \boldsymbol{B}$. As will be shown in the following sections, it is important to consider both $\boldsymbol{B}_{\text{aniso}}$ and $\boldsymbol{B}_{\text{int}}$ together for dealing with the Zeeman behavior of magnetization in magnetic crystals.

On the other hand, for the antisymmetric tensor $K_{mn} = -K_{nm}$, the spin–spin correlation energy can be calculated as

$$\begin{aligned}\mathsf{H}_{mn} &= \sum_{m.n}\sum_{i,j} S_{m,i}K_{mn}S_{n,j} = \sum_{m.n}\sum_{i,j}\frac{1}{2}\left(S_{m,i}K_{mn}S_{n,j} + S_{n,j}K_{nm}S_{m,i}\right)\\ &= \frac{1}{2}\sum_{m.n}\sum_{i,j} K_{mn}\left(S_{m,i}S_{n,j} - S_{m,j}S_{n,i}\right).\end{aligned}$$

Defining a vector $\boldsymbol{D} = \left(\frac{K_{mn}}{2}, \frac{K_{mn}}{2}, \frac{K_{mn}}{2}\right)$, this expression implies a new type of spin–spin couplings, which can be written as

$$\mathsf{H}_{mn} = \boldsymbol{D}.(\boldsymbol{S}_m \times \boldsymbol{S}_n), \tag{11.22}$$

which is known as Dzyaloshinsky–Moriya interaction [3, 4]. Not always significant in magnetic crystals, a coupling as (11,22) is required in unusual magnetic crystals called *antisymmetric magnets*, such as $\alpha\, Fe_2O_3$ (α-hematite). However, we shall not pursue antisymmetric crystals, because such couplings exist not only in rare examples, but the problem is too specific to include in the present discussion.

Designating an applied magnetic field by $\boldsymbol{B}_0$, the external work by $\boldsymbol{B}_0$ on a magnetic crystal is expressed by $-\boldsymbol{M}.\boldsymbol{B}_0$, for which we can consider a torque $\boldsymbol{T}_0$ for rotating the vector $\boldsymbol{M}$, namely, $\boldsymbol{T}_0 = \boldsymbol{M} \times \boldsymbol{B}_0$, which strains the lattice structure. Nonetheless, a magnetized crystal is strained by $\boldsymbol{T}_0$ or $\boldsymbol{M}$ so that traditional Weiss field $\boldsymbol{B}_{int}$ in strained crystals should be revised as

$$\boldsymbol{B}_{int} = \lambda\boldsymbol{M} + \boldsymbol{B}_{aniso} \quad \text{where} \quad -\boldsymbol{M}.\boldsymbol{B}_{aniso} = \Delta U_{mag}. \tag{11.23a}$$

Here, $\boldsymbol{B}_{aniso}$ is the effective field due to ΔU_{mag}, and $\lambda\boldsymbol{M}$ represents the Weiss field for $\boldsymbol{B}_0 = 0$.

The ferromagnetic phase below T_c shows a complicated domain pattern. As illustrated in Fig. 11.9, movable domains by $\boldsymbol{B}_0$ are not simply binary, but there are 45° walls between magnetized domains in perpendicular directions, that is, $\boldsymbol{M}_1 \perp \boldsymbol{M}_2$, in addition to ordinary 180° walls between oppositely magnetized domains, that is, $\boldsymbol{M}_1 \parallel -\boldsymbol{M}_1$. Inside domain walls, the lattice should be significantly strained by rotating magnetization, hence signified by matrix elements $K^{(1)}_{n,ij}$ and the related potential $\Delta U^{(1)}_{mag}$. Despite of the strains, domain walls are movable by a field $\boldsymbol{B}$ in modest strength, pushing perpendicular magnetization toward surfaces. Such a pattern can be explained in terms of the surface/volume ratio in domain structure, which however remains only qualitative.

11.8 Antiferromagnetic and Ferrimagnetic States

Insulating oxides, fluorides, and sulfates of transition elements exhibit a variety of antiferromagnetic and ferrimagnetic states, depending on the nature of electron exchange mechanisms in crystals. Neutron scattering experiments show diffraction patterns of crystal planes of spins below critical temperatures, providing evidence for ordered spins in two sublattices. Figure 11.10 shows the unit structure in MnF_2 crystals, indicating that Mn^{2+} ions at the body-center positions and those at the corners form sublattices of spins oriented in different directions. These spin sublattices, A and B, are magnetically specified by vectors $\boldsymbol{M}_A$ and $\boldsymbol{M}_B$, respectively, which cannot be independent, but interacting via electron exchange at nearest distances. We consider Weiss fields $\boldsymbol{B}^{(A)}_{int}$ and $\boldsymbol{B}^{(B)}_{int}$ in sublattices A and B, which are expressed as

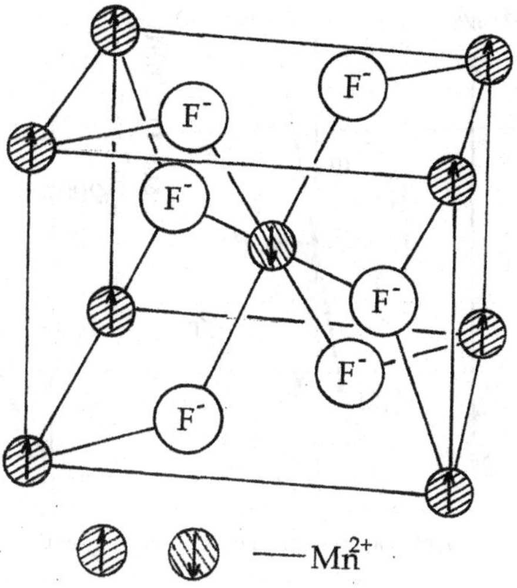

Fig. 11.10 Unit cell structure in antiferromagnetic MnF_2.

$$\boldsymbol{B}_{\mathrm{int}}^{(\mathrm{A})} = \lambda \boldsymbol{M}_{\mathrm{A}} + \lambda' \boldsymbol{M}_{\mathrm{B}} \quad \text{and} \quad \boldsymbol{B}_{\mathrm{int}}^{(\mathrm{B})} = \lambda \boldsymbol{M}_{\mathrm{B}} + \lambda' \boldsymbol{M}_{\mathrm{A}}.$$

In the presence of a weak external field $\boldsymbol{B}_{\mathrm{o}}$, sublattice magnetizations $\boldsymbol{M}_{\mathrm{A}}$ and $\boldsymbol{M}_{\mathrm{B}}$ become parallel to $\boldsymbol{B}_{\mathrm{o}} - \lambda \boldsymbol{M}_{\mathrm{B}}$ and $\boldsymbol{B}_{\mathrm{o}} - \lambda \boldsymbol{M}_{\mathrm{A}}$ in equilibrium, respectively. Figure 11.11a illustrates for two cases of $\boldsymbol{B}_{\mathrm{o}} \parallel \boldsymbol{M}_{\mathrm{A}} - \boldsymbol{M}_{\mathrm{B}}$ and $\boldsymbol{B}_{\mathrm{o}} \perp \boldsymbol{M}_{\mathrm{A}} - \boldsymbol{M}_{\mathrm{B}}$.

In the perpendicular case, we can see the relation

$$2\lambda \left|\boldsymbol{M}_{\mathrm{A,B}}\right| \sin\varphi = \left|\boldsymbol{B}_{\mathrm{o}}\right|,$$

or the susceptibility is given by

$$\chi_{\perp} = \frac{2\left|\boldsymbol{M}_{\mathrm{A.B}}\right| \sin\varphi}{\left|\boldsymbol{B}_{\mathrm{o}}\right|} = \frac{1}{\lambda - \lambda'} \tag{11.23b}$$

where

$$\lambda = \frac{2}{\mathrm{N}g^2\beta^2}\left\{J\left(\frac{\boldsymbol{G}}{2}\right) - J(0)\right\}.$$

On the other hand, in the parallel case,

$$M_{\mathrm{A}} = \frac{\mathrm{N}g\beta S}{2}\mathrm{B}_S\left\{\frac{g\beta S(B_{\mathrm{o}} - \lambda M_{\mathrm{B}} - \lambda' M_{\mathrm{A}})}{\mathrm{k_B}T}\right\} \quad \text{and}$$

$$M_{\mathrm{B}} = \frac{\mathrm{N}g\beta S}{2}\mathrm{B}_S\left\{\frac{g\beta S(B_{\mathrm{o}} - \lambda M_{\mathrm{A}} - \lambda' M_{\mathrm{B}})}{\mathrm{k_B}T}\right\},$$

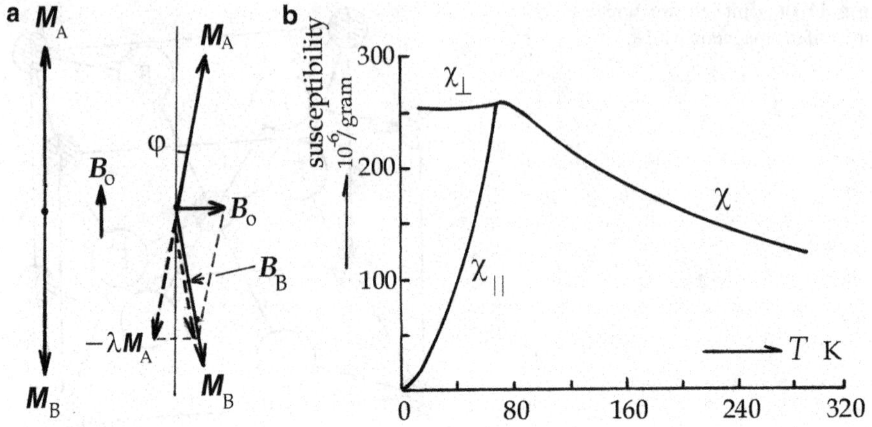

Fig. 11.11 (**a**) For calculating $\chi_\parallel$ and $\chi_\perp$ at 0 K. (**b**) Magnetic susceptibility of MnF_2.

from which we calculate

$$\chi_\parallel = \lim_{B_o \to 0} \frac{M_A - M_B}{B_o}. \tag{11.24}$$

From this, it is clear that for $T \to 0$, $M_A = M_B$, and hence $\chi_\parallel = 0$, while we should have $\chi_\parallel = \chi_\perp$ at the critical point, as numerically illustrated in Fig. 11.11b.

11.9 Fluctuations in Ferromagnetic and Antiferromagnetic States

In ferromagnetic and antiferromagnetic states, the magnetization vector $\boldsymbol{M}$ is in thermodynamic equilibrium with the crystal lattice. As in equilibrium with the strain energy, the fluctuations in $\boldsymbol{M}$ is not of random character but anisotropic with respect to symmetry axes. Such fluctuation modes can be studied by *magnetic resonance*, where an external magnetic field $\boldsymbol{B}_o$ is applied on the sample crystal. Conventional magnetic resonance is an experimental method at $\boldsymbol{G} = 0$, but here the resonance at $|\boldsymbol{G}| = \frac{1}{2}$ can be detected by neutron inelastic scatterings in magnetic crystals with an applied field $\boldsymbol{B}_o$; nonetheless, we call such experiments also as magnetic resonance.

In a magnetized crystal strained by an applied field $\boldsymbol{B}_o$, the magnetic strain potential ΔU_{mag} can be represented by $\boldsymbol{B}_{aniso}$, as defined by (11.21b). Hence, when $\boldsymbol{B}_o$ is in parallel to a symmetry axis, we have $\boldsymbol{B}_o \parallel \boldsymbol{B}_{aniso}$, around which $\boldsymbol{M}$ is in Larmor precession. In scattering spectra with $\boldsymbol{B}_o$, a magnetic resonance can take place at a frequency determined by $\boldsymbol{B}_o$ plus $\boldsymbol{B}_{aniso}$.

11.9.1 Ferromagnetic Resonance

A ferromagnetic sample crystal needs to be prepared in ellipsoidal form in order to characterize it by a uniform magnetization, whereas an antiferromagnetic crystal has no macroscopic magnetization as a whole. Typical experimental arrangements for a ferromagnetic and antiferromagnetic resonance are illustrated Fig. 11.12a, b, respectively.

Inside of a ferromagnetic crystal in ellipsoidal shape, we consider a uniform $\boldsymbol{B}$ that is expressed by components $B_i = B_0 - N_i M_i$, where $i = x, y, z$; we assume $\boldsymbol{B}_0 \parallel z$-axis, and N_x, N_y, and N_z are *demagnetization factors* of an ellipsoid. We apply then an oscillating field $\boldsymbol{B}_1$ in the xy-plane to observe magnetic resonance absorption of $\boldsymbol{B}_1$ radiation. In this case, by the torque $M \times \boldsymbol{B}$, the perpendicular components M_x and M_y are rotated, as described in the following. We write

$$B_x = B_0 - N_x M_x,\ B_y = B_0 - N_y M_y, \quad \text{and} \quad B_z = B_0 - N_z M_z;$$

hence the equations of motion of $\boldsymbol{M}$ are

$$\begin{aligned} \dot{M}_x &= \gamma(M_y B_z - M_z B_y) = \gamma\{B_0 + (N_y - N_z)M_z\}M_y, \\ \dot{M}_y &= \gamma(M_z B_x - M_x B_z) = -\gamma\{B_0 - (N_z - N_x)M_z\}M_x, \end{aligned} \tag{i}$$

and

$$\dot{M}_z = 0,$$

where the constant γ represents $g\beta/\hbar$ of an atomic magnetic moment. From the last equation, we have $M_z = M$, which is constant of time. Equations in (i) have a steady-state solution, if

$$\begin{vmatrix} i\omega & \gamma\{B_0 + (N_y - N_z)M\} \\ -\gamma\{B_0 - (N_z - N_x)M\} & i\omega \end{vmatrix} = 0. \tag{ii}$$

Therefore, the ferromagnetic resonance frequency in the applied field B_0 can be obtained from (ii) by solving

$$\omega_0^2 = \gamma^2\{B_0 + (N_y - N_z)M\}\{B_0 + (N_x - N_z)M\}. \tag{iii}$$

For a spherical sample, $N_x = N_y = N_z$ so that

$$\omega_0 = \gamma B_0.$$

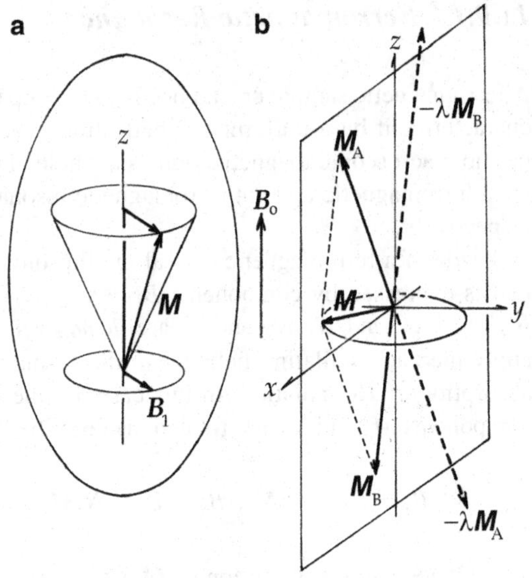

Fig. 11.12 (**a**) Larmor precession in a ferromagnetic crystal in ellipsoidal shape. (**b**) Larmor precession in an antiferromagnetic crystal.

If B_o is applied in perpendicular and parallel directions to a flat-plate specimen, we have $N_x = N_y = 0, N_z = 1$, and $N_x = N_z = 0,\ N_y = 1$, respectively, for which the resonance frequencies are obtained as

$$\omega_o = \gamma(B_o - M) \quad \text{and} \quad \omega_o = \gamma\sqrt{B_o(B_o + M)}.$$

Values of γ are normally expressed as $g\beta$, where experimental values of g are reported as 2.10, 2.18, and 2.21 for metallic specimens of Fe, Co, and Ni, respectively. These ω_o are called ferromagnetic resonance frequencies, representing Larmor frequencies of $\boldsymbol{M}$ in the effective field $B = \boldsymbol{B}_o - (N)\boldsymbol{M}$.

11.9.2 Antiferromagnetic Resonance

While ferromagnetic resonance experiments are performed with external fields $\boldsymbol{B}_o$ and $\boldsymbol{B}_1$, the antiferromagnetic resonance is studied by neutron inelastic scatterings at Brillouin zone boundaries. Nevertheless, theoretically, we can interpret magnetic fluctuations by analogy with magnetic resonance experiments. As illustrated in Fig. 10.2, magnetic fluctuations expressed as $\Delta\varepsilon = \hbar\Delta\omega$ can be displayed in intensity spectra around the center $\varepsilon_o = \hbar\omega_o$, from which the resonance frequency ω_o can be identified.

Consider an antiferromagnetic crystal that is characterized by two sublattice magnetizations $\boldsymbol{M}_A$ and $\boldsymbol{M}_B$. Primarily antiparallel near the transition point, but these are strongly perturbed by internal exchange fields $-\lambda\boldsymbol{M}_B$ and $-\lambda\boldsymbol{M}_A$ plus the anisotropic fields $\boldsymbol{B}_{\text{aniso}}$ and $-\boldsymbol{B}_{\text{aniso}}$, respectively. The anisotropic field is taken in the direction z of easy magnetization axis, representing lattice strains that are responsible for magnetic fluctuations in the motion of $\boldsymbol{M}_{\text{A}}$ and $\boldsymbol{M}_{\text{B}}$. In the absence of external fields, $\boldsymbol{M}_{\text{A}}$ and $\boldsymbol{M}_{\text{B}}$ are in precession around the effective fields $\pm\boldsymbol{B}_{\text{aniso}}$. Hence, we can write

$$\boldsymbol{B}_{\text{A}} = -\lambda\boldsymbol{M}_{\text{B}} + \boldsymbol{B}_{\text{aniso}} \quad \text{and} \quad \boldsymbol{B}_{\text{B}} = -\lambda\boldsymbol{M}_{\text{A}} - \boldsymbol{B}_{\text{aniso}},$$

as illustrated in Fig. 11.12b. Assuming that $M_{\text{A}z} = +M$ and $M_{\text{B}z} = -M$, equations of motion for the x- and y-components are

$$\dot{M}_{\text{A}x} = \gamma\{M_{\text{A}y}(\lambda M + B_{\text{aniso}}) - M(-\lambda M_{\text{B}y})\}$$
$$\dot{M}_{\text{A}y} = \gamma\{M(-\lambda M_{\text{B}x}) - M_{\text{A}x}(\lambda M + B_{\text{aniso}})\}$$

and

$$\dot{M}_{\text{B}x} = \gamma\{M_{\text{B}y}(-\lambda M - B_{\text{aniso}}) - (-M)(-\lambda M_{\text{A}y})\}$$
$$\dot{M}_{\text{B}y} = \gamma\{(-M)(-\lambda M_{\text{A}x}) - M_{\text{B}x}(-\lambda M - B_{\text{aniso}})\}.$$

Writing $M_{\text{A}+} = M_{\text{A}x} + \text{i}M_{\text{A}y}$; $M_{\text{B}+} = M_{\text{B}x} + \text{i}M_{\text{B}y}$ and assuming that all of these M's fluctuate as proportional to $\exp(-\text{i}\omega t)$, the above equations become

$$-i\omega M_{\text{A}+} = i\gamma\{M_{\text{A}+}(B_{\text{aniso}} + B_{\text{ex}}) + M_{\text{B}+}B_{\text{ex}}\}$$
$$-i\omega M_{\text{B}+} = i\gamma\{M_{\text{B}+}(B_{\text{aniso}} + B_{\text{ex}}) + M_{\text{A}+}B_{\text{ex}}\},$$

where $B_{\text{ex}} = \lambda M$. These equations have a solution, if

$$\begin{vmatrix} \gamma(B_{\text{aniso}} + B_{\text{ex}}) - \omega & \gamma B_{\text{ex}} \\ \gamma B_{\text{ex}} & \gamma(B_{\text{aniso}} + B_{\text{ex}}) + \omega \end{vmatrix} = 0.$$

Accordingly, the resonance frequency is given by

$$\omega_{\text{o}}^2 = \gamma^2 B_{\text{aniso}}(B_{\text{aniso}} + 2B_{\text{ex}}). \tag{iv}$$

However, the condition $B_{\text{ex}} > B_{\text{aniso}}$ is common among antiferromagnets studied in practice so that

$$\omega_{\text{o}} \approx \gamma\sqrt{B_{\text{aniso}}B_{\text{ex}}}$$

is an adequate formula to identify the fluctuation mode. For a typical antiferromagnet MnF_2, estimated values were $B_{\text{ex}} = 540\,\text{kG}$ and $B_{\text{aniso}} = 8.8\,\text{kG}$ at 0 K, with which the resonant frequency ω_{o} was predicted as 280GHz. Experimentally, the resonance was found at 261GHz.

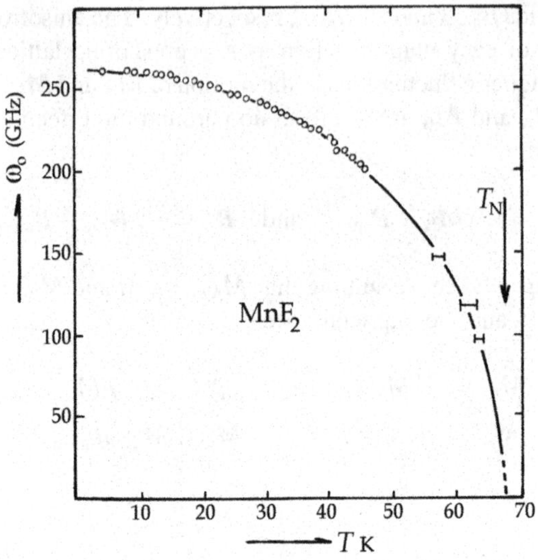

Fig. 11.13 Antiferromagnetic resonance frequency versus temperature for MnF_2.

In fact, ω_o was temperature-dependent in antiferromagnetic crystals, as shown in Fig. 11.13, implying that such magnetic fluctuations represent a soft mode with a diminishing frequency toward $T_N = 67K$.

Exercise 11

1. Discuss the reason why we consider the magnetic spin is primarily independent of the crystal lattice.
2. Review the Pauli exclusion principle quantum-mechanically in terms of indistinguishable electrons.
3. Discuss differences between magnetic symmetry and molecular correlation symmetry in crystals.
4. Physically, what is the direction of easy magnetization? Can a crystal be magnetized in other directions?
5. What is the significance of magnetic resonance experiments?

References

1. C. Kittel, *Introduction to Solid State Physics*, 5th edn. (Wiley, New York, 1976)
2. F. Reif, *Fundamentals of Statistical and Thermal Physics* (McGraw-Hill, New York, 1965)
3. T. Moriya, *Magnetism I* (Acad. Press, New York, 1963), p. 85
4. J. Kanamori, *Magnetism I* (Acad. Press, New York, 1963), p. 128

Chapter 12
Phonon and Electron Statistics in Metals

Discovered by Kamerlingh-Onnes in 1911, superconductivity in metals was a peculiar phenomenon, but is now known as due to phase transition of conduction electrons in periodic lattice at low temperatures. Related to lattice excitations, the transition is a basic occurrence in many-electron systems in crystals. In this chapter, we discuss statistical properties of many-phonon and many-electron systems.

Quantum mechanically, condensed electrons are mutually correlated in like manner as phonons in crystals. Prescribed by the Pauli principle, the correlation energy of electrons is characterized by spin–spin interactions $\boldsymbol{s}_i \cdot \boldsymbol{s}_j$, analogous to $\boldsymbol{\sigma}_i \cdot \boldsymbol{\sigma}_j$ for pseudospins. Similar phase transitions may therefore be expected, but electrons are quantum-mechanical particles behaving different from phonons.

12.1 Phonon Statistics Part 2

In Chap. 2, phonons were discussed as *particles* characterized by the energy $\hbar\omega_k$ and momentum $\hbar k$, when lattice vibrations are quantized. Quantized vibrations can thus be equivalent to a large number of phonons, behaving like independent particles. In classical language, phonons are described as independent and collision-free, but in quantum field theory, they are correlated via overlapped wavefunctions.

In Chap. 2, we discussed that a normal mode of lattice vibrations is indexed by the wavevector k for the Hamiltonian H_k and obtained the relation (2.14) to define operators b_k and $b_k^\dagger$. The total Hamiltonian and total number of phonons are then expressed by

$$\mathrm{H} = \sum_k \mathrm{H}_k = \sum_k \hbar\omega_k \left(b_k^\dagger b_k + \frac{1}{2} \right) \quad \text{and} \quad \mathrm{N} = \sum_k b_k^\dagger b_k, \tag{12.1}$$

respectively. Thermodynamic properties of a phonon gas can therefore be determined by a system of normal modes k with energies $\varepsilon_k = n_k \hbar\omega_k$ consisting of $n_k = b_k^\dagger b_k$

M. Fujimoto, *Thermodynamics of Crystalline States*,
DOI 10.1007/978-1-4614-5085-6_12,

phonons at a given temperature T, moving freely in all directions of k. Noted that ω_k and k are characteristic frequencies and wavenumber of the modes, for which the number of phonons n_k can be arbitrary, while the total $\mathrm{N} = \sum_k n_k$ is left to be determined thermodynamically.

As defined in Sect. 2.2, the operators $b_k^\dagger$ and b_k are defined to produce the energy ε_k plus and minus one phonon energy $\hbar\omega_k$, being referred to as the creation and annihilation operators, respectively. Taking a commutation relation $[b_{k'}, b_k^\dagger] = \delta_{k',k}$, we note that these operators commute as $b_{k'}b_k^\dagger = b_{k'}^\dagger b_k$ for $k' \neq k$, signifying that the lattice is invariant for exchanging phonons $\hbar\omega_k$ and $\hbar\omega_{k'}$ between the states k and k'. In terms of energy and momentum, two phonons $(\hbar\omega_k, \hbar k)$ and $(\hbar\omega_{k'}, \hbar k')$ are therefore *unidentifiable particles* in a crystal, where each phonon is characterized by the rest energy $\hbar\omega_0$ at $|k| = 0$. Considering a large number of identical particles of $\hbar\omega_0$, the vibration field is equivalent to a gaseous material described by the Bose–Einstein statistics, as discussed in Sect. 2.7.

For a small $|k|$ compared with reciprocal lattice spacing, the lattice can be considered as a continuous medium. In the field theory, classical displacements q_i and the conjugate momenta p_i can be expressed by continuous variables, $q(r,t)$and $p(r,t)$, propagating along the direction of k. For a symmetric field, these variables are represented by Fourier series in one dimension, which are functions of the *phase* of propagation, that is, $\phi = k \cdot r - \omega_k t$, where $\boldsymbol{r}$ and t are arbitrary position and time in a crystal. Correspondingly, the quantized field can be expressed by the wavefunction $\Psi(\boldsymbol{r},t)$and its conjugate momentum $\pi(\boldsymbol{r},t) = \rho\frac{\partial\Psi(\boldsymbol{r},t)}{\partial t}$, where ρ is the effective mass density. Normalizing in a volume Ω in periodic structure, these field variables can be expressed as

$$\Psi(\boldsymbol{r},t) = \frac{1}{\sqrt{\Omega}}\sum_k \sqrt{\frac{\hbar}{2\rho\omega_k}}\left\{b_k \exp i(\boldsymbol{k}\cdot\boldsymbol{r} - \omega_k t) + b_k^\dagger \exp i(-\boldsymbol{k}\cdot\boldsymbol{r} - \omega_k t)\right\} \quad (12.2)$$

and

$$\pi(\boldsymbol{r},t) = \frac{i}{\sqrt{\Omega}}\sum_k \sqrt{\frac{\hbar\rho\omega_k}{2}}\left\{-b_k \exp i(\boldsymbol{k}\cdot\boldsymbol{r} - \omega_k t) + b_k^\dagger \exp i(-\boldsymbol{k}\cdot\boldsymbol{r} - \omega_k t)\right\},$$

where $b_k^\dagger = b_{-k}$. Using the normalizing condition

$$\int_\Omega \Psi(\boldsymbol{r},t)\delta(\boldsymbol{r}-\boldsymbol{r}')\mathrm{d}\Omega = \Psi(\boldsymbol{r}',t)$$

for the field $\Psi(\boldsymbol{r},t)$, we can show the following commutation relations for these field variables Ψ and π, that is,

$$\pi(\boldsymbol{r},t)\Psi(\boldsymbol{r}',t) - \Psi(\boldsymbol{r}',t)\pi(\boldsymbol{r},t) = \frac{\hbar}{i}\delta(\boldsymbol{r}-\boldsymbol{r}'),$$

$$\Psi(\boldsymbol{r},t)\Psi(\boldsymbol{r}',t) - \Psi(\boldsymbol{r}',t)\Psi(\boldsymbol{r},t) = 0,$$

and

$$\pi(\boldsymbol{r},t)\pi(\boldsymbol{r}',t) - \pi(\boldsymbol{r}',t)\pi(\boldsymbol{r},t) = 0.$$

Writing the Hamiltonian density as $\mathrm{H} = \frac{\pi^2}{2\rho} + \frac{\kappa}{2}(\mathrm{grad}\Psi)^2$ where κ is a restoring constant, and (12.2) are solutions of the Heisenberg equation $-i\hbar\frac{\partial\Psi(r,t)}{\partial t} = \mathrm{H}\Psi(\boldsymbol{r},t)$, as verified by using the *Lagrangian formalism*. Nevertheless, we shall not discuss the formal theory in this book; interested readers are referred to a standard book, for example, *Quantenfeldtheorie des Festkörpers* by Haken [1].

In thermodynamics of phonons, we consider that energy levels $n_k\hbar\omega_k, \ldots$ are like excited levels that are occupied by n_k phonons at a given temperature T. Considering the correlations among phonons, we must multiply the isothermal probability $\exp\left(-\frac{n_k\hbar\omega_k}{\mathrm{k_B}T}\right)$ by the statistical weight λ^{n_k}, where λ is the *adiabatic probability* for adding one phonon to this level. Due to *phonon–phonon correlations*, phonon numbers n_k can be changed, as if driven by an adiabatic potential, which is expressed equivalently by the probability λ. In fact, λ is equivalent to the *chemical potential* μ, if writing $\lambda = \exp\frac{\mu}{\mathrm{k_B}T}$, in open *thermodynamic systems* [2]; in this case, phonon gas is open to the correlation energy. Hence, we write the partition function of H_k as

$$Z_k = \sum_{n_k} \lambda^{n_k} \exp\left(-\frac{n_k\hbar\omega_k}{\mathrm{k_B}T}\right) = \sum_{n_k}\left\{\lambda\exp\left(-\frac{\hbar\omega_k}{\mathrm{k_B}T}\right)\right\}^{n_k}.$$

This expression is an infinite power series of a small quantity $x = \lambda\exp\left(-\frac{\hbar\omega_k}{\mathrm{k_B}T}\right) < 1$, so that Z_k can be written simply as $\frac{1}{1-x}$, summing over $n_k = 0, 1, 2, \ldots, \infty$; thus, we obtain

$$Z_k = \frac{1}{1 - \lambda\exp\left(-\dfrac{\hbar\omega_k}{\mathrm{k_B}T}\right)}.$$

The partition function for the whole crystal is therefore determined by the product $Z = \Pi_k Z_k$, with which the thermal average of the number of phonons can be calculated as

$$\langle n\rangle = \frac{1}{Z}\sum_k n_k\lambda^{n_k}\exp\left(-\frac{n_k\hbar\omega_k}{\mathrm{k_B}T}\right) = \lambda\frac{\partial}{\partial\lambda}\ln Z = \lambda\frac{\partial}{\partial\lambda}\sum_k \ln Z_k.$$

Writing $\langle n\rangle = \sum_k \langle n_k\rangle$, we obtain

$$\langle n_k\rangle = \lambda\frac{\partial}{\partial\lambda}\ln Z_k = -x\frac{\mathrm{d}}{\mathrm{d}x}\ln(1-x) = \frac{x}{1-x} = \frac{1}{\frac{1}{\lambda}\exp\dfrac{\hbar\omega_k}{\mathrm{k_B}T} - 1}.$$

Returning to μ, the average number of phonons $\langle n_k \rangle$ in the k-state is therefore expressed by

$$\langle n_k \rangle = \frac{1}{\exp\dfrac{\hbar\omega_k - \mu}{k_B T} - 1}. \tag{12.3}$$

In the limit of $T \to \frac{\hbar\omega_k - \mu}{k_B} = 0$, we have $\langle n_k \rangle \to \infty$, meaning that if $\mu = \varepsilon_k$, almost all phonons occupy the first level $\varepsilon_k = \hbar\omega_k$ at T. This phenomenon is known as the *Bose–Einstein condensation.* On the other hand, if $\mu = 0$, there is no condensation, signifying that phonons are independent particles. Evidenced by experimental results, the Bose–Einstein condensation does occur at very low temperatures, indicating that phonons are correlated particles.

12.2 Conduction Electrons in Metals

12.2.1 The Pauli Principle

If characterized as a charged mass, the electron is a classical particle; however, it is a quantum particle in a multi-electron system, as signified by mutual correlations with other electrons. Electrons in atomic orbital obey the Pauli principle, permitting two electrons with antiparallel spins to occupy atomic states together with unpaired electrons. An electron is signified by an internal degree of freedom, taking two opposite directions $\sigma = \pm 1$ of an intrinsic spin vector. Two electrons coupled with antiparallel *spins* are not distinguishable from each other, which can be combined by a force that is different from the electrostatic repulsion. Although the Coulomb interaction cannot be ignored, spin correlations are more significant at close distance.

In a multi-electron system, a state of two unidentifiable electrons can generally be described by a wavefunction $\Psi(\boldsymbol{r}_1, \sigma_1;\ \boldsymbol{r}_2, \sigma_2)$, where $\boldsymbol{r}_1$ and $\boldsymbol{r}_2$ are their positions and σ_1 and σ_2 express their intrinsic states. Defining an operator $\mathcal{P}$ to exchange electrons 1 and 2 in such a two-particle state [3], we write

$$\mathcal{P}\Psi(\boldsymbol{r}_1, \sigma_1;\ \boldsymbol{r}_2, \sigma_2) = \Psi(\boldsymbol{r}_2, \sigma_2;\ \boldsymbol{r}_1, \sigma_1).$$

Operating $\mathcal{P}$ once again on both sides, the state should return to the original one, and hence,

$$\mathcal{P}^2\Psi(\boldsymbol{r}_1, \sigma_1;\ \boldsymbol{r}_2, \sigma_2) = \mathcal{P}\Psi(\boldsymbol{r}_2, \sigma_2;\ \boldsymbol{r}_1, \sigma_1) = \Psi(\boldsymbol{r}_1, \sigma_1;\ \boldsymbol{r}_2, \sigma_2),$$

from which we obtain

$$\mathcal{P}^2 = 1 \quad \text{or} \quad \mathcal{P} = \pm 1\,. \tag{12.4}$$

Therefore, the wavefunction of two unidentifiable electrons should be either *symmetric* or *antisymmetric* combination of wavefunctions for their exchange, which can be specified by $\mathcal{P} = +1$or by $\mathcal{P} = -1$, respectively, and the operator P is called the *parity*.

For such a pair of electrons, the parity $\mathcal{P}$ and the Hamiltonian H are commutable, that is, $[\mathrm{H}, \mathcal{P}] = 0$, so that the eigenfunction of two electrons can be given by either symmetrical or antisymmetrical combinations of $\Psi(\boldsymbol{r}_1, \sigma_1; \boldsymbol{r}_2, \sigma_2)$ and $\Psi(\boldsymbol{r}_2, \sigma_2; \boldsymbol{r}_1, \sigma_1)$, that is,

$$\frac{1}{\sqrt{2}}\{\Psi(\boldsymbol{r}_1, \sigma_1; \boldsymbol{r}_2, \sigma_2) \pm \Psi(\boldsymbol{r}_2, \sigma_2; \boldsymbol{r}_1, \sigma_1)\}.$$

Phonons and electrons are examples of different parities, $\mathcal{P} = +1$ and $\mathcal{P} = -1$, respectively; accordingly, the parity and spin represent *intrinsic* properties of these particles. Phonons in crystals and electromagnetic *photons* are both characterized by zero spins and $\mathcal{P} = +1$, whereas electrons and protons are particles of spin $\pm\frac{1}{2}$ and $\mathcal{P} = -1$. In general, *elementary particles* have spins $\sigma = 1\mathrm{n}$, where n is either *even* or *odd* integers, corresponding to the parity +1 or $-$ 1, respectively. Such intrinsic properties are quantum mechanical and absent in classical particles. We will discuss a large number of conduction electrons in metals, which are nearly free particles under normal thermodynamic conditions, but correlated quantum mechanically at very low temperatures.

Consider electrons in nearly free motion in normal metals. In a metallic crystal of cubic volume $V = L^3$ with periodic boundary conditions, we assume *one-particle state* for each electron in the first approximation, which is expressed by a plane wave

$$\Psi_{k,\sigma} = (\exp i\boldsymbol{k}\cdot\boldsymbol{r})\chi(\sigma) \tag{12.5}$$

with energy eigenvalues $\varepsilon(\boldsymbol{k}) = \frac{\hbar^2 k^2}{2m}$, where $\boldsymbol{k}$ is the wavevector and $\chi(\sigma)$ is the spin function. Due to the Pauli principle, each of these energy states specified by $\boldsymbol{k}$ and σ can be occupied by a single electron and a pair of electrons with antiparallel spins; these energy levels in a metal are thereby filled up to a level $\boldsymbol{k} = \boldsymbol{k}_\mathrm{F}$, called the *Fermi level*. If the number of electrons is odd, only the top level $\varepsilon(\boldsymbol{k}_\mathrm{F})$ is occupied by one electron with arbitrary spin, while all other levels $\varepsilon(\mathbf{k})$ for $k<k_\mathrm{F}$ are filled with two electrons with antiparallel spins at $T = 0$ K. Electrons near the Fermi level ε (k_F) can then be excited by an applied electric field $\boldsymbol{E}$, causing electrical conduction; on the other hand, all other electrons stay intact at levels $\varepsilon(\boldsymbol{k})$ below $\varepsilon(\boldsymbol{k}_\mathrm{F})$. In this Sommerfeld model, total number of states can be determined by the spherical volume of radius $k_\mathrm{F}$in the reciprocal space, times two, that is, $\frac{4\pi k_\mathrm{F}^3}{3} \times 2$, where 2 is the spin degeneracy. On the other hand, a small cube of volume $\left(\frac{2\pi}{L}\right)^3$ in the reciprocal space is occupied by one state only. Denoting the total number of electron per volume by N, we have the relation $N = 2 \times \frac{4\pi k_\mathrm{F}^3}{3}\Big/\left(\frac{2\pi}{L}\right)^3$. Letting $L^3 = V$ the volume

of a crystal, $\rho_0 = \frac{N}{V}$ is the number density of electrons, and the Fermi energy can be expressed as

$$\varepsilon_F = \varepsilon(\boldsymbol{k}_F) = \frac{\hbar^2}{2m}\left(3\pi^2\rho_0\right)^{\frac{2}{3}}. \tag{12.6}$$

12.2.2 *The Coulomb Interaction*

In Sommerfeld model, Coulomb interactions between electrons were ignored, as justifiable as the matter of approximation. In addition, normal crystals are electrically neutralized by ionic charges. Nevertheless, Thomas and Fermi [4] considered that Coulomb interactions are insignificant in a multi-electron system, as verified in their theory by the screening effect.

Placing a charge $-e$ at the origin of coordinates in a metal, they showed that there is no significant electrostatic effect due to other charges. Considering the static potential energy $V_0 = \frac{1}{4\pi\varepsilon_0}\frac{e^2}{r}$, the kinetic energy $\varepsilon(\boldsymbol{k}_F)$ of an electron is perturbed, that is, $\varepsilon(\boldsymbol{k}_F{}') = \varepsilon(\boldsymbol{k}_F) + V_0$, resulting in the effective density $\rho' = \frac{1}{3\pi^2\hbar^3}\left(2mV_0 + p_F^2\right)^{\frac{3}{2}}$ where $p_F = \hbar k_F$ is the Fermi-level momentum. With this density ρ', we can write the Poisson equation $\nabla^2 V' = -\frac{\rho'-\rho_0}{\varepsilon_0}$, where ε_0 is the dielectric constant of the metal, whereas the Coulomb potential V_0 should satisfy the Laplace equation $\nabla^2 V_0 = 0$. We can calculate the density difference approximately as

$$\rho' - \rho_0 = \frac{(2m)^{\frac{3}{2}}}{3\pi^2\hbar^3\varepsilon_0}\left\{\left(V_0 + \frac{p_F^2}{2m}\right)^{\frac{3}{2}} - \left(\frac{p_F^2}{2m}\right)^{\frac{3}{2}}\right\} \simeq \frac{p_F^3}{3\pi^2\hbar^3\varepsilon_0}\left\{\left(1 + \frac{V_0}{p_F^2/2m}\right)^{\frac{3}{2}} - 1\right\}$$
$$\simeq \frac{4mp_F}{\pi\hbar^3}V_0\left\{1 + \mathrm{O}\left(\frac{V_0}{p_F^2/2m}\right)\right\}.$$

Assuming that $V' = V_0\{1 + \mathrm{O}(\frac{V_0}{p_F^2/2m})\}$ for a small ratio $\frac{V_0}{p_F^2/2m} = \frac{V_0}{\varepsilon_F}$, the Poisson equation can be reduced to

$$\frac{1}{r^2}\frac{\mathrm{d}}{\mathrm{d}r}\left(r^2\frac{\mathrm{d}V'}{\mathrm{d}r}\right) = -\kappa V',$$

where $\kappa = \frac{4mp_F}{\pi\hbar^3\varepsilon_0}$. This equation gives a solution expressed by

$$V' \sim \exp\left(-\kappa r^2\right),$$

implying that the charge $-e$ at $r = 0$ is *screened* within an effective length $\kappa^{-\frac{1}{2}}$. Such a screening length depends on the value of $\kappa \geq k_F$ and is shorter than the nearest neighbor distance in a typical metal. In spite of the crude assumption of $V_0 < \varepsilon_F$, we may consider that Coulomb interactions are screened in metals and hence ignored, supporting Sommerfeld *free electron* model.

12.2.3 The Bloch Theorem in Equilibrium Crystals

In a stable crystal at a given *p–T* condition, conduction electrons cannot be entirely free from the lattice potential,

$$V(\boldsymbol{r}) = V(\boldsymbol{r} + n_1\boldsymbol{a}_1 + n_2\boldsymbol{a}_2 + n_3\boldsymbol{a}_3),$$

characterized by lattice periodicity, where (n_1, n_2, n_3) are integers along the symmetry axes. The Fourier transform of $V(\boldsymbol{r})$ can be expressed as

$$V(\boldsymbol{r}) = \sum_G V_G \exp i\boldsymbol{G}\cdot\boldsymbol{r}, \tag{12.7a}$$

where $G = h\boldsymbol{a}_1{}^* + k\boldsymbol{a}_2{}^* + l\boldsymbol{a}_3{}^*$ is a translation vector in the reciprocal lattice, corresponding to the lattice translation $\boldsymbol{R} = n_1\boldsymbol{a}_1 + n_2\boldsymbol{a}_2 + n_3\boldsymbol{a}_3$, as defined by (3.7).

In adiabatic approximation, the electronic motion is perturbed by the lattice potential, resulting in a modulated wavefunction

$$\Psi_k(\boldsymbol{r}, \sigma) = u_k(\boldsymbol{r}) \exp i\boldsymbol{k}\cdot\boldsymbol{r}\, \chi(\sigma), \tag{12.7b}$$

where the amplitude $u_k(\boldsymbol{r})$is a periodic function in the lattice. However, such a function as called Bloch function is not uniquely determined, as seen from the relation

$$\Psi_{k,\sigma}(\boldsymbol{r}, \sigma) = \{u_k(\boldsymbol{r}) \exp i(\pm\boldsymbol{G}\cdot\boldsymbol{r})\} \exp i\{(\boldsymbol{k} \mp \boldsymbol{G})\cdot\boldsymbol{r}\}\chi(\sigma), \tag{12.7c}$$

which is held at any lattice point $\boldsymbol{r} = \boldsymbol{R}_\mathrm{n}$ for any $\boldsymbol{k}$ satisfying the relation

$$\boldsymbol{k} \mp \boldsymbol{G} = \boldsymbol{k}; \tag{12.8}$$

thereby, (12.7c) is another Bloch function. Along the normal direction of a *crystal plane* (*h,k,l*), (12.8) can be interpreted as a Bragg diffraction of the plane wave $\exp i\boldsymbol{k}\cdot\boldsymbol{r}$ by such planes of $\pm G$, reflecting and interfering in phase for a constructive diffraction pattern. Such diffraction is originated from *elastic* collisions of electrons and periodic lattice. Therefore, the Bloch wave in the lattice behaves like a free wave modified as if reflected elastically from Brillouin-zone boundaries. Corresponding to (12.8), the electron energy is specified by $\varepsilon(\boldsymbol{k} \mp \boldsymbol{G}) = \varepsilon(\boldsymbol{k})$, from which we obtain $\pm 2\boldsymbol{k}\cdot\boldsymbol{G} + \boldsymbol{G}^2 = 0$; hence, $\boldsymbol{k} = \boldsymbol{G}/2$ indicates zone boundaries. Therefore, for the Bloch waves, the reciprocal space can be divided into many of these first zones surrounded by planes $\pm\boldsymbol{G}/2$. In this scheme, it is sufficient to have one zone, to deal with the energy of an independent electron in a periodic lattice. Figure 12.1 shows schematically the effect of a periodic lattice in one dimension.

In fact, at zone boundaries, two waves $\exp i\left(\pm\frac{\boldsymbol{G}}{2}\cdot\boldsymbol{r}\right)$ are no longer independent, as perturbed by the lattice potential, resulting in splitting of the degenerate energy

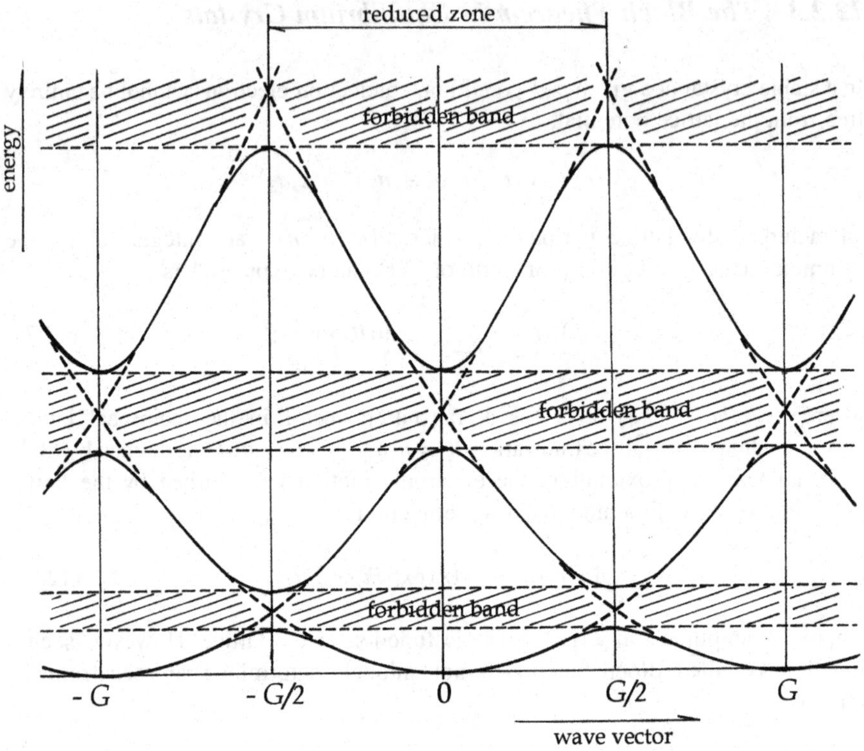

Fig. 12.1 Energy band structure of a Bloch electron in one-dimensional crystal.

into $\varepsilon\left(\frac{G}{2}\right) \pm V_{\frac{G}{2}}$. Therefore, the perturbed energy $\varepsilon(\boldsymbol{k})$ forms a band structure characterized by forbidden energy gap $2V_{\frac{G}{2}}$, as shown in the figure. In addition, the energy band theory shows that the mass for a Bloch electron should be calculated as an effective mass $m^* = \frac{1}{\hbar^2}\frac{d^2\varepsilon}{dk^2}$, and

$$\varepsilon(\boldsymbol{k}) = \sum_{i,j} \frac{\hbar^2}{m^*} k_i k_j + \text{const.}$$

Therefore, electrons in a metal can be described primarily as free particles, although modulated by the periodic lattice.

12.3 Many-Electron Systems

Although described as nearly free particles, at very low temperature, we cannot ignore spin–spin correlations among electrons in metals, as governed by the Pauli principle. Due to such correlations, thermal properties of conduction electrons cannot be determined by the Boltzmann statistics.

We consider a system of N electrons that are primarily independent free particles. The wavefunction Ψ of the system can be constructed by one-electron wavefunction $\Psi_{k,\sigma}$ expressed by (12.5) with energies $\varepsilon(k) = \frac{\hbar^2 k^2}{2m^*}$. Counting these electrons by 1, 2,. . ., N, the wavefunction of N electrons can be constructed with a linear combination of products of $\Psi_{k,\sigma}$in the form of Slater determinant, that is,

$$(1,2,.....,\mathrm{N}) = \begin{vmatrix} \Psi_{k_1,\sigma}(1) & \Psi_{k_2,\sigma}(2) & \Psi_{k_3,\sigma}(3) & \ldots\ldots & \Psi_{k_\mathrm{N},\sigma}(\mathrm{N}) \\ \Psi_{k_1,\sigma}(2) & \Psi_{k_2,\sigma}(3) & \Psi_{k_3,\sigma}(4) & \ldots\ldots & \Psi_{k_\mathrm{N},\sigma}(\mathrm{N}-1) \\ \Psi_{k_1,\sigma}(3) & \Psi_{k_2,\sigma}(4) & \Psi_{k_3,\sigma}(5) & \ldots\ldots & \Psi_{k_\mathrm{N},\sigma}(\mathrm{N}-2) \\ \ldots\ldots & \ldots\ldots & \ldots\ldots & \ldots\ldots & \ldots\ldots \\ \Psi_{k_1,\sigma}(\mathrm{N}) & \Psi_{k_2,\sigma}(\mathrm{N}-1) & \Psi_{k_3,\sigma}(\mathrm{N}-2) & \ldots\ldots & \Psi_{k_\mathrm{N},\sigma}(1) \end{vmatrix},$$

to satisfy the Pauli principle.

In the presence of a lattice potential $V(\mathrm{n})$, the eigenstate of one electron should be determined by Schrödinger equation

$$-\frac{\hbar^2}{2m}\Delta\Psi_{k',\sigma} + V\Psi_{k',\sigma} = \varepsilon(k').$$

In this approximation, $\Psi(1,2,\ldots,\mathrm{N})$ is not the eigenfunction of the whole system, but the above determinant composed of the perturbed one-electron functions $\Psi_{k,\sigma}(n_k)$ can represent the perturbed state of N electrons.

For indistinguishable electrons, the number $n_k = 1,2,\ldots,\mathrm{N}$in the determinant $\Psi(1,2,\ldots,\mathrm{N})$ can be considered to signify if an electron is present or absent in each of one-electron states $\Psi_{k,\sigma}(n_k)$, by letting $n_k = 1$ or $n_k = 0$, respectively [5]. For example, by $(0_k, 1_k, \ldots, 1_k)$, we mean that the first state is empty, and all others are occupied by one electron each.

Creation and annihilation operators $a_k^\dagger$ and a_k can be defined for electrons to express the following properties:

$$a_k^\dagger \Psi_{k,\sigma}(0_k) = \Psi_{k,\sigma}(1_k). \tag{12.9a}$$

$$a_k^\dagger \Psi_{k,\sigma}(1_k) = 0. \tag{12.9b}$$

$$a_k \Psi_{k,\sigma}(0_k) = 0. \tag{12.9c}$$

$$a_k \Psi_{k,\sigma}(1_k) = \Psi_{k,\sigma}(0_k). \tag{12.9d}$$

Among these, particularly (12.9b) manifests the Pauli principle; namely, no more than one particle can be in any state specified by k and σ. It follows from (12.9b) and (12.9c) that

$$\left(a_k^\dagger\right)^2 = (a_k)^2 = 0.$$

To exchange two electrons in the system, we can use two creation operators $a_k^\dagger$ and $a_{k'}^\dagger$ and write

$$a_k^\dagger a_{k'}^\dagger\{\Psi_1(0_1)\Psi_2(0_2)\ldots\Psi_{k,\sigma}(0_k)\ldots\Psi_{k',\sigma'}(0_{k'})\ldots\}$$
$$=\Psi_1(0)\Psi_2(0)\ldots\Psi_{k,\sigma}(1_k)\ldots\Psi_{k',\sigma'}(1_{k'})\ldots.$$

The antisymmetric nature for two electrons can then be expressed as

$$a_k^\dagger a_{k'}^\dagger\Psi(\ldots,0_k,\ldots,0_{k'},\ldots)=-a_{k'}^\dagger a_k^\dagger\Psi(\ldots,0_{k'},\ldots,0_k,\ldots). \tag{12.10}$$

Therefore, exchanging two creation operators is equivalent to exchanging $\Psi_{k,\sigma}$ and $\Psi_{k',\sigma'}$ in the Slater determinant. It is noted that (12.9a–12.9d) and (12.10) are satisfied if these operators obey the commutation rules

$$[a_k,\ a_{k'}^\dagger]_+=a_k a_{k'}^\dagger+a_{k'}^\dagger a_k=\delta_{k,k'}, \tag{12.11}$$

$$[a_k^\dagger,\ a_{k'}^\dagger]_+=0,\quad [a_k,\ a_{k'}]_+=0.$$

Like in the phonon case, the number operator in the state $\Psi_{k,\sigma}$ can be defined by

$$n_k=a_k^\dagger a_k \tag{12.12}$$

so that the total number of electrons is expressed as $\mathrm{N}=\sum_k n_k$. We can confirm from the commutating relations in (12.11) that

$$n_k\Psi_{k,\sigma}(0_k)=a_k^\dagger a_k\Psi_{k,\sigma}(0_k)=0$$

and

$$n_k\Psi_{k,\sigma}(1_k)=a_k^\dagger\{a_k\Psi_{k,\sigma}(1_k)\}=a_k^\dagger\Psi_{k,\sigma}(0_k)=\Psi_{k,\sigma}(1_k),$$

indicating that eigenvalues of the number operator n_k are 0 and 1, respectively.

The Hamiltonian for non-interacting electrons can be expressed as

$$\mathrm{H}=\sum_k\varepsilon(\boldsymbol{k})n_k=\sum_k\varepsilon(\boldsymbol{k})a_k^\dagger a_k, \tag{12.13}$$

where k in the above one-dimensional theory can be replaced by a vector $\boldsymbol{k}$, for which $\varepsilon(\boldsymbol{k})$ is the eigenvalue of one-electron Hamiltonian in three dimensions.

Despite of other perturbations in general, (12.13) can be used for the many-electron system, specifying the eigenvalue as $\varepsilon(\boldsymbol{k}')$ with a perturbed wavevector $\boldsymbol{k}'$.

The parity of electrons $\mathcal{P} = -1$ is scalar, and the correlation energy between electrons at $\boldsymbol{r}_i$ and $\boldsymbol{r}_j$ can be described by a scalar potential $V(\boldsymbol{r}_i - \boldsymbol{r}_j)$, representing spatial correlations. Total correlations are expressed as

$$\mathrm{H}_{\mathrm{int}} = \frac{1}{2}\sum_{i \neq j} V(|\boldsymbol{r}_i - \boldsymbol{r}_j|), \tag{12.14}$$

which is determined by matrix elements between one-electron states in Slater determinants. For nearly free electrons, $\mathrm{H}_{\mathrm{int}}$ may have nonzero matrix elements if the potential $V(|\boldsymbol{r}_i - \boldsymbol{r}_j|)$ is a periodic function associated with lattice excitations. Assuming that

$$V(|\boldsymbol{r}_i - \boldsymbol{r}_j|) = \sum_{\pm \boldsymbol{Q}} V_{\boldsymbol{Q}} \exp\{-i\boldsymbol{Q}\cdot|\boldsymbol{r}_i - \boldsymbol{r}_j|\} \quad \text{or}$$
$$V_{\boldsymbol{Q}} = \frac{1}{\Omega}\int_{\Omega} \mathrm{d}^3|\boldsymbol{r}_i - \boldsymbol{r}_j| \exp i\boldsymbol{Q}\cdot|\boldsymbol{r}_i - \boldsymbol{r}_j|, \tag{12.15}$$

such matrix elements can be written as

$$\langle \boldsymbol{k}, \boldsymbol{k}'|\mathrm{H}_{\mathrm{int}}|\boldsymbol{k}'', \boldsymbol{k}'''\rangle = \frac{1}{2\Omega}\iint_{\Omega} \mathrm{d}^3\boldsymbol{r}_i \mathrm{d}^3\boldsymbol{r}_j V(|\boldsymbol{r}_i - \boldsymbol{r}_j|) \exp i$$
$$\times \{(\boldsymbol{k} - \boldsymbol{k}'')\cdot\boldsymbol{r}_i + (\boldsymbol{k}' - \boldsymbol{k}''')\cdot\boldsymbol{r}_j\}.$$

Therefore, $\mathrm{H}_{\mathrm{int}}$ related to (12.15) gives rise to a secular perturbation if we have the relations

$$\boldsymbol{k} - \boldsymbol{k}' = \boldsymbol{Q} \quad \text{and} \quad \boldsymbol{k}'' - \boldsymbol{k}''' = -\boldsymbol{Q} \tag{12.16}$$

or

$$\boldsymbol{k} - \boldsymbol{k}' = \boldsymbol{k}''' - \boldsymbol{k}'',$$

where $\pm\boldsymbol{Q}$ are phonon vectors in crystals. This implies that an electron emits a phonon $\hbar\boldsymbol{Q}$ on collision with the lattice, which is absorbed by the lattice; hence, the two-electron collision process (12.16) appears as if elastic scatterings. For this process, the matrix element of $\mathrm{H}_{\mathrm{int}}$ can be expressed as

$$\langle \boldsymbol{k}, \boldsymbol{Q}|\mathrm{H}_{\mathrm{int}}|\boldsymbol{k}'', -\boldsymbol{Q}\rangle = \frac{1}{2}\sum_{k,k'',Q} V_Q a^{\dagger}_{k+Q} a_k a^{\dagger}_{k''-Q} a_{k''}. \tag{12.17}$$

In a static periodic lattice potential, we noticed that electrons can be *scattered* elastically as Bragg diffraction; in contrast, such scatterings in dynamic lattice may occur from a modulated structure. As discussed later, Fröhlich considered phonon–electron interactions similar to the above argument.

12.4 Fermi–Dirac Statistics for Conduction Electrons

In a metallic crystal, energy levels $\varepsilon(\boldsymbol{k})$ of one conduction electron are practically continuous, which are thermally accessible in thermodynamic environment. Obeying the Pauli principle, the one-electron state Ψ_k is either vacant or occupied by one electron in specified spin state, that is, $\Psi_k(0)$ or $\Psi_k(1)$ with thermal probabilities 1 or $\exp\left(-\frac{\varepsilon(\boldsymbol{k})}{\mathrm{k_B}T}\right)$, corresponding to excitation energy zero or ε, respectively. Considering the spin–spin correlations among electrons, the latter should be modified by multiplying the factor λ as in the case of phonons; we can write the Gibbs sum

$$\mathrm{Z}_k = 1 + \lambda \exp\left(-\frac{\varepsilon(\boldsymbol{k})}{\mathrm{k_B}T}\right).$$

For electrons, the energy $\varepsilon(\boldsymbol{k})$ can be occupied by only one electron in an arbitrary spin state, so that $n_k = 1$. Therefore, the thermodynamic probability for a one-particle state εto be occupied by 1 electron is expressed as

$$P(\varepsilon, 1) = \frac{1}{\exp\dfrac{\varepsilon-\mu}{\mathrm{k_B}T} + 1}. \tag{12.18}$$

Electrons obey the probability (12.18) at a given temperature T, which is called the Fermi–Dirac distribution.

Exercise 12

1. Consider a gas of free electrons, which will not condense even at $T = 0\,\mathrm{K}$. However, we consider spin–spin correlations for such a gas embedded in a crystal. Why? Discuss on the physical reason for the difference.
2. The Bose–Einstein distribution is determined by the probability for an energy level $\varepsilon = n\varepsilon_0$ to be occupied by n particles of energy $\varepsilon_0 = \hbar\omega$, which is expressed by

$$P_{\mathrm{BE}}(\varepsilon, n) = \frac{1}{\exp\dfrac{\varepsilon-\mu}{\mathrm{k_B}T} - 1}.$$

In contrast, the Fermi–Dirac distribution is given by

$$P_{\mathrm{FD}}(\varepsilon, 1) = \frac{1}{\exp\dfrac{\varepsilon-\mu}{\mathrm{k_B}T} + 1}.$$

These distributions become identical if $\exp\frac{\varepsilon-\mu}{\mathrm{k_B}T} \gg 1$, that is,

$$P_{BE}(\varepsilon, n) \approx P_{\mathrm{FD}}(\varepsilon, 1) \approx \exp\frac{\mu-\varepsilon}{\mathrm{k_B}T}.$$

This is the Gibbs factor in Boltzmann–Gibbs statistics. Discuss these results in terms of the criterion for quantum and classical particles.

3. Show that a pressure of a Fermi electron gas in the ground state is given by

$$p = \frac{(3\pi^2)^{\frac{2}{3}}}{5}\frac{\hbar^2}{m}\left(\frac{N}{V}\right)^{\frac{5}{3}}.$$

Find an expression for the entropy of a Fermi electron gas, assuming $\mathrm{k_B}T \ll \varepsilon_{\mathrm{F}}$.

References

1. H. Haken, *Quantenfeldtheorie des Festkörpers* (B. G. Teubner, Stuttgart, 1973)
2. C. Kittel, H. Kroemer, *Thermal Physics* (Freeman, San Francisco, 1980)
3. B. Kursunoglu, *Modern Quantum Theory* (Freeman, San Francisco, 1962)
4. L.D. Landau, E.M. Lifshitz, *Quantum Mechanics*, vol. 3 of Theoretical Physics (Pergamon Press, London 1958)
5. C. Kittel, *Quantum Theory of Solids* (Wiley, New York, 1963)

Chapter 13
Superconducting Metals

The superconducting state of a metal is characterized not only by near-zero electrical resistance but also by *perfect diamagnetism* of a charged liquid, which was felt somewhat unusual on discovery, because of the unknown charge carriers. Presumably, it is a common occurrence in metallic crystals at low temperatures, exhibiting electromagnetic properties of charge carriers in superconducting states. It is now known to be related to a phase transition from normal to superconducting states characterized by the critical temperatures T_c. In this chapter, superconductivity and early theories are reviewed, prior to discussing the theory by Bardeen, Cooper, and Schieffer in Chap. 14. These theories were able to identify the charge carrier, wherefrom we realize the significant role played by the lattice for superconducting transitions.

13.1 Superconducting States

13.1.1 Near-Zero Electrical Resistance

The normal electric conduction in metals originates from drifting electrons, as described by the Ohm law for a current I and applied voltage V. Energy loss during drifting can be attributed to inelastic collisions between electrons and lattice; the electrical resistance defined by

$R = V/I$ obeys normally Matthiessen's rule [1] expressed by

$$R = R_{\text{ideal}} + R_{\text{res}}.$$

Here, R_{ideal} is due to scatterings by the lattice, whereas the residual resistance R_{res} is caused by impurities and other imperfections in the lattice. The normal resistance R_{ideal} exhibits temperature dependence proportional to T^5 at temperatures below the Debye's temperature Θ_D, whereas R_{res} is virtually

M. Fujimoto, *Thermodynamics of Crystalline States*,

DOI 10.1007/978-1-4614-5085-6_13,

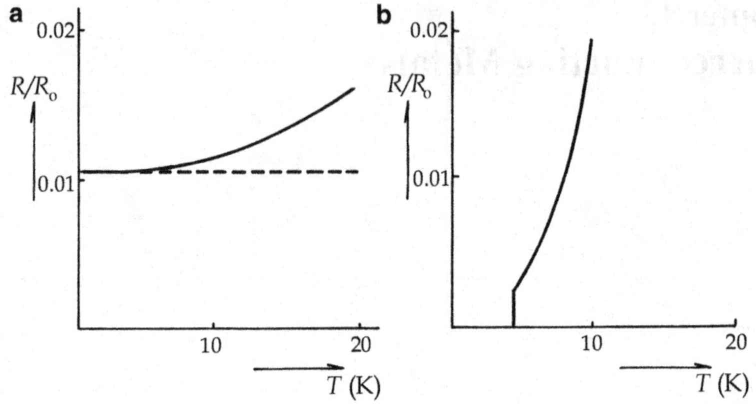

Fig. 13.1 (**a**) Electrical resistance in normal metals. (**b**) Electrical resistance of superconducting mercury.

temperature independent. Figure 13.1a, b compares a typical normal resistance of platinum with superconducting mercury. The Matthiessen's rule is clear from Fig. 13.1a, but the electric resistance of Hg shows, in contrast, an abrupt drop down to near zero at a specific temperature T_c. This was observed as a peculiar property of a metal, judging from a normal conductivity shown in Fig. 13.1a. The near-zero resistance of Hg below T_c implies that the corresponding current becomes infinite, if the Ohm law is applied; the metal can then be characterized by infinite conductivity. If so, using a superconducting coil, a *persistent* current could be obtained, which was actually confirmed by Collins, who estimated the value of resistance almost zero, that is, $R_{super} < R_{normal} \times 10^{-15}$.

Early attempts to obtain a high magnetic field with a superconducting solenoid were nevertheless a failure, and thereby, we discovered that the superconductivity was destroyed by its own magnetic field. Later on, it became clear that at temperatures below T_c, there is a *critical magnetic field* H_c to destroy the superconductivity. Such a critical field H_c was found to be a function of T, showing a parabola-like temperature dependence as shown in Fig. 13.2a. The observed $H_c - T$ curve was approximately parabola as given by

$$H_c = H_o\left(1 - \frac{T^2}{T_c^2}\right), \tag{13.1}$$

which was found valid to temperatures T sufficiently close to 0 K, although the lowest available temperature was limited to about 1.7 K at that time of experiment. Here H_o was determined by extrapolating H_c to the absolute zero, which is however relatively a weak field, as evaluated from Fig. 13.2b. Inspired by such a $H_c - T$ curve, Fig. 13.2b implies a phase diagram between normal and superconducting phases, while the origin for H_c is not immediately clear.

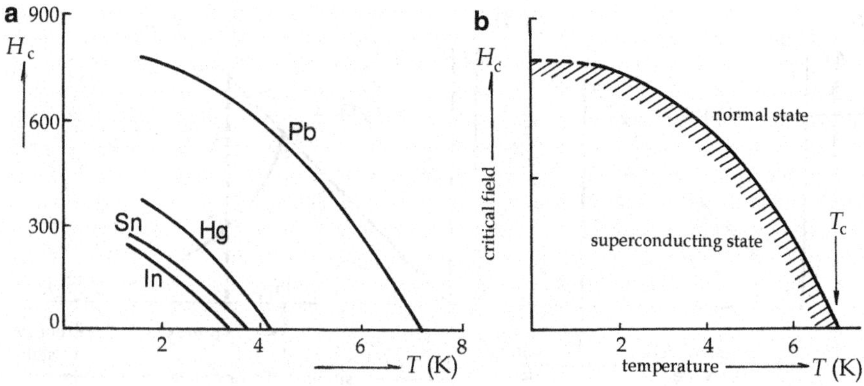

Fig. 13.2 $H_c - T$ phase diagram of superconducting metals. (**a**) Observed results. (**b**) Idealized phase diagram.

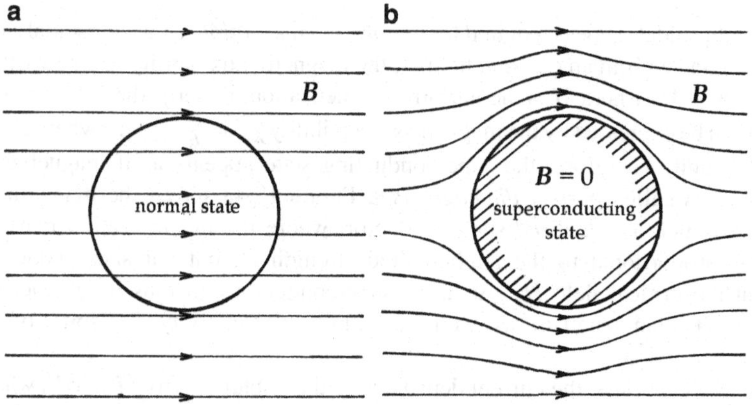

Fig. 13.3 The Meissner effect in an idealized superconducting sphere. (**a**) Normal state. (**b**) Superconducting state.

13.1.2 The Meissner Effect

Meissner and Ochsenfeld (1933) discovered the effect of a magnetic field to a superconducting state. If a conductor in ellipsoidal shape was cooled down below the transition temperature in a magnetic field, they found that all of the field lines were pushed out of the body, as illustrated by Fig. 13.3. Such a magnetic effect, known as the *Meissner effect*, idealizes the superconducting state by *zero flux density* inside the conductor, that is,

$$\boldsymbol{B} = 0. \tag{13.2}$$

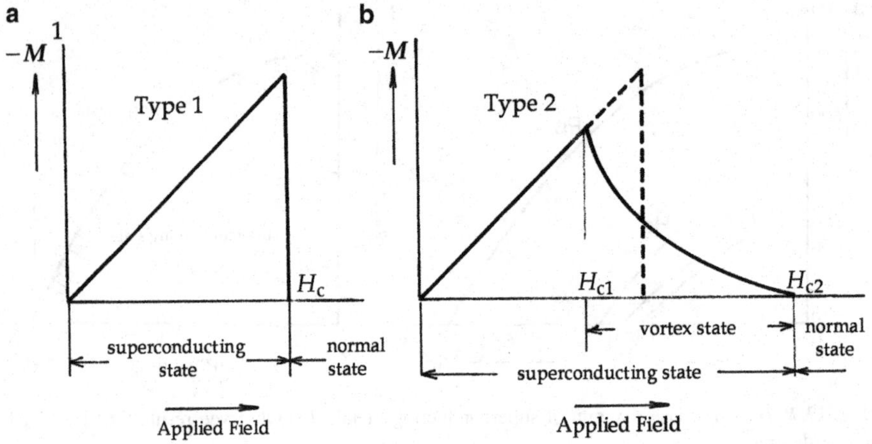

Fig. 13.4 Magnetization curves for Meissner effect in superconductors. (**a**) Type 1. (**b**) Type 2.

An ellipsoidal shape is required for *uniform magnetization*. In an ellipsoidal sample of superconductor in an external field $\boldsymbol{H}$, the magnetic flux density can be written as $\boldsymbol{B} = \mu_o \boldsymbol{H} + \boldsymbol{M}$, where $\boldsymbol{M}$ is the uniform magnetization; thereby, the Meissner effect (13.2) can be expressed by the magnetic susceptibility $\chi_m = \frac{M}{H} = -\mu_o$, where $\mu_o = \frac{1}{4\pi}$ in MKS unit. Therefore, the superconducting state appears as if magnetized by H, which is called *perfect diamagnetism*. Figure 13.4a shows the magnetization curve for such an *idealized* superconductor, where the magnetization rises from zero as proportional to the applied field strength H, but vanishes suddenly to zero at the critical field H_c. An ideal superconductor can thus be characterized by perfect diamagnetism as well as infinite conductivity, in spite of their unknown relation.

If using Ohm law, the current density should be obtained from $\boldsymbol{j} = \sigma \boldsymbol{E}$, where $\boldsymbol{E}$ is an applied electric field. In this case, in order to obtain infinite σ, the electric field $\boldsymbol{E}$ should be zero. We can therefore write $\frac{\partial \boldsymbol{B}}{\partial t} = 0$ from the Maxwell equations for $\boldsymbol{E} = 0$, and hence, $\boldsymbol{B}$ should be constant of time. This suggests that the magnetic flux density $\boldsymbol{B}$ penetrated into a conductor at temperatures above T_c, which should be *frozen* as specified by $\boldsymbol{B} = \text{const.}(\neq 0)$ in the superconducting state. In contrast, the Meissner effect specifies that $\boldsymbol{B} = 0$ in the superconductor, and the Ohm law cannot support the superconducting current density $\boldsymbol{j}_s$. Therefore, the Meissner effect appears to be independent from $\boldsymbol{j}_s$, whose origin was nevertheless clarified later by London's electromagnetic theory, as will be discussed in Sect. 13.3.

The Meissner effect can be understood in terms of magnetization energy, meaning that an ellipsoidal superconductor is characterized by the energy density $\mu_o \boldsymbol{M} \cdot \boldsymbol{H}$, which changes as a function of temperature T below T_c. Hence, at such a given temperature as $T < T_c$, we can enhance the Meissner effect at rate of either $-\boldsymbol{M} \cdot d\boldsymbol{H}$ or $\boldsymbol{H} \cdot d\boldsymbol{M}$, which can be added to the Gibbs function in thermodynamic argument.

The magnetization $\boldsymbol{M}$ due to Meissner effect is distributed in general inside of a non-ellipsoidal crystal, in which the critical field H_c cannot be uniform. On the other hand, pure metals are usually studied with specimens in long cylindrical shape, where H_c takes a single value. An idealized superconductor characterized by a magnetization curve in Fig. 13.4a is referred to as Type 1. However, there are some superconductors of alloy compounds, which are characterized by intrinsically distributed H_c, which are called Type 2. In Fig. 13.4b, shown is a typical magnetization curve of a Type 2 superconductor, which is signified by two critical fields H_{c1} and H_{c2}, as indicated in the figure. The region between H_{c1} and H_{c2} is generally called the *vortex region*. Type 2 superconductors can be considered for important applications, if H_{c2} is sufficiently high, for the superconductivity is extended in a vortex region to relatively high magnetic fields.

13.1.3 Thermodynamic Equilibrium Between Normal and Superconducting Phases

We consider a Type 1 superconducting state of a metal specified by magnetization $\boldsymbol{M}$ due to the Meissner effect. As inferred from Fig. 13.2b, the superconducting state is in mixed normal and superconducting phases in equilibrium. In contrast to the nonmagnetic normal phase, the superconducting phase is characterized by a magnetization $\boldsymbol{M}$ interacted with a magnetic field $\boldsymbol{H} = -\frac{\boldsymbol{M}}{\mu_0}$, so that the Meissner effect can be expressed by an external work $\boldsymbol{H} \cdot \mathrm{d}\boldsymbol{M}$ per unit superconducting volume. Therefore, under a constant volume condition, the first law of thermodynamics can be expressed for the superconducting phase as

$$\mathrm{d}U = H \cdot \mathrm{d}M + T\,\mathrm{d}S.$$

Defining the Gibbs function by

$$G(H,T) = U - TS - HM, \tag{13.3}$$

the thermal equilibrium can be determined by the inequality $\mathrm{d}G \leq 0$ against external variables T and H.

The condition for two phases to coexist at a temperature T and critical field H_c is determined by densities of Gibbs functions g_s and g_n defined by relations $g_s V_s = G_s$, $g_n V_n = G_n$, and $V_s + V_n = V$, to give an equal value in thermal contact. We therefore write

$$g_s(H_c,T) = g_n(H_c,T) \tag{13.4}$$

on the boundary curve in Fig. 13.2b. For the normal phase, g_n has nothing to do with H_c, so that $g_n(H_c,T)$ in (13.4) can be replaced by $g_n(0,T)$.

At T_c characterized by $H_c = 0$, we have

$$g_n(0, T_c) = g_s(0, T_c), \tag{13.5}$$

signifying that the transition is continuous at $H = 0$. On the other hand, if $H \neq 0$ for $T < T_c$, the Gibbs function $G_s(H_c, T)$ should be calculated as

$$G_s(H_c, T) = G_s(0, T) - V_s \int_{H_c}^{0} M(H)\mathrm{d}H = G_n(0, T) - V_s\left(\frac{1}{2}\mu_0 H_c^2\right)$$

or

$$g_s(H_c, T) = g_n(0, T) - \frac{1}{2}\mu_0 H_c^2. \tag{13.6}$$

Accordingly, the transition is discontinuous at temperatures below T_c.

The latent heat L per volume can then be calculated with the thermodynamic relation $S_n - S_s = -\left\{\frac{\partial}{\partial T}(G_n - G_s)\right\}_V$, resulting in

$$L = -\mu_0 T H_c \frac{\mathrm{d}H_c}{\mathrm{d}T}. \tag{13.7}$$

Referring to Ehrenfest's classification, such a transformation below T_c is first order, changing between normal and superconducting phases by applying magnetic field H in the range $0 \leq H \leq H_c$. In contrast, the transition at T_c is second order, where $L = 0$, and hence, $H_c = 0$ from (13.7).

The specific heat exhibits a discontinuity at T_c, due to a sudden change in the magnetization, which can be calculated at a constant volume V as

$$c_s - c_n = T\left\{\frac{\partial(s_s - s_n)}{\partial T}\right\}_V = \mu_0 T\left\{H_c \frac{\mathrm{d}^2 H_c}{\mathrm{d}T^2} + \left(\frac{\mathrm{d}H_c}{\mathrm{d}T}\right)^2\right\}. \tag{13.8a}$$

Since $H_c = 0$ at T_c, the discontinuity at T_c is given by

$$(c_s - c_n)_{T_c} = \mu_0 T_c \left(\frac{\mathrm{d}H_c}{\mathrm{d}T}\right)^2, \tag{13.8b}$$

which is known as Rutger's formula.

The temperature dependence of H_c has a significant implication. In the above argument, it is particularly so when the absolute zero is approached. According to the classical *third law* of thermodynamics, we should expect $s_n - s_s \to 0$ in the limit of $T \to 0$. This condition is fulfilled in (13.8b), if

$$\lim_{T\to 0}\frac{dH_c}{dT}=0, \tag{13.9}$$

implying that the $H_c - T$ curve has a horizontal tangent at $T = 0$; therefore, the equilibrium curve in Fig. 13.2b is not exactly parabolic.

13.2 Long-Range Order in Superconducting States

The superconducting state below T_c is composed of two volumes of normal and superconducting states, where the volume ratio depends on the magnitude of applied magnetic field. According to Sect. 13.1, the transition at T_c is second order with no latent heat, but accompanying a discontinuous specific heat. Below T_c, the superconducting state consists of normal and superconducting phases, whose volumes vary adiabatically with an external magnetic field; hence, the transition is first order. It is logical to define an order parameter, as in Landau theory, to prescribe the ordering process in a superconducting state. We can sketch such a *mesoscopic process* in terms of temperature and applied magnetic field, for which Landau theory can be employed to discuss the validity of mean-field concepts in a superconducting state.

In a superconducting state, we can postulate an order parameter defined as $\eta = 0$ at T_c and $\eta = 1$ for complete order at $T = 0$ K. For the density $g(\eta, T)$ of Gibbs function, we can write

$$g(0,T_c)=-\frac{1}{2}\gamma T_c^2, \tag{i}$$

where γT_c represents the specific heat of conduction electrons at T_c, as derived from Sommerfeld formula $c_{el} = \gamma T$ for an electron gas. On the other hand, $g(1, 0)$ should represent the order energy of condensed electrons per unit volume at $T = 0$, which is given by

$$g(1,0)=-\frac{1}{2}\mu_o H_o^2. \tag{ii}$$

Unlike binary systems, inversion symmetry can be disregarded in multi-electron system, so that the Gibbs function can be expanded into a series with even and odd power terms. That is,

$$g(\eta,T)=g_o+\alpha(H,T)\eta+\frac{1}{2}\beta(H,T)\eta^2+\cdots. \tag{iii}$$

Gorter and Casimir wrote $g(\eta, T)$ in an alternative form

$$g(\eta,T)=g_o-\frac{\mu_o H_c^2}{2}\eta-\frac{\gamma T^2}{2}\sqrt{1-\eta}. \tag{13.10}$$

Here, the factor $\sqrt{1-\eta}$ of the third term in (13.10) is somewhat arbitrarily chosen though their $g(\eta, T)$ in (13.10) was adequate for the requirements (i) and (ii). In Landau function (iii), the coefficients α, β, ... can be arbitrary, so that (13.10) has no conflict with (iii). At any rate, Gorter–Casimir's formula is a useful proposition to explain observed specific heat and $H_c(T)$ for $T < T_c$.

Using (13.10), thermal equilibrium can be determined from $\left(\frac{\partial g}{\partial \eta}\right)_{H,T} = 0$, that is,

$$\frac{\gamma T^2}{4\sqrt{1-\eta}} = \frac{\mu_o H_o^2}{2}.$$

At $T = T_c$, we should have $\eta = 0$, and hence, $\frac{\gamma T_c^2}{4} = \frac{\mu_o H_o^2}{2}$. In this way, the order parameter can be expressed as

$$\eta = 1 - \left(\frac{T}{T_c}\right)^4. \tag{13.11}$$

Substituting η into (13.10) and differentiating two times with respect to temperature, we can derive the specific heat:

$$c_s = \frac{3\gamma T^3}{T_c^2}. \tag{13.12}$$

In this model, the result $c_s \propto T^3$ shows the same temperature dependence as c_{Debye} for lattice vibrations, but these are distinguishable by different proportionality factors.

Using (13.12) and the Sommerfeld formula $c_n = \gamma T$ for a normal conductor, we can derive the parabolic expression $H_c = H_o\left(1 - \frac{T^2}{T_c^2}\right)$, representing temperature dependence in mean-field approximation; the empirical formula (13.1) is thus compatible with the Gorter–Casimir theory. Nonetheless, it is significant to realize that such assumptions as (iii) and (13.10) agree conceptually with an internal adiabatic potential that drives ordering processes.

Figure 13.5a shows experimental data of specific heat of gallium metal. Experimentally, normal and superconducting phases are separated in the presence of a magnetic field with strength lower than H_c in mixed volumes like domains in binary magnets. To analyze the temperature dependence of the specific heat of the superconducting phase, $c_s/\gamma T_c$ versus T_c/T were plotted against the applied magnetic field in the range between 0 and 300 gauss. Deviating slightly from the linear relation, we can consider that the mean-field approximation does not fully support the experimental result, while normal phase is primarily a free-electron gas. Nevertheless, such anomalies are particularly significant at extremely low temperatures. Figure 13.5b in semilog plot shows clearly an anomaly in logarithmic character, that is,

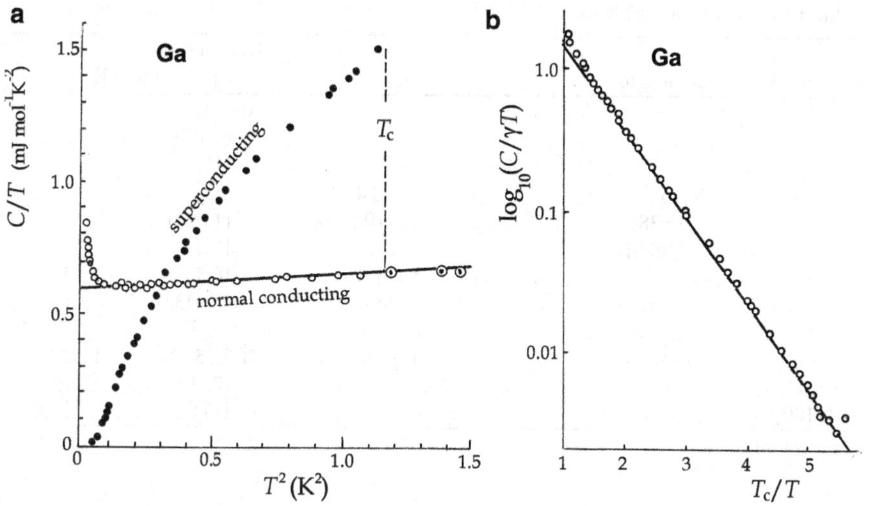

Fig. 13.5 Specific heat measured in superconducting gallium. (**a**) C/T versus T^2. (**b**) A semilog plot of $C/\gamma T$ versus T_c/T.

$$c_s \sim \exp\left(-\frac{b}{T}\right) \tag{13.13}$$

with a constant b, which is referred to as a *logarithmic anomaly* [2].

The exponential behavior of c_s in (13.13) allows the temperature dependence to be interpreted in terms of a Boltzmann factor $\exp\left(-\frac{E_g}{k_B T}\right)$. Assuming $E_g = k_B b$ to represent an *energy gap* in the superconducting spectrum, we can consider that such an exponential c_s indicates the presence of an energy gap $E_g \propto b$ between normal electrons and unidentified superconducting charge carriers. Writing as $E_g = \frac{\Delta}{2}$, the parameter Δ represents the *electron–electron* coupling in the theory of superconducting transition in Chap. 14. The logarithmic anomaly is therefore a significant evidence for the essential gap Δ in the theory of superconducting states. Bardeen, Cooper, and Schrieffer derived the expression of the gap $\Delta(T)$ in their theory. Using their formula, values of $\Delta(0)$ were estimated by extrapolating measured plots to $T = 0$ K for various superconductors, which are listed in Table 13.1, including observed T_c and $H_c(0)$ as well, where $H_c(0)$ was previously denoted as H_0.

13.3 Electromagnetic Properties of Superconductors

Superconductivity due to infinite electrical conductivity in Ohm law was re-characterized by perfect diamagnetism. These two phenomena were considered as consistent with the Maxwell theory of electromagnetism, but found as otherwise,

Table 13.1 Superconducting metals[a]

Δ(0) (in eV) extrapolated to 0 K			(T_c in K)	[H_o (in Gauss) extrapolated to 0 K]	
				Al 3.4 (1.140) [105]	
	V 16 (5.38) [1420]		Zn 2.4 (0.875) [53]	Ga 3.3 (1.091) [51]	
	Nb 30.5 (9.50) [1980]	Mo 2.7 (0.92) [95]	Cd 1.5 (0.56) [30]	In 10.5 (3.0435) [293]	Sn 11.5 (3.722) [309]
La 19 (6.00) [1100]	Ta 14 (4.438) [830]		Hg 16.5 (4.153) [412]	Tl 7.35 (2.39) [171]	Pb 27.3 (7.193) [803]

[a]Data compiled from [2]

because of charge carriers of unknown type. Normal and superconducting phases were separated by an external magnetic field, where the latter is characterized by the Meissner effect and the former phase exhibits a usual conduction if described by the Ohm law. We therefore assume that superconducting charge carriers are particles of charge e' and mass m', different from ordinary electrons of e and m.

Normal and superconducting current densities, $\boldsymbol{J}_n$ and $\boldsymbol{J}_s$, are considered as determined by an applied electric field as

$$\boldsymbol{J}_n = \sigma \boldsymbol{E} \tag{13.14a}$$

and

$$\Lambda \frac{\partial \boldsymbol{J}_s}{\partial t} = \boldsymbol{E}, \tag{13.14b}$$

respectively. Equation (13.14a) is the Ohm law that is applicable to normal conduction electrons, whereas (13.14b) is equivalent to the current density $\boldsymbol{J}_s = n_s e' \boldsymbol{v}_s$ of n_s superconducting particles in accelerating motion, that is,

$$m' \frac{d\boldsymbol{v}_s}{dt} = e' \boldsymbol{E}.$$

Here, superconducting charge carriers are assumed as particles of unknown charge e' and mass m'; hence, the parameter Λ in (13.14b) can be expressed as

$$\Lambda = \frac{m'}{n_s e'^2}. \tag{13.15}$$

Noting that the electric field $\boldsymbol{E}$ for the Ohm law is an *irrotational vector*, that is, curl $\boldsymbol{E} = 0$, the field $\boldsymbol{E}$ for (13.14b), on the other hand, should be *rotational* as characterized by curl $\boldsymbol{E} \neq 0$. Equation (13.5b) suggests that $\boldsymbol{J}_s$ should be attributed to such an *inductive* $\boldsymbol{E}$, and we can write the Maxwell equations for a superconducting state of *supercurrent* density $\boldsymbol{J}_s$, that is,

$$\text{curl}\,\boldsymbol{E} = -\frac{\partial \boldsymbol{B}}{\partial t} \quad \text{and} \quad \text{curl}\,\boldsymbol{B} = \mu_0\left(\boldsymbol{J}_s + \varepsilon_0\frac{\partial \boldsymbol{E}}{\partial t}\right).$$

Combining the first equation with (13.14b), we obtain

$$\text{curl}\left(\Lambda\frac{\partial \boldsymbol{J}_s}{\partial t}\right) = -\frac{\partial \boldsymbol{B}}{\partial t}. \tag{13.16}$$

Assuming $\frac{\partial E}{\partial t} = 0$ for a steady current $\boldsymbol{J}_s$, the Maxwell equation can be reduced to curl $\boldsymbol{B} = \mu_0\boldsymbol{J}_s$, indicating that $\boldsymbol{B}$ is distributed in the superconductor. Combining this and (13.16), we can write

$$\frac{\partial \boldsymbol{B}}{\partial t} = -\text{curl}\left(\frac{\Lambda}{\mu_0}\text{curl}\frac{\partial \boldsymbol{B}}{\partial t}\right) = -\lambda^2\text{curl curl}\frac{\partial \boldsymbol{B}}{\partial t} = \lambda^2\nabla^2\frac{\partial \boldsymbol{B}}{\partial t},$$

where the factor λ is a constant defined as

$$\lambda^2 = \frac{\Lambda}{\mu_0}.$$

The time derivatives, $\dot{\boldsymbol{J}}_s, \dot{\boldsymbol{E}}_s$, and $\dot{\boldsymbol{B}}_s$, represent local properties in a superconductor, but not the bulk body that exhibits the Meissner effect. We notice that (13.16) is for $\dot{\boldsymbol{J}}_s$, and the corresponding $\dot{\boldsymbol{E}}_s$ should determine $\dot{\boldsymbol{B}}_s$ that penetrates to a depth λ from the superconducting surface. Writing the above equations, for simplicity, for $\dot{B}_z$ in the z direction perpendicular to the surface, we have

$$\lambda^2\nabla^2\dot{B}_z - \dot{B}_z = 0, \tag{13.17}$$

whose solution is given by $\dot{B}_z = \dot{B}_0\exp\left(-\frac{z}{\lambda}\right)$, where $z = \lambda$ represents the effective penetration depth of $\dot{B}_z$, and hence, $\dot{B}_z = 0$ for $z > \lambda$. However, this result is by no means the same as $B_z = 0$, so that the Meissner effect does not arise from Maxwell theory directly.

To obtain the Meissner effect in the region $0 < z < \lambda$, London revised (13.16) and (13.17) as

$$\text{curl}(\Lambda\boldsymbol{J}_s) = -\boldsymbol{B} \tag{13.18}$$

and

$$\boldsymbol{B} = \lambda^2\nabla^2\boldsymbol{B}. \tag{13.19}$$

Equation (13.18) gives the relation between $\boldsymbol{J}_s$ and $\boldsymbol{B}$, while the constant λ in (13.19) describes the penetration depth. Using these expressions, we may describe that the Meissner effect is *incomplete* in the layer of thickness λ, and we have complete Meissner effect $\boldsymbol{B} = 0$ beyond $z = \lambda$. Equation (13.19) may be oversimplified, giving however a correct order magnitude of λ, estimated as about 10^{-5} cm.

The London equation (13.18) can be modified by using the vector potential $\boldsymbol{A}$ defined by $\boldsymbol{B} = \mathrm{curl}\boldsymbol{A}$ derived from the basic relation $\mathrm{div}\boldsymbol{B} = 0$ that is compatible with $\mathrm{curl}\,\boldsymbol{E} \neq 0$. Using the vector potential $\boldsymbol{A}$, (13.17) can be expressed as

$$\mathrm{curl}(\Lambda \boldsymbol{J}_s + \boldsymbol{A}) = 0. \tag{13.20a}$$

If the superconducting body is *simply connected*, as illustrated by the curve C_1 in Fig. 13.6, we can write

$$\Lambda \boldsymbol{J}_s + \boldsymbol{A} = \nabla\chi' \tag{13.20b}$$

where χ' is an arbitrary scalar function that satisfies $\nabla^2\chi' = 0$, hence giving unique values of $\boldsymbol{J}_s$ and $\boldsymbol{A}$ at any point inside C_1. On the other hand, for a *multiply connected* body, (13.20a) should be expressed in integral form to obtain a relation between the *persistent surface supercurrent* and the vector potential $\boldsymbol{A}$. Leaving the latter case aside, in a simply connected body, we can choose the *gauge* from $\mathrm{div}\boldsymbol{A} = 0$ for a time-independent case, to write another vector potential as $\boldsymbol{A} + \nabla\chi$ with an arbitrary χ. This allows the choice $\chi = -\chi'$. Using such a gauge χ', called the *London gauge*, (13.20b) can be simplified as

$$\boldsymbol{J}_s = -\frac{1}{\Lambda}\boldsymbol{A} = -\frac{1}{\mu_0\lambda^2}\boldsymbol{A}. \tag{13.21}$$

With (13.21), called London equation, the Meissner effect is obtained for a simply connected metal.

Equation (13.21) indicates that each of n_s superconducting carriers gains a momentum by $\frac{m'}{e'}\boldsymbol{j}_s$, where $n_s\boldsymbol{j}_s = \boldsymbol{J}_s$ represents the total supercurrent density. Therefore, for a system of independent superconducting particles, we write the Hamiltonian

$$\mathsf{H}_{\text{system}} = \sum_i \frac{1}{2m'}\{\boldsymbol{p}_s(\boldsymbol{r}_i) - e'\boldsymbol{A}(\boldsymbol{r}_i)\}^2, \tag{13.22}$$

in which the momentum $\boldsymbol{p}_s(\boldsymbol{r}_i)$ and position $\boldsymbol{r}_i$ constitute a field that can interact with the electromagnetic field. In this case, assuming the field of $(\boldsymbol{p}_s(\boldsymbol{r}_i), \boldsymbol{r}_i)$ as continuous variables, we can define the Hamiltonian density H by $\mathsf{H}_{\text{system}} = \int_{\text{volume}} \mathsf{H}\,\mathrm{d}^3\boldsymbol{r}$, that is,

$$\mathsf{H} = \frac{1}{2m'}\{\boldsymbol{p}_s(\boldsymbol{r}) - e'\boldsymbol{A}(\boldsymbol{r})\}^2,$$

which is called *one-particle Hamiltonian*.

The vector potential $\boldsymbol{A}(\boldsymbol{r})$ is invariant under a gauge transformation, $\boldsymbol{A}(\boldsymbol{r}') = \boldsymbol{A}(\boldsymbol{r}) + \nabla\chi(\boldsymbol{r})$, where the scalar function $\chi(\mathbf{r})$ can be arbitrarily selected from solutions of $\nabla^2\chi(\boldsymbol{r}) = 0$. However, the gauge invariance of the Hamiltonian $\mathcal{H}$ must be fabricated to be consistent with the invariance of electromagnetic field. In quantum theory, considering $\boldsymbol{p}_s(\boldsymbol{r})$ as a differential operator $-i\hbar\nabla$, the Schrödinger equation $\mathrm{H}\psi(\boldsymbol{r}) = E\psi(\boldsymbol{r})$ with the energy eigenvalue E can be modified to be invariant under the gauge transformation. We postulate that the equation

$$\frac{1}{2m'}\{-i\hbar\nabla_{r'} - e'\boldsymbol{A}(\boldsymbol{r}')\}^2\psi(\boldsymbol{r}') = E\psi(\boldsymbol{r}')$$

is transformed to

$$\frac{1}{2m'}\{-i\hbar\nabla_r - e'\boldsymbol{A}(\boldsymbol{r}) - e'\nabla_r\chi(\boldsymbol{r})\}^2\psi(\boldsymbol{r}') = E\psi(\boldsymbol{r}').$$

We notice that the latter can be satisfied with a transformed function $\Psi(\boldsymbol{r}') = \exp\frac{ie'\chi(r)}{\hbar}\psi(\boldsymbol{r})$.

The function $\psi(\boldsymbol{r}')$ can therefore be characterized by a phase shift $\frac{e'\chi(r)}{\hbar}$, which arises from the gauge $\chi(\boldsymbol{r})$.

On the other hand, the momentum $\boldsymbol{p}(\boldsymbol{r}) = -i\hbar\nabla_r$ is a field operator related to the speed $\boldsymbol{v}(\boldsymbol{r})$ of a superconducting carrier particle, and hence, we can write $m'\boldsymbol{v}(\boldsymbol{r}) = \boldsymbol{p}_s(\boldsymbol{r}) - e'\boldsymbol{A}(\boldsymbol{r})$. The total momentum $\boldsymbol{P}_s(\boldsymbol{r}) = n_s\boldsymbol{p}_s(\boldsymbol{r})$ can then be written as

$$\boldsymbol{P}_s(\boldsymbol{r}) = e'\Lambda\boldsymbol{J}_s(\boldsymbol{r}) + e'\boldsymbol{A}(\boldsymbol{r}). \tag{13.23}$$

Hence, (13.20a) and (13.20b) become

$$\mathrm{curl}\boldsymbol{P}_s(\boldsymbol{r}) = 0 \quad \text{and} \quad \boldsymbol{P}_s(\boldsymbol{r}) = n_s\nabla\chi'(\boldsymbol{r}),$$

where $\chi'(\boldsymbol{r})$ is a scalar function in a simply connected conductor. Nevertheless, considering $-n_s\chi'$ as the transformation gauge, the London equation (13.21) is equivalent to

$$\boldsymbol{P}_s(\boldsymbol{r}) = 0. \tag{13.24}$$

Thus, the London equation is equivalent to (13.24), implying a long-range order of momenta of superconducting charge carriers.

In multiply connected conductor, $\chi'(\boldsymbol{r})$ cannot take a unique value at $\boldsymbol{r}$. Figure 13.6 sketches a superconducting domain surrounded by a normal phase shaded in the figure. In a practical metal, the body may have a void, shaded and bordered by Σ, which can be in normal phase or an empty space. Inside the closed curve C_1, the metal is superconducting, whereas C_2 may contain a different phase inside as shown. Such C_1 and C_2 represent simply connected and multiply connected

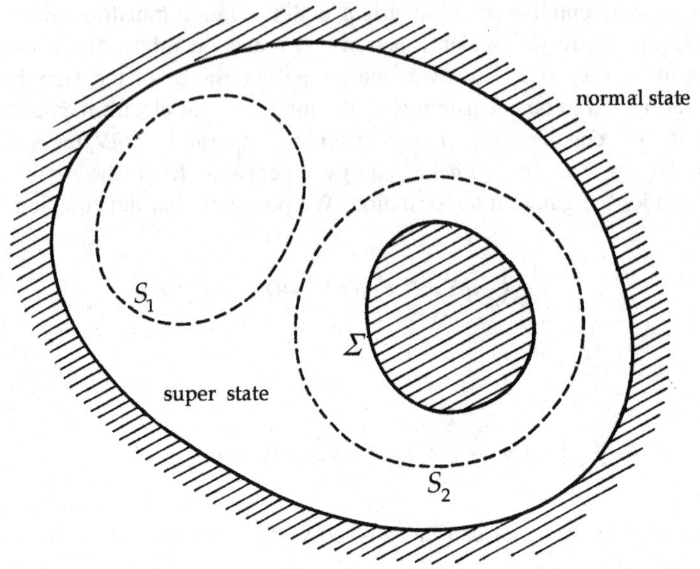

Fig. 13.6 A superconducting body surrounded by a normal conductor. A closed surface S_1 is filled by superconducting state completely, whereas the surface S_2 enclosing a void or normal state Σ. These spaces in S_1 and S_2 are singly and multiply connected.

spaces, respectively. Equation (13.23) is a formula at an arbitrary local point $\boldsymbol{r}$; hence, in integral form for C_1, we have $\int_{S_{1+}} \boldsymbol{p}_s \cdot \mathrm{d}\boldsymbol{S}_{1+} = -\int_{S_{1-}} \boldsymbol{p}_s \cdot \mathrm{d}\boldsymbol{S}_{1-}$, where $\boldsymbol{S}_{1+}$ and $\boldsymbol{S}_{1-}$ are surfaces above and below the page, respectively, both subtending the curve C_1 in common. For such a simply connected conductor, fluxes of $\boldsymbol{p}_s$ threading in and out of C_1 are equal and opposite, leading to (13.24).

A persisting current was an important finding, flowing along a superconducting coil for considerably long time of practically about a year or so. Evidenced by such persisting currents, the Meissner effect in a simply connected body can be rephrased for a multi-connected body by trapping the flux of magnetic field lines threading Σ . Choosing the curve C_2 including Σ as a doubly connected space, two integrals $\oint_{S_2} \boldsymbol{p}_s \cdot \mathrm{d}S_2$ and $\oint_{\Sigma} \boldsymbol{p}_s \cdot \mathrm{d}\Sigma$ can be calculated, but the former vanishes because of (13.24), so that only the latter contributes the total integral. In this case, it is convenient to use the magnetic induction vector $\boldsymbol{B}$, and for the trapped flux inside Σ, we write the induction law as

$$-\frac{\partial}{\partial t}\oint_{\Sigma} \boldsymbol{B} \cdot \mathrm{d}S_{\Sigma} = \oint_{\Sigma} (\mathrm{curl}\boldsymbol{E}) \cdot \mathrm{d}S_{\Sigma} = \oint_{\Sigma} \boldsymbol{E} \cdot \boldsymbol{dl}_{\Sigma},$$

where we have ignored the surface penetration of $\boldsymbol{B}$, for simplicity, and used the Stokes theorem to derive the last expression. Here, $\boldsymbol{S}_{\Sigma}$ represents a closed surface subtending the curve Σ, and $\mathrm{d}\boldsymbol{l}_\Sigma$ is the line element along Σ. Using (13.14b) to replace $\boldsymbol{E}$ by $\boldsymbol{J}_\mathrm{s}$, we obtain

$$\frac{\partial}{\partial t}\left(\oint_\Sigma \boldsymbol{B}\cdot\mathrm{d}\boldsymbol{S}_\Sigma + \Lambda\oint_\Sigma \boldsymbol{J}_\mathrm{s}\cdot\mathrm{d}\boldsymbol{l}_\Sigma\right) = 0. \tag{13.25a}$$

Defining the total flux $\Phi = \oint_\Sigma \boldsymbol{B}\cdot\mathrm{d}\boldsymbol{S}_\Sigma + \Lambda\oint_\Sigma \boldsymbol{J}_\mathrm{s}\cdot\mathrm{d}\boldsymbol{l}_\Sigma$ through the curve Σ, (13.25a) is written as

$$\frac{\partial\Phi}{\partial t} = 0 \quad \text{or} \quad \Phi = \text{const.} \tag{13.25b}$$

The first surface integral in (13.25a) can be replaced by $\int_\Sigma (\mathrm{curl}\boldsymbol{A})\cdot\mathrm{d}\boldsymbol{S}_\Sigma = \oint_\Sigma \boldsymbol{A}\cdot\mathrm{d}\boldsymbol{l}_\Sigma$, and hence, the total flux is expressed as

$$\Phi = \oint_\Sigma (\boldsymbol{A} + \Lambda\boldsymbol{J}_\mathrm{s}).\mathrm{d}\boldsymbol{l}_\Sigma. \tag{13.26}$$

For a simply connected body, we have shown in (13.20a) and (13.20b) that the vector $\boldsymbol{A} + \Lambda\boldsymbol{J}_\mathrm{s}$ can be expressed as $\nabla\chi'$, and the London equation (13.21) was obtained by choosing the London gauge, resulting in $\Phi = 0$. In a multiply connected body, on the contrary $\Phi = \text{const.}$ because of a trapped magnetic flux in Σ, for which the function χ' cannot be uniquely determined, indicating that the supercurrent density $\boldsymbol{J}_\mathrm{s}$ returns to any point on the passage repeatedly. Mathematically, the line integral in (13.25a) can be restricted to one circulation along the path Σ, where $\psi(\boldsymbol{r}') = \psi(\boldsymbol{r})$ on returning to $\boldsymbol{r}' = \boldsymbol{r}$, accompanying a phase shift $\frac{e'\chi}{\hbar} = 2\pi$. Therefore, the London gauge can be selected as $\chi' = n_\mathrm{s}\frac{2\pi\hbar}{e'}$, and in such a superconductor the trapped flux in a hole Σ can be expressed as

$$\Phi = \oint_\Sigma \nabla\chi'(\boldsymbol{r})\cdot\mathrm{d}\boldsymbol{l}_\Sigma = n_\mathrm{s}\frac{2\pi\hbar}{e'}. \tag{13.27}$$

Experimentally, it was found that Φ is determined by $e' = 2e$, indicating that each of the superconducting charge carriers consists of two electrons. Therefore, the quantized flux Φ is expressed by an integral multiple of $\Phi_\mathrm{o} = \frac{h}{2e} = 2.0678 \times 10^{-15}\,\mathrm{T\,m^2}$, namely, $\Phi = n_\mathrm{s}\Phi_\mathrm{o}$ where the unit flux Φ_o is called a *fluxoid*. Owing to trapped flux in voids in practical conductors, supercurrents can exist on their surfaces, even after removing the applied magnetic field. This explains a persisting current observed in a superconducting coil. Practical superconductors are thus signified by such trapped magnetic flux in their superconducting phases, in addition to Ohm's currents in their normal phases.

13.4 The Ginzburg–Landau Equation and the Coherence Length

In the forgoing, we discussed that the superconducting phase is characterized by an ordered momentum $\boldsymbol{P}_s(\boldsymbol{r})$ in (13.23), while in a multiply connected conductor, the flux Φ of an applied field is trapped in voids as in (13.27). The electromagnetic relations in Sect. 13.3 are local, referring to an arbitrary point in the surface layer of depth λ. Inside the layer, for distributed $\boldsymbol{J}_s(\boldsymbol{r})$ and $\boldsymbol{A}(\boldsymbol{r})$, their spatial correlations should be taken into account to determine the thermodynamic properties. It is logical to consider the wavefunction $\psi(\boldsymbol{r})$ to represent distributed order variables in terms of the distributed density as

$$n_s = \psi^*(\boldsymbol{r})\psi(\boldsymbol{r}). \tag{13.28}$$

Considering (13.28) for the local order parameter $\eta(\boldsymbol{r})$ expressed for the Gibbs function density $g(\eta)$ given by (iii) in Sect. 13.2, Ginzburg and Landau [2] wrote Gibbs' free-energy density for the superconducting state as

$$g_s = g_n + \frac{1}{2m'}|(\boldsymbol{p}_s - e'\boldsymbol{A})\psi|^2 + \alpha|\psi|^2 + \frac{1}{2}\beta\left(|\psi|^2\right)^2 - \frac{1}{2\mu_o}\boldsymbol{M}^2. \tag{13.29}$$

Here on the right side, the second term represents the *kinetic energy* of the order parameter, the third and fourth are the adiabatic potential as in Landau theory of a binary system, and the last term is the magnetic field energy density. As interpreted in Chap. 7, we consider that the third and fourth terms represent the adiabatic potential associated with lattice deformation. In contrast to (iii) in Sect. 13.2, the Ginzburg–Landau function (13.29) describes a superconducting transition in detail, as it includes the kinetic energy. It is particularly important to determine the role of long-range order in the superconducting domain coexisting with normal conducting domains.

In thermal equilibrium, the total Gibbs function $G_s = \int_V g_s \mathrm{d}V$ should be minimized against arbitrary variations of ψ and $\boldsymbol{A}$. Hence, from the relation $\delta G_s = 0$, we obtain

$$\delta g_s = \left(\frac{\partial g_s}{\partial \psi}\right)_{\boldsymbol{A}} \delta\psi + \left(\frac{\partial g_s}{\partial \boldsymbol{A}}\right)_{\psi} \cdot \delta\boldsymbol{A} = 0.$$

Hence, from $\left(\frac{\partial g_s}{\partial \psi}\right)_{\boldsymbol{A}} = 0$ and $\left(\frac{\partial g_s}{\partial \boldsymbol{A}}\right)_{\psi} = 0$, we can derive the equations

$$\frac{1}{2m'}(-i\hbar\nabla - e'A)^2\psi + (\alpha + \beta|\psi|^2)\psi' = 0 \tag{13.30}$$

and

$$J_s(r) = -\frac{i\hbar e'}{2m'}(\psi^*\nabla\psi - \psi\nabla\psi^*) - \frac{e'^2}{m'}\psi^*\psi A, \tag{13.31}$$

respectively. Equation (13.30) is called the Ginzburg–Landau equation, and (13.31) provides the formula for the supercurrent density.

In the absence of an applied field, that is, $A = 0$, Equation (13.30) is identical to (7.1a) that is a nonlinear equation of propagation in an adiabatic potential $\Delta U = \alpha\psi + \beta|\psi|^2\psi$. For simplicity, we write one-dimension wavefunction as $\psi(x)$ at a given time t, for which (13.30) can be expressed as

$$\frac{\hbar^2}{2m'}\frac{d^2\psi}{dx^2} + \alpha\psi + \beta|\psi|^2\psi = 0, \tag{13.32}$$

where we consider $\alpha < 0$ and $\beta > 0$ for $T < T_c$. In fact, the differential equation of this type was already discussed in Sect. 7.1, and hence, the previous results can be used directly for the present argument. We are in fact interested in such a solution expressed in a form $\psi = \psi_0 f(\phi)$, where $\phi = kx$ represents the phase of propagation.

We set the boundary conditions $\psi = 0$ at $x = 0$ and $\psi =$ const. at $x = \infty$, signifying the threshold and complete superconducting order, respectively. Near at $x = 0$, we can assume ψ is very small, so that (13.32) for the threshold is approximately expressed as

$$\frac{\hbar^2}{2m'}\frac{d^2\psi}{dx^2} + \alpha\psi = 0. \tag{13.33}$$

This is a simple harmonic oscillator equation, whose solution for the lowest energy is given as $\psi \sim \exp i(x/\xi)$, where $\xi = \sqrt{\frac{\hbar^2}{2m'|\alpha|}}$.

For $x \to \infty$, on the other hand, we consider that the nonlinear term $\beta|\psi|^2\psi$ in (13.32) is significant for a finite amplitude ψ_0. Letting $\left(\frac{d\psi}{dx}\right)_{x\to\infty} = \psi_0$ in the nonlinear equation of (13.32), we obtain $\psi_0 = \sqrt{|\alpha|/\beta}$ and the solution

$$\psi(x) = \sqrt{\frac{|\alpha|}{\beta}}\tanh\frac{x}{\sqrt{2}\xi}.$$

Here, the parameter ξ signifies an approximate distance for the density $n_s = |\psi|^2$ to reach a plateau in the direction x, which is a significant measure for the supercurrent in Type 1 superconductor. By definition, ξ depends on $m'|\alpha|$ that is an intrinsic constant of the superconductor and is not the same as the penetration depth λ of an applied field.

Using $|\psi|^2 = \frac{|\alpha|}{\beta}$ in (13.29), we obtain the relation

$$g_s - g_n = -\frac{\alpha^2}{2\beta},$$

which should be representing an ordered state of the superconductor. Since this relation should be identical to (13.6), we have $\frac{\alpha^2}{2\beta} = \frac{1}{2}\mu_0 H_c^2$; hence,

$$H_c = \sqrt{\frac{\alpha^2}{\mu_0 \beta}}. \tag{13.34}$$

The corresponding supercurrent density (13.31) can be expressed as

$$\boldsymbol{J}_s(\boldsymbol{r}) = -\frac{e'^2}{m'}|\psi|^2 \boldsymbol{A},$$

which is London equation (13.20), and therefore, the penetration depth λ can be calculated as

$$\lambda = \sqrt{\frac{m'\beta}{\mu_0 e'^2 \alpha}}. \tag{13.35}$$

On the other hand, (13.30) can be written for the transition threshold as

$$\frac{1}{2m'}\left(-i\hbar\nabla - e'\boldsymbol{A}\right)^2\psi = \alpha\psi.$$

If the magnetic field B applied parallel to the y axis, this equation is expressed as

$$-\frac{\hbar}{2m'}\left(\frac{\partial^2}{\partial x^2} + \frac{\partial^2}{\partial z^2}\right)\psi + \frac{1}{2m'}\left(i\hbar\frac{\partial}{\partial y} + e'B\right)^2\psi = \alpha\psi.$$

Setting $\psi = \psi_0(x)\exp i\left(k_y y + k_z z\right)$, we obtain

$$\frac{1}{2m'}\left\{-\hbar^2\frac{d^2}{dx^2} + \hbar^2 k_z^2 + \left(\hbar k_y - e'Bx\right)^2\right\}\psi_0 = \sigma\psi_0,$$

which can be modified as

$$\frac{1}{2m'}\left\{-\hbar^2\frac{d^2}{dx^2} + e'^2B^2x^2 - 2\hbar k_y e'Bx\right\}\psi_0 = \left(\alpha - \frac{\hbar^2\left(k_y^2 + k_z^2\right)}{2m'}\right)\psi_0.$$

Using a coordinate transformation $x - \frac{\hbar k_y e' B}{2m'} \rightarrow x'$, this equation can be expressed as a harmonic oscillator equation, that is,

$$\left(-\frac{\hbar^2}{2m'}\frac{d^2}{dx'^2} - \frac{1}{2}m'\omega_L^2 x'^2\right)\psi_o = \left(\alpha - \frac{\hbar^2 k_z^2}{2m'}\right)\psi_o,$$

where $\omega_L = \frac{e'B}{m'}$. Therefore, the threshold H_{c2} in a Type 2 superconductor is determined by the lowest eigenvalue of the above equation, that is, α, if $k_z = 0$. That is, from the relation $\frac{1}{2}\hbar\omega_L = \alpha$, we can solve this equation for $B = \mu_o H_{c2}$. Combining the relations (13.33), (13.36), and (13.34) for ξ, λ, and α vs. H_c, respectively, we can derive the equation

$$H_{c2} = \sqrt{2}\kappa H_c \quad \text{where} \quad \kappa = \frac{\lambda}{\xi}. \tag{13.36}$$

When $\kappa > \frac{1}{\sqrt{2}}$, we have $H_{c2} > H_c$, characterizing a Type 2 superconductor. The ratio κ is thus a significant parameter to determine the type of superconductors: Type 1 or 2 depending on the value of κ, either $\kappa < \frac{1}{\sqrt{2}}$ or $\kappa > \frac{1}{\sqrt{2}}$, respectively.

It is interesting that we can write H_{c2} in terms of the quantized flux $\Phi = n_s\Phi_o$, where $\Phi_o = \frac{2\pi\hbar}{e'}$. Combining relations (13.36) and (13.34) with the unit flux Φ_o, we can state that

$$H_{c2} = \frac{2m'\alpha}{e'\hbar} \cdot \frac{e'\Phi_o}{2\pi\hbar} \cdot \frac{\hbar^2}{2m'\xi^2} = \frac{\Phi_o}{2\pi\xi^2}. \tag{13.37}$$

We note that H_{c2} is equal to the unit flux per area $2\pi\xi^2$ that signifies the ordered area at the superconducting threshold.

Exercise 13

1. Consider a superconducting plate of large area and thickness δ, where the penetration of a magnetic field can be along the normal direction x. Show that

$$B(x) = B_o \frac{\cosh\frac{x}{\lambda}}{\cosh\frac{\delta}{2\lambda}}.$$

   The effective magnetization $M(x)$ in the plate is given by $B(x) - B_o = M(x)$. Derive the equation $M(x) = -B_o\frac{\delta^2 - 4x^2}{8\lambda^2}$.
2. In a superconductor at temperatures in the range $0 < T < T_c$, we consider the normal and supercurrents, j_n and j_s, obeying the Ohm law and the London equation, respectively.

Consider a plane electromagnetic wave $(\boldsymbol{E}, \boldsymbol{B}) \propto \exp i(\boldsymbol{k} \cdot \boldsymbol{r} - \omega t)$ propagating through the superconductor. Using the Maxwell equations, derive the dispersion relation between $\boldsymbol{k}$ and ω that can be expressed as

$$c^2k^2 = \frac{\sigma\omega}{\varepsilon_o} i - \frac{c^2}{\lambda^2} + \omega^2.$$

Here, σ and λ are the normal conductivity and superconducting penetration depth, respectively.

3. Consider an infinitely long cylindrical superconductor, where the magnetic induction B has a cylindrical symmetry. Write the penetration equation $\boldsymbol{B} - \lambda^2 \nabla^2 \boldsymbol{B} = 0$ with cylindrical coordinates (ρ, θ, z) and show

$$B(\rho) = \frac{\Phi_0}{2\pi\lambda^2} \ln \frac{\lambda}{\rho} \quad \text{for} \quad \xi < \rho < \lambda$$

$$B(\rho) = \frac{\Phi_0}{2\pi\lambda^2} \sqrt{\frac{\pi\lambda}{2\rho}} \exp \frac{-\rho}{\lambda} \quad \text{for} \quad \lambda < \rho.$$

Here

$$\Phi_0 = 2\pi \int_0^\infty B(\rho)\rho d\rho$$

is the total flux.

References

1. C.G. Kuper, *An Introduction to the Theory of Superconductivity* (Oxford Univ. Press, New York, 1968)
2. C. Kittel, *Introduction to Solid State Physics*, 6th edn. (Wiley, New York, 1986)

Chapter 14
Theories of Superconducting Transitions

Apart from exotic electromagnetic phenomena, observed critical temperatures T_c of superconducting transitions depend on the isotopic mass M composing the lattice, providing evidence for the lattice to be involved in the transition mechanism. On the other hand, microscopic electron–electron correlations are essential in the multi-electron system; in addition, Fröhlich proposed electron–lattice interactions. Based on his proposal, Cooper elaborated an electron-pair model for superconducting charge carriers. Bardeen, Cooper, and Schrieffer then established the theory of superconducting transitions, constituting the objective in this chapter. The superconducting transition in metals can now be interpreted in terms of Cooper pairs, which is analogous to pseudospin clusters for structural phase changes in insulating crystals.

14.1 The Fröhlich's Condensate

Electron scatterings are the essential mechanism for interactions with the lattice in the superconducting state. It is noted that electron scatterings by a harmonic lattice are elastic, where the energy and momentum are conserved. In such elastic scatterings, electrons behave like *free particles* with no loss of energy, so that the lattice remains as *rigid*. In contrast, the phonon-modulated lattice is strained, where electron scatterings can be inelastic collisions, exchanging energy and momentum with the lattice. At low temperatures, lattice vibrations in acoustic mode are slow in timescale of observation, so that the lattice distortion is detectable, as ionic displacements occur in finite magnitude.

Denoting an ionic displacement by $\boldsymbol{u}$ from a regular lattice point $\boldsymbol{r}_o$, the displaced lattice coordinate can be expressed as $\boldsymbol{r} = \boldsymbol{r}_o + \boldsymbol{u}$; thereby the lattice potential $V(\boldsymbol{r})$ is modified in the first order as

$$V(\boldsymbol{r}) = V(\boldsymbol{r}_o) + \boldsymbol{u} \cdot \boldsymbol{\Delta} V(\boldsymbol{r}_o), \tag{14.1a}$$

M. Fujimoto, *Thermodynamics of Crystalline States*,

DOI 10.1007/978-1-4614-5085-6_14,

where $V(\boldsymbol{r}_o)$ represents the rigid lattice that is characterized by the space group. In a modulated lattice, on the other hand, the displacement $\boldsymbol{u}(\boldsymbol{r})$ is expressed by Fourier series

$$\boldsymbol{u}(\boldsymbol{r}) = \sum_{\boldsymbol{q}} \boldsymbol{u}_{\boldsymbol{q}} \exp i(\boldsymbol{q} \cdot \boldsymbol{r} - \omega t), \tag{14.1b}$$

where $\boldsymbol{q}$ is the wavevector of a lattice excitation and the lattice symmetry holds no longer in principle. In Chap. 2, lattice vibrations at small $\boldsymbol{q}$ were discussed as a vibration *field*, which is quantum mechanically equivalent to a large number of phonons $(\hbar\omega, \hbar\boldsymbol{q})$. Assuming that the lattice is constructed with mass points M, the Fourier transform $\boldsymbol{u}_{\boldsymbol{q}}$ is expressed with phonon creation and annihilation operators, $b_q^\dagger$ and b_q, as

$$\boldsymbol{u}_{\boldsymbol{q}} = \frac{\boldsymbol{q}}{q}\sqrt{\frac{2M\omega}{\hbar}}\left(b_q^\dagger - b_q\right). \tag{14.1c}$$

The Hamiltonian of electrons and the hosting lattice can be written as

$$\mathsf{H} = \mathsf{H}_{\mathrm{el}} + \mathsf{H}_{\mathrm{L}} + \mathsf{H}_{\mathrm{int}} + \mathsf{H}_{\mathrm{coul}}, \tag{14.2}$$

where the electronic term is given by (12.13), that is,

$$\mathsf{H}_{\mathrm{el}} = \sum_{\boldsymbol{k}} \varepsilon(\boldsymbol{k}) a_{\boldsymbol{k}}^\dagger a_{\boldsymbol{k}}, \tag{14.3a}$$

composing of many *one-electron modes* specified by the wavevector $\boldsymbol{k}$ and kinetic energy $\frac{\hbar^2 k^2}{2m}$ where m is the effective mass. The lattice term is given by the phonon energy

$$\mathsf{H}_{\mathrm{L}} = \sum_{\boldsymbol{q}} (\hbar\omega_q) b_q^\dagger b_q. \tag{14.3b}$$

The Coulomb interaction term $\mathsf{H}_{\mathrm{coul}}$ is added in (14.2), however ignorable if justified by the shielding effect in a conductor of packed charges. We assume the Bloch function for each electron to behave like a free particle in the lattice in the first approximation. Hence, the *one-electron wavefunction* is expressed as $\psi_k(\boldsymbol{r}) = \psi_o \exp i(\boldsymbol{k} \cdot \boldsymbol{r} + \varphi_k)$, where the phase constants φ_k are randomly distributed in the $\boldsymbol{k}$-space.

The interaction term $\mathsf{H}_{\mathrm{int}}$ arising from the term $\boldsymbol{u} \cdot \nabla V(\boldsymbol{r}_o)$ in (14.1a) is signified by such a matrix element as

$$\int_{\Omega} \langle u_q' | \psi_k'(\boldsymbol{r}) \boldsymbol{u} \cdot \nabla V(\boldsymbol{r}_o) \psi_k(\boldsymbol{r}) | u_q \rangle \mathrm{d}^3\boldsymbol{r}. \tag{14.4}$$

Here $\langle u_q'|$ and $|u_q\rangle$ are phonon wavefunctions before and after scatterings. Because of the sinusoidal nature of the wavefunction of electrons, the integral $\int_\Omega \psi_{k'}{}^*(\boldsymbol{r})\boldsymbol{u}\cdot\nabla V(\boldsymbol{r}_0)\psi_k(\boldsymbol{r})\mathrm{d}^3\boldsymbol{r}$ does not vanish if $\boldsymbol{k}'-\boldsymbol{k}=\boldsymbol{q}$, which is detectable at low frequencies if observed at $\boldsymbol{r}_0$ in sufficiently long timescale of observation. In this case, $\mathsf{H}_{\mathrm{int}}$ can be expressed as proportional to a delta function $\delta(\boldsymbol{k}'-\boldsymbol{k}-\boldsymbol{q}+\boldsymbol{G})$, where $\boldsymbol{G}$ is any reciprocal vector of the lattice. Hence, in (14.4) we assume $\boldsymbol{G}=0$ for simplicity, ignoring all other $\boldsymbol{G}\neq 0$.

The matrix elements (14.4) have undetermined factors $\exp i\Delta\varphi_{\mathrm{k}}$ due to the phase difference $\Delta\varphi$ in the scattering process in a crystal; however, it can be maximized at $\Delta\boldsymbol{k}=\boldsymbol{q}$, if we have a phasing process $\Delta\varphi_{\mathrm{k}}\rightarrow 0$. Following Born and Huang, we assume such a phasing process for minimizing structural strains as $\Delta\varphi\rightarrow 0$; $\mathsf{H}_{\mathrm{int}}$ can then represent *condensates* of electrons in phase with the lattice excitations, achieving $\Delta\boldsymbol{k}=\boldsymbol{q}$ at $\Delta\varphi_k=0$. Regarding the phonon states, in this case, the above matrix element can be off-diagonal, that is, $\langle u_q'|\ldots|u_q\rangle\neq 0$, where the time-dependent $\mathsf{H}_{\mathrm{int}}$ is thermally stabilized by phasing at a given temperature, forming stable condensates in the modified lattice. Abrupt rise in the specific heat at T_{c} can be attributed to such a phasing process.

Following Fröhlich [1], $\mathsf{H}_{\mathrm{int}}$ at a given time can be expressed as

$$\mathsf{H}_{\mathrm{int}}=-\int_\Omega \mathrm{d}^3\boldsymbol{r}\{\rho(\boldsymbol{r})\boldsymbol{u}(\boldsymbol{r})\cdot\nabla V(\boldsymbol{r})\} \tag{14.5}$$

where $\rho=\sum_{\mathrm{i}}\rho_q\exp(-i\boldsymbol{q}\cdot\boldsymbol{r}_{\mathrm{i}})$ represents the modulated electron density to interact with $\boldsymbol{u}(\boldsymbol{r})$ in phase. The Fourier transform, $\rho_q=\sum_{\mathrm{i}}\rho(\boldsymbol{r}_{\mathrm{i}})\exp(i\boldsymbol{q}\cdot\boldsymbol{r}_{\mathrm{i}})/\Omega$, is then expressed as $\rho(\boldsymbol{r})\exp(i\boldsymbol{q}\cdot\boldsymbol{r})/\Omega$ for a small $|\boldsymbol{q}|$, and we can write

$$\rho_q=\sum_k a_{k+q}^{\dagger}a_k \tag{14.6}$$

in the second quantization scheme. The energy $\mathsf{H}_{\mathrm{int}}$ can therefore be expressed in terms of ρ_q as

$$\mathsf{H}_{\mathrm{int}}=i\sum_q D_q\left(\rho_q b_q^{\dagger}-\rho_q^{\dagger}b_q\right), \tag{14.7a}$$

where D_q is determined by the phonon spectra as

$$D_q^2=\frac{4}{9}\frac{C^2\hbar\omega_q}{NM\Omega v_0^2} \tag{14.7b}$$

under isothermal conditions. Here, $C=\frac{\hbar^2}{2m}\int_\Omega|\nabla\Psi_0|^2\mathrm{d}^3\boldsymbol{r}$ is constant, if Ψ_0 is a Bloch function, where v_0 is the velocity of the phonon mode of $\boldsymbol{q}$. If otherwise, considering H_{int} as a perturbation, we can leave D_q generally just as a temperature-dependent constant.

In the second quantization scheme, the Hamiltonian H represents the total ordering process, where H_{int} represents a perturbation from condensates, which we shall call Fröhlich's condensate. The phasing can be attributed to synchronizing electron scatterings at a given temperature, which are driven by the adiabatic potential of dynamic lattice displacements.

14.2 The Cooper Pair

Due to *off-diagonal* elements of displacements accompanied with distributed $\Delta\varphi$, Fröhlich's condensates in crystals cannot be independent of each other at low temperatures. We assume a thermodynamic process for the two condensates to bound together signified by inversion symmetry $\boldsymbol{q} \rightleftarrows -\boldsymbol{q}$. However, for a randomly selected Fröhlich's condensates i and j, the vectors $\boldsymbol{q}$ and $-\boldsymbol{q}$ are specified with uncertain phases $\Delta\varphi_i$ and $\Delta\varphi_j$; for such pairs to be considered as *identical* charge carriers, these phases should be synchronized with $\boldsymbol{u}$ in a slow passage on decreasing temperature. While keeping the factor D_q as temperature-dependent factor in (14.7a), we consider such a phasing process for a stable pair of electrons, by analogy to pseudospin clusters in binary systems.

We write the Hamiltonian for identical pairs of electrons as $\tilde{\mathsf{H}}$, which should be different from H for independent Fröhlich's condensates, but $\tilde{\mathsf{H}}$ is derivable from H by a canonical transformation with an action variable S that represents the thermal process. We write the Hamiltonian (14.2) as

$$\mathsf{H} = \mathsf{H}_o + \lambda \mathsf{H}' \qquad \text{(i)}$$

where $\lambda\mathsf{H}'$ replaces H_{int} for convenience. The canonical transformation

$$\tilde{\mathsf{H}} = \exp(-S)\mathsf{H} \exp S \qquad \text{(ii)}$$

converts H to $\tilde{\mathsf{H}}$, for which the action S can be attributed to an adiabatic potential for minimal lattice strains.

Expanding $\tilde{\mathsf{H}}$ into a power series, we have

$$\tilde{\mathsf{H}} = \mathsf{H} + [\mathsf{H}, S] + \frac{1}{2}[[\mathsf{H}, S], S] + \cdots$$

Using (i) for H, this expression can be written for a small S as

$$\tilde{\mathsf{H}} = \mathsf{H}_o + \lambda\mathsf{H}' + [\mathsf{H}_o, S] + [\lambda\mathsf{H}', S].$$

If S is determined by such a way that

$$\lambda \mathrm{H}' + [\mathrm{H_o}, S] = 0, \tag{iii}$$

we obtain

$$\tilde{\mathrm{H}} = \mathrm{H_o} + [\lambda \mathrm{H}', S]. \tag{iv}$$

Since $\mathrm{H_o}$ is considered as diagonal, from (iii) we can write matrix elements of S as

$$\langle 1_q|S|0_q\rangle = -iD_q \sum_k \frac{a^\dagger_{k-q}a_k}{\varepsilon_k - \varepsilon_{k-q} - \hbar\omega_q} \quad \text{and}$$
$$\langle 0_q|S|1_q\rangle = iD_q \sum_{kk'} \frac{a^\dagger_{k'+q}a_{k'}}{\varepsilon_{k'} - \varepsilon_{k'+q} + \hbar\omega_q} \tag{v}$$

between two phonon states $|1_{\boldsymbol{q}}\rangle$ and $|0_{\boldsymbol{q}}\rangle$, expressing occupied and unoccupied states by $\hbar\omega_{\boldsymbol{q}}$. With these expressions, $\tilde{\mathrm{H}}$ can be expressed

$$\tilde{\mathrm{H}} = \mathrm{H_o} + \frac{D_q^2}{2}\sum_{k,k'} a^\dagger_{k'+q}a_{k'}a^\dagger_{k-q}a_k\left(\frac{1}{\varepsilon_k - \varepsilon_{k-q} - \hbar\omega_q} - \frac{1}{\varepsilon_{k'} - \varepsilon_{k'+q} + \hbar\omega_q}\right). \tag{vi}$$

In (v), scatterings $\boldsymbol{k} \to \boldsymbol{k} - \boldsymbol{q}$ and $\boldsymbol{k}' + \boldsymbol{q} \to \boldsymbol{k}'$ can be considered as *emission* and *absorption* of a phonon $(\hbar\omega_q, \boldsymbol{q})$ between phonon states $\boldsymbol{u}_q$ and $\boldsymbol{u}_{-q}$, respectively. These terms in (vi) show singular behaviors, if $|\Delta\boldsymbol{k}| = |\boldsymbol{q}|$ and $-\Delta\boldsymbol{k} \approx \Delta\boldsymbol{k}'$, which are achieved after thermal phasing. Such a scattering process is illustrated by a vector diagram in Fig. 14.1a.

Considering that the off-diagonal elements $a^\dagger_{k-q}a_k$ and $a^\dagger_{k'+q}a_{k'}$ of the electron density matrices are associated with the coupling process, $\tilde{\mathrm{H}}$ represents the pair of electrons as a bound object, for which the momentum conservation restricts the coupling to a specific scattering $\boldsymbol{k}' = -\boldsymbol{k}$. In this case, illustrated in Fig. 14.1b, the binding energy of two electrons can be determined from (vi) as

$$\tilde{\mathrm{H}} = D_q^2 \sum_{q,k} \frac{\hbar\omega_q}{(\varepsilon_{k-q} - \varepsilon_k)^2 - (\hbar\omega_q)^2} a^\dagger_{k-q}a_k a^\dagger_{-k+q}a_{-k}, \tag{14.8a}$$

which can be written as

$$\tilde{\mathrm{H}} = -\sum_{k,q} V(\boldsymbol{k}, \boldsymbol{q}) a^\dagger_{k-q}a_k a^\dagger_{-k+q}a_{-k}, \tag{14.8b}$$

where

$$-V(\boldsymbol{k}, \boldsymbol{q}) = D_q^2 \frac{\hbar\omega_q}{(\varepsilon_{k-q} - \varepsilon_k)^2 - (\hbar\omega_q)^2}. \tag{14.8c}$$

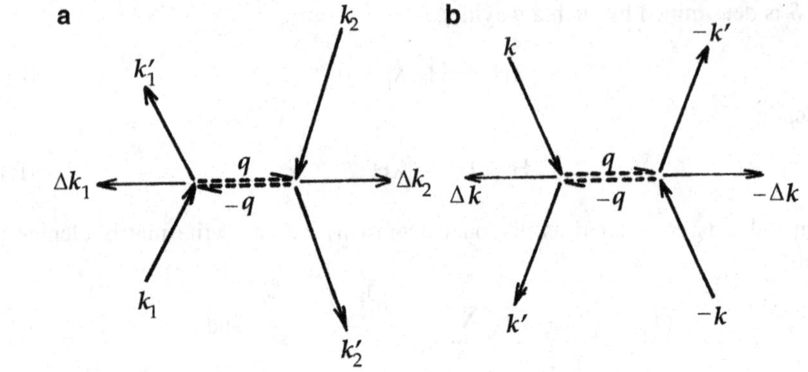

Fig. 14.1 (**a**) A coupled Fröhlich's condensates. (**b**) A Cooper pair.

The interaction $\tilde{H}$ is attractive for $V(\boldsymbol{k}, \boldsymbol{q})>0$, signifying that such an electron pair can be a charge carrier $2e$ in the superconducting state. In this case, from (14.8c), we should have

$$\left|\varepsilon_{k\pm q} - \varepsilon_k\right| < \hbar\omega_q; \tag{14.8d}$$

otherwise, $\tilde{H}$ is positive and repulsive. By definition, the constant D_q is proportional to $\hbar\omega_q$, and (14.8d) implies that one-electron energies ε_k are modulated by an amount $\Delta\varepsilon_k$ smaller than $\hbar\omega_q$, to keep $\tilde{H}$ as attractive. In the limit of $\Delta\varepsilon_k \to \hbar\omega_q$, we obtain an intense coupling of two electrons (14.8c).

However, $\boldsymbol{k}'$and $\boldsymbol{k}$ are distributed with the scattering geometry $\mathbf{k}' = \boldsymbol{k} \pm \boldsymbol{q}$ for a given $\boldsymbol{q}$, thereby modulating the spherical Fermi surface ε_F by $\pm\hbar\omega_q$. Therefore, $\tilde{H}$ is responsible for thermal instability of the Fermi surface for the particular scatterings $\Delta\boldsymbol{k} = -\Delta\boldsymbol{k}'$. Figure 14.2 shows schematically a Fermi sphere in the $\boldsymbol{k}$-space modulated by $\pm\hbar\omega_q$, leading to a *singularity* when $\Delta\varepsilon_k = \hbar\omega_q$. With singular scatterings in Fig. 14.1b, Cooper [2] proposed that this specific interaction produces a *pseudo-particle* of two electrons, called *Cooper pairs,* playing the essential role in superconducting states. Supported by experimental evidence, Cooper pairs are identical pseudo-particles, whose number is undetermined but considerable for the superconducting phase. Bardeen, Cooper, and Schrieffer [3] elaborated the theory of superconducting transitions, considering Cooper pairs as charge carriers.

It is realized that a Cooper pair can be treated as a single particle of charge $e' = 2e$ and mass $m' = 2m$, whose motion can be discussed relative to the lattice. Placed at the fixed center of mass, a Cooper pair is a single particle in free space; however, with respect to the fixed lattice, an adiabatic potential must be considered.

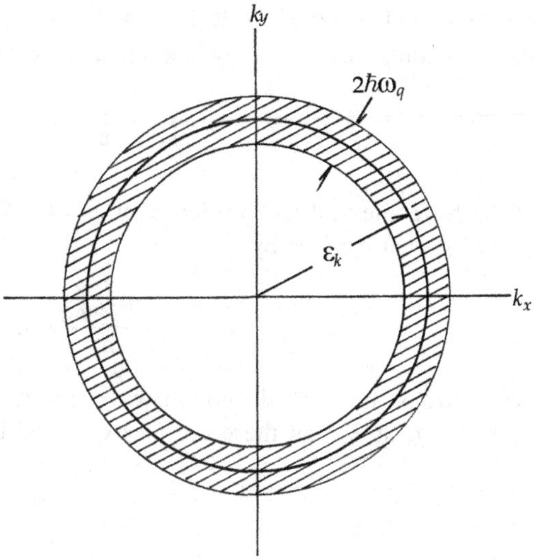

Fig. 14.2 A one-electron energy surface $\varepsilon_\mathbf{k}$ modulated by Debye phonons in $k_x k_y$ space.

It is noted that the Cooper pair can be characterized by inversion $\boldsymbol{q} \rightleftarrows -\boldsymbol{q}$, for which a binary order variable can be defined. In this case, a superconducting transition can be described via inversion of a binary transition for two Fröhlich's condensates. In fact, Anderson showed that such an order variable can be defined from the Bardeen–Cooper–Schrieffer theory.

14.3 Critical Anomalies and the Superconducting Ground State

14.3.1 Critical Anomalies and a Gap in a Superconducting Energy Spectrum

As characterized by $\Delta\boldsymbol{k}' = -\Delta\boldsymbol{k}$, the Cooper pair was described with respect to the center-of-mass coordinate system, where $\boldsymbol{k}' + \boldsymbol{k} = 0$. However, with respect to the lattice, $\boldsymbol{k}' + \boldsymbol{k}$ cannot be equal to zero, for which an adiabatic potential is responsible. Considering such pairs to emerge at the threshold of a superconducting phase, the transition can be characterized by inversion symmetry $\boldsymbol{q} \rightleftarrows -\boldsymbol{q}$.

The transition is initiated by the potential $-\nabla V(\boldsymbol{r}_\mathrm{o})$ that emerges at T_c, where critical anomalies appear as related to fluctuations, as discussed in Sect. 6.3. It is noted that the Cooper pair is associated with the symmetric mode of fluctuations $\frac{1}{\sqrt{2}}\left(u_q + u_{-q}\right)$; hence, we have the antisymmetric mode $\frac{1}{\sqrt{2}}\left(u_q - u_{-q}\right)$ for uncoupled electrons. The wavefunction of the Cooper pair is symmetrically modulated by the lattice wave $u_{\pm q} \sim \exp(\pm i\boldsymbol{q} \cdot \boldsymbol{r})$, so that the modulated wavefunction of an electron pair is given by $\frac{1}{\sqrt{2}}\psi_k(u_q + u_{-q})$. Therefore, the corresponding kinetic energy is modulated between ε_{k+q} and ε_{k-q}, and

$$\varepsilon_{k\pm q} - \varepsilon_k = \pm\frac{\hbar^2 kq}{m}.$$

In this context, the Fermi level for $k = k_F$ is split into two, showing an *energy gap* at T_c that is given by

$$2\frac{\hbar^2 k_F q_o}{m} = E_g,$$

where q_o is the critical phonon wavevector. The coherence length defined by $\xi_o = 1/2q_o$ at T_c can therefore be expressed by

$$\xi_o = \frac{\hbar^2 k_F}{mE_g}. \tag{14.9}$$

Clearly, the lattice modulation is responsible for the coherence length ξ_o, however not defined as related to the lattice in Ginzburg–Landau theory. The critical singularity is due to the discontinuity in forming Cooper pairs, similar to clusters in structural transitions.

For a Cooper pair, we consider scattering processes signified by $\boldsymbol{k} \to \boldsymbol{k}' = \boldsymbol{k} + \boldsymbol{q}$ and $-\boldsymbol{k} \to -\boldsymbol{k}' = -\boldsymbol{k} - \boldsymbol{q}$, with respect to the center-of-mass coordinates. Therefore, in the relative coordinate frame of reference, the wavefunctions $\psi_k(\boldsymbol{r})$ and $\psi_{k'}(\boldsymbol{r})$ must be functions of $\boldsymbol{r} = \boldsymbol{r}_1 - \boldsymbol{r}_2$ with respect to the fixed center of mass, that is, $\boldsymbol{r}_o = 0$. We can write Schrödinger equation

$$\left\{\frac{\hbar^2}{2m}\left(\boldsymbol{k}^2 + \sum_{\boldsymbol{k}'}\boldsymbol{k}'^2\right) + \tilde{\mathrm{H}}\right\}\psi_{k,k'}(r) = \lambda\psi_{\boldsymbol{k},\boldsymbol{k}'}(\boldsymbol{r}),$$

where λ is the eigenvalue and $\psi_{k,k'}(r) = \sum_{\boldsymbol{k}} \alpha_k \exp i\boldsymbol{k}\cdot\boldsymbol{r}$. Denoting the unperturbed energy by $E_{\boldsymbol{k}}$, we have the secular equation

$$(E_{\boldsymbol{k}} - \lambda)\alpha_k + \sum_{k'} \alpha_{k'}\left\langle \boldsymbol{k}, -\boldsymbol{k}\middle|\tilde{\mathrm{H}}\middle|\boldsymbol{k}', -\boldsymbol{k}'\right\rangle = 0,$$

where $\boldsymbol{k} = \boldsymbol{k}' - \boldsymbol{q}$ and $-\boldsymbol{k} = -\boldsymbol{k}' + \boldsymbol{q}$. The perturbation $\tilde{\mathrm{H}}$ in (13.8b) is related with distributed $\boldsymbol{k}'$ deviated from $\boldsymbol{k}_F$ in a small range limited by the Debye cutoff frequency ω_D. In this case, the second term in the above equation is assumed as a constant C in the first approximation, so that we obtain $\alpha_k = \frac{C}{E_k - \lambda}$. As the second approximation, the summation is replaced by an integral over the corresponding energy $E_{k'}$, that is,

$$(E_k - \lambda)\alpha_k = -V \int_{E_F}^{E_k} \alpha_{k'} \rho_{k'} dE_{k'},$$

where $E_{k'} - E_F = \Delta < \hbar\omega_D$. Assuming $\rho_{k'} \approx \rho_F$, we obtain

$$\frac{1}{\rho_F V} = \int_{E_F}^{E_{k'}} \frac{dE_{k'}}{E_{k'} - \lambda} = \ln\frac{E_{k'} - \lambda}{E_F - \lambda} = \ln\frac{E_{k'} - E_F + \Delta}{\Delta}.$$

If $E_{k'} - E_F = \hbar\omega_D$, we have the expression

$$\Delta = \frac{2\hbar\omega_D}{\exp\frac{1}{\rho_F V} - 1}, \tag{14.10}$$

representing the binding energy of a Cooper pair with respect to the Fermi level.

14.3.2 Order Variables in Superconducting States

Bardeen, Cooper, and Schrieffer published the theory of superconducting transitions in 1957, assuming that Cooper pairs are dominant charge carriers in the superconducting phase. Hereafter, their theory is called the BCS theory.

We write the Hamiltonian for interacting electrons as expressed by

$$\mathsf{H} = \sum_k \left(\varepsilon_k a_k^\dagger a_k + \varepsilon_{-k} a_{-k}^\dagger a_{-k} \right) - \sum_{k',k} V(\boldsymbol{k}, -\boldsymbol{k}) a_{k'}^\dagger a_{-k'}^\dagger a_{-k'} a_k$$

where $\varepsilon_k = \varepsilon_{-k}$ is eigenvalues of a single electron at $\boldsymbol{k}$ and $-\boldsymbol{k}$ on the Fermi surface. Here one-electron energy is degenerated ε_k, referring to the Fermi energy as zero. The above Hamiltonian can be reexpressed by

$$\mathsf{H} = \sum_{k,k'} \varepsilon_k a_k^\dagger a_k - \sum_{k',k;q} V(\boldsymbol{k}', \boldsymbol{k}; \boldsymbol{q}) a_{k'+q}^\dagger a_{k-q}^\dagger a_{k'} a_k. \tag{i}$$

The wavefunction can be expressed as $\psi(\ldots, n_k, \ldots, n_{k'}, \ldots)$, where n_k, $n_{k'}$ take values either 1 or 0, signifying that one-particle states $\boldsymbol{k}$ and $\boldsymbol{k}'$ are either occupied or unoccupied, respectively [4]. On the other hand, in BCS theory the Cooper pair is described by a two-particle wavefunction $\psi(n_k, n_{k'})$, and hence using number operators $n_k = a_k^\dagger a_k$ and $n_{-k}^\dagger = a_{-k}^\dagger a_{-k}$, we can rewrite (i) for the BCS Hamiltonian H_{BCS} as

$$\mathsf{H}_{BCS} = -\sum_k (1 - n_k - n_{-k})\varepsilon_k - V \sum_{k',k} a_{k'}^\dagger a_{-k'}^\dagger a_{-k} a_k, \tag{ii}$$

where V is a constant. Applying (ii) to $\psi(n_k, n_{-k})$ in the two-particle subspace, we have paired states for $n_k = n_{-k}$, corresponding to $\boldsymbol{k}$ and $-\boldsymbol{k}$, where either both of one-electron states are occupied or unoccupied, and hence,

$$(1 - n_k - n_{-k})\psi(1_k, 1_{-k}) = -\psi(1_k, 1_{-k}) \quad \text{and}$$
$$(1 - n_k - n_{-k})\psi(0_k, 0_{-k}) = \psi(0_k, 0_{-k}).$$

Therefore, expressing these paired states by column matrices $\begin{pmatrix} 0 \\ 1 \end{pmatrix}$ and $\begin{pmatrix} 1 \\ 0 \end{pmatrix}$, respectively, the operator $1 - n_k - n_{-k}$ can be expressed by a matrix

$$1 - n_k - n_{-k} = \begin{pmatrix} 1 & 0 \\ 0 & -1 \end{pmatrix} = \sigma_{k,z},$$

which is the z-component of Pauli's matrix $\boldsymbol{\sigma}$. Further noting that

$$a_k^\dagger a_{-k}^\dagger \psi(1_k 1_{-k}) = 0 \quad \text{and} \quad a_k^\dagger a_{-k}^\dagger \psi(0_k 0_{-k}) = \psi(1_k 1_{-k}),$$

the operator $a_k^\dagger a_{-k}^\dagger$ can be assigned to the x- and y-components of Pauli's spin $\boldsymbol{\sigma}$. Considering that

$$\sigma_{k,x} = \begin{pmatrix} 0 & 1 \\ 1 & 0 \end{pmatrix} \quad \text{and} \quad \sigma_{k,y} = \begin{pmatrix} 0 & -i \\ i & 0 \end{pmatrix},$$

we can define $\sigma_k^+ = \sigma_{k,x} + i\sigma_{k,y} = \begin{pmatrix} 0 & 2 \\ 0 & 0 \end{pmatrix}$ and $\sigma_k^- = \sigma_{k,x} - i\sigma_{k,y} = \begin{pmatrix} 0 & 0 \\ 2 & 0 \end{pmatrix}$. Then, we obtain the relations

$$a_k^\dagger a_{-k}^\dagger = \frac{1}{2}\sigma_k^- \quad \text{and} \quad a_{-k} a_k = \frac{1}{2}\sigma_k^+.$$

Using these results,

$$\mathsf{H}_{BCS} = -\sum_k \varepsilon_k \sigma_{k,z} - \frac{V}{4}\sum_{k',k} \sigma_{k'}^- \sigma_k^+$$
$$= -\sum_k \varepsilon_k \sigma_{k,z} - \frac{V}{4}\sum_{k',k} \left(\sigma_{k',x}\sigma_{k,x} + \sigma_{k',y}\sigma_{k,y}\right). \tag{iii}$$

In this argument, $\sigma_{k,z}$ is an operator for creating or annihilating paired states $\pm \boldsymbol{k}$. However, writing (iii) as a scalar product of vectors σ_k and a field $\boldsymbol{F}_k$ defined by $\left(-\frac{V}{2}\sum_{k'}\sigma_{k',x}, -\frac{V}{2}\sum_{k'}\sigma_{k',y}, \varepsilon_k\right)$, that is,

$$\mathsf{H}_{\text{BCS}} = -\sum_{\boldsymbol{k}} \sigma_k \cdot \boldsymbol{F}_k. \tag{iv}$$

This expression represents the interaction energy of classical vectors σ_k in the field $\boldsymbol{F}_k$, which corresponds to a Weiss field due to other $\sigma_{k'}$ in the system. In order for (iv) to be minimum, σ_k should be parallel to $\boldsymbol{F}_k$, characterizing that these components of the vector $\boldsymbol{\sigma}$ are in phase with F_k.

In (iv), the direction of such classical vectors σ_k can be expressed by an angle θ_k from the z-axis, representing the effective sinusoidal phase. Assuming that σ_k is in the x–z plane for simplicity, we write the relation $\sigma_k \parallel \boldsymbol{F}_k$ as

$$\frac{F_{k,x}}{F_{k,z}} = \frac{\sigma_{k,x}}{\sigma_{k,z}} = \frac{\frac{V}{2}\sum_{k'}\sigma_{k',x}}{\varepsilon_k} = \tan\theta_k. \tag{v}$$

In fact, V is a function of $\boldsymbol{k}'$ and $\boldsymbol{k}$, depending on the distance between σ_k and $\sigma_{k'}$, which can however be considered as constant in the critical region determined by a small $|\boldsymbol{k}|$ and ε_k. In addition, $V(\boldsymbol{k}', \boldsymbol{k})$ in (14.8c) has a well-defined singularity at $\theta_k = 0$, where we can write $\sigma_{k',x} = \sigma_{k'\text{o}} \sin\theta_{k'}$. Noticing that the amplitude $\sigma_{k'\text{o}}$ depends only on $|\boldsymbol{k}'|$, we can write $V_k{}' = \sum_{k'} V(\boldsymbol{k}', \boldsymbol{k})\sigma_{|k'\text{o}|}$ and obtain

$$\tan\theta_k = \frac{V_k{}'}{2\varepsilon_k}\sum_{k'}\sin\theta_{k'}. \tag{vi}$$

Such a phase angle θ_k as determined by (vi) expresses the mesoscopic wave of a superconducting cluster. Nevertheless, we pay attention to the singular behavior of $\tan\theta_k$ at $\theta_k = 0$, which represents a boundary wall between $\sigma_{k\text{o}} = +1$ and -1, like a domain wall in magnetic ordering. Therefore, by writing $\tan\theta_k = \frac{\Delta_k}{\varepsilon_k}$, the singularity of at $\theta_k = 0$ can be attributed to the parameter Δ_k, which satisfies the relation $\Delta_k = \frac{V_k{}'}{2}\sum_{k'}\frac{\Delta_k}{\sqrt{\Delta_k^2+\varepsilon_{k'}^2}}$, or

$$1 = \frac{V_k{}'}{2}\sum_{k'}\frac{1}{\sqrt{\Delta_k^2 + \varepsilon_{k'}^2}}. \tag{vii}$$

Here, replacing the summation $\sum_{k'}\ldots$ by integration over the distributed ε_k in a region between $-\hbar\omega_{\text{D}}$ and $+\hbar\omega_{\text{D}}$, where ω_{D} is Debye frequency, this relation can be calculated as

$$1=\frac{V'\rho_{\mathrm{F}}}{2}\int_{-\hbar\omega_{\mathrm{D}}}^{+\hbar\omega_{\mathrm{D}}}\frac{d\varepsilon}{\sqrt{\Delta^2+\varepsilon^2}}=V'\rho_{\mathrm{F}}\sinh^{-1}\frac{\hbar\omega_{\mathrm{D}}}{\Delta},$$

omitting indexes k, k' for simplicity, where ρ_{F} is the density of states at the Fermi level $\varepsilon=\varepsilon_F$. For $V'\rho_{\mathrm{F}}=1$, we can derive the expression

$$\Delta=\frac{\hbar\omega_{\mathrm{D}}}{\sinh\frac{1}{V'\rho_{\mathrm{F}}}}\cong 2\hbar\omega_{\mathrm{D}}\exp\left(-\frac{1}{V'\rho_{\mathrm{F}}}\right), \qquad \text{(viii)}$$

showing that the energy gap Δ is positive, if $V'>0$. In this case, the Weiss field $\boldsymbol{F}$ is by no means static, while interpretable as a field for inverting the spin $\sigma_{\boldsymbol{k}}$ between $\sigma_{\boldsymbol{k}z}=\pm1$, for which a work required is $2|\boldsymbol{F}|=2\sqrt{\varepsilon_k^2+\Delta_k^2}$. The minimum work is 2Δ at the Fermi level that is obtained in the limit of $\varepsilon_k\to0$, representing the energy gap in a superconductor at $T\le T_{\mathrm{c}}$.

The BCS Hamiltonian $\mathsf{H}_{\mathrm{BCS}}$ is responsible for generating Cooper pairs at the Fermi level, so that the ground state of a superconductor can be specified by (iv) plus the energy for creating electron pairs. Assuming $\sigma_{ky}=0$, the ground state can be characterized by

$$\begin{aligned}E_{\mathrm{o}}&=-\sum_{k}\varepsilon_k\sigma_k\mathrm{o}\cos\theta_k-\frac{V}{4}\sum_{k',k}\sigma_k{}'x\sigma_kx+\sum_{\text{pair}}2\varepsilon_k\\&=-\sum_{k}\varepsilon_k\sigma_{k\mathrm{o}}\left(\cos\theta_k+\frac{1}{2}\sin\theta_k\tan\theta_k\right)+\sum_{\text{pair}}2\varepsilon_k,\end{aligned}$$

where the second term can be simplified by (vii) as

$$\sum_{k}\varepsilon_k\sin\theta_k\tan\theta_k=\sum_{k}\frac{\Delta^2}{\sqrt{\varepsilon_k^2+\Delta^2}}=\frac{2\Delta^2}{V'}.$$

Replacing the sum $\sum_k\ldots$ by integration, we obtain

$$E_{\mathrm{o}}=2\rho_{\mathrm{F}}\int_0^{\hbar\omega_{\mathrm{D}}}\left(\varepsilon-\frac{\varepsilon^2}{\sqrt{\varepsilon^2+\Delta^2}}\right)d\varepsilon-\frac{\Delta^2}{V'}.$$

Assuming $\Delta\ll\hbar\omega_{\mathrm{D}}$, we can evaluate E_{o} as

$$E_{\mathrm{o}}=\rho_{\mathrm{F}}(\hbar\omega_{\mathrm{D}})^2\left(1-\sqrt{1+\left(\frac{\Delta}{\hbar\omega_{\mathrm{D}}}\right)^2}\right)\approx-\frac{1}{2}\rho_{\mathrm{F}}\Delta^2.$$

In thermodynamics, we have to consider the energy gap Δ as a function of temperature. Near the critical temperature T_c, the interaction potential $V_k{}' = \sum_{k'} V(\boldsymbol{k}', \boldsymbol{k})\sigma_{|k'_o|}$ can be calculated as a statistical average $V\sum_{k'}\langle\sigma_{k'z}\rangle$, where

$$\langle\sigma_{k'z}\rangle = \frac{\exp\frac{(+1)F_{k'z}}{k_B T} - \exp\frac{(-1)F_{k'z}}{k_B T}}{\exp\frac{(+1)F_{k'z}}{k_B T} + \exp\frac{(-1)F_{k'z}}{k_B T}} = \tanh\frac{F_{k'z}}{k_B T}.$$

Hence, the relation (vi) can be written as

$$\tan\theta_k = \frac{V}{2\varepsilon_k}\sum_{k'}\tanh\frac{F_{k'z}}{k_B T}\sin\theta_{k'} = \frac{\Delta}{\varepsilon_k},$$

where $F_{k'z}$ is in energy unit, so we replaced it by $\sqrt{\varepsilon_{k'}^2 + \Delta_{k'}^2}$. If the transition threshold is characterized by $\Delta = 0$, this equation can be utilized to specify the critical temperature T_c as

$$1 = V\sum_{k'}\frac{1}{2\varepsilon_{k'}}\tanh\frac{\varepsilon_{k'}}{k_B T}.$$

Replacing $\sum_{k'}\ldots$ by integration again, we obtain

$$\frac{2}{V\rho_F} = \int_{-\hbar\omega_D}^{+\hbar\omega_D}\frac{d\varepsilon}{\varepsilon}\tanh\frac{\varepsilon}{2k_B T} = 2\int_0^{\frac{\hbar\omega_D}{2k_B T_c}}\frac{\tanh\xi}{\xi}d\xi, \tag{ix}$$

which is the BCS formula for T_c. By graphical integration, these authors showed that T_c is given as

$$k_B T_c = 1.14\hbar\omega_D\exp\left(-\frac{1}{V\rho_F}\right). \tag{x}$$

Combined with (viii), the energy gap is related with T_c as $2\Delta = 3.5\,T_c$; the experimental values of $\frac{2\Delta}{T_c}$ obtained from Sn, Al, Pb, and Cd were 3.5, 3.4, 4.1, and 3.3, respectively. In fact, Δ is a function of temperature T. Figure 14.3 shows the theoretical BCS curve of $\frac{\Delta}{k_B T_c}$ versus $\frac{T}{T_c}$, compared with experimental results. The agreement is quite reasonable, supporting the assumption they made for the interaction potential $V_k{}'$. The isotope effect expressed by $M^{0.5}T_c = \text{const.}$ follows directly from the BCS results (x), since the Debye frequency ω_D is proportional to $M^{-0.5}$.

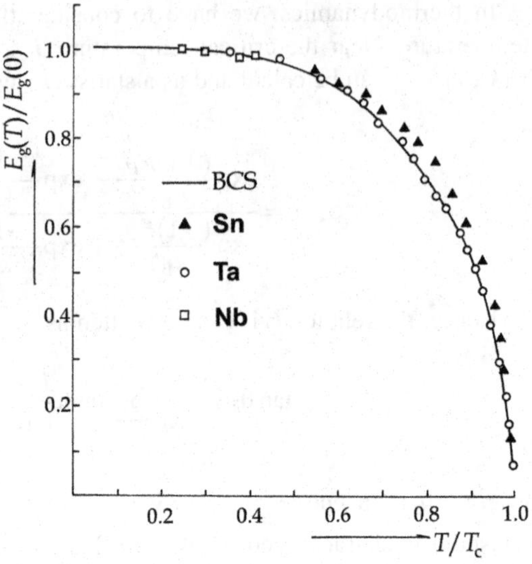

Fig. 14.3 The energy gap E_g as a function of temperature T. The BCS formula compared with experimental results.

14.3.3 BCS Ground States

Supported by the isotope effect, the BCS Hamiltonian describes the interacting electron–lattice system in a metal. Here, we formulate the equation of motion for thermodynamics of superconducting transition.

Writing the BCS Hamiltonian (i) as $\mathsf{H}_{\text{BCS}} = \sum_k \mathsf{H}_k$ where

$$\mathsf{H}_k = \varepsilon_k \left(a_k^\dagger a_k + a_{-k}^\dagger a_{-k} \right) - V \sum_{k'} a_{k'}^\dagger a_{-k'}^\dagger a_{-k} a_k,$$

the operators $a_k^\dagger$ and a_k are dynamical variables, whose time variations are determined by

Heisenberg's equations

$$i\hbar \frac{\partial a_k^\dagger}{\partial t} = \left[a_k^\dagger, \mathsf{H}_k \right] \quad \text{and} \quad i\hbar \frac{\partial a_k}{\partial t} = [a_k, \mathsf{H}_k].$$

Using the identity relations $a_k a_k = 0$ and $a_k^\dagger a_k^\dagger = 0$ in these equations, we derive

$$i\hbar \dot{a}_k = \varepsilon_k a_k - a_{-k}^\dagger \left(V \sum_{k'} a_{-k'} a_{k'} \right)$$

and

$$i\hbar\dot{a}^{\dagger}_{-k} = -\varepsilon_k a^{\dagger}_{-k} - a_k\left(V\sum_{k'} a^{\dagger}_{k'}a^{\dagger}_{-k'}\right).$$

We can define the quantity $\Delta_k = V\sum_{k'} a_{-k'}a_{k'}$ and its complex conjugate $\Delta_k^* = V\sum_{k'} a^{\dagger}_{k'}a^{\dagger}_{-k'}$, indicating interactions between Fröhlich's condensates for $|\varepsilon_k|<\hbar\omega_D$. If $|\varepsilon_k|>\hbar\omega_D$ on the other hand, we have to take $\Delta_k = \Delta_k^* = 0$ for no electron pairs to be formed. Using Δ_k, these equations of motion are linearized as

$$i\hbar\dot{a}_k = \varepsilon_k a_k - \Delta_k a^{\dagger}_{-k} \quad \text{and} \quad i\hbar\dot{a}_{-k} = -\varepsilon_k a^{\dagger}_{-k} - \Delta_k{}^* a_k. \tag{i}$$

We can solve (i) for a_k and a_{-k} that are proportional to $\exp\left(-i\frac{\lambda_k}{\hbar}t\right)$, if we can find λ_k to satisfy the determinant equation

$$\begin{vmatrix} \lambda_k - \varepsilon_k & \Delta_k \\ \Delta_k & \lambda_k + \varepsilon_k \end{vmatrix} = 0 \quad \text{or} \quad \lambda_k^2 = \varepsilon_k^2 + \Delta_k\Delta_k{}^*. \tag{ii}$$

The real λ_k represents an eigenvalue for a Cooper pair. The eigen-operators for a Cooper pair can be determined by linear combinations of these one-electron operators a_k and a_{-k}:

$$\begin{aligned} \alpha_k &= u_k a_k - v_k a^{\dagger}_{-k}, \quad \alpha_{-k} = u_k a_{-k} + v_k a^{\dagger}_k, \\ \alpha^{\dagger}_k &= u_k a^{\dagger}_k - v_k a_{-k}, \quad \alpha^{\dagger}_{-k} = u_k a^{\dagger}_{-k} + v_k a_k. \end{aligned} \tag{iiia}$$

The reverse relations are

$$\begin{aligned} a_k &= u_k \alpha_k + v_k \alpha^{\dagger}_{-k}, \quad a_{-k} = u_k \alpha_{-k} - v_k \alpha^{\dagger}_k, \\ a^{\dagger}_k &= u_k \alpha^{\dagger}_k + v_k \alpha_{-k}, \quad a^{\dagger}_{-k} = u_k \alpha^{\dagger}_{-k} - v_k \alpha_k. \end{aligned} \tag{iiib}$$

These relations (iiia) and (iiib) are known as the *Bogoliubov transformation* [5]. The coefficients u_k and v_k are real for the symmetric and antisymmetric combinations of one-electron operators, respectively, with regard to inversion $\boldsymbol{k} \to -\boldsymbol{k}$, that is, $u_k = u_{-k}$ and $v_k = -v_{-k}$, normalized as $u_k^2 + v_k^2 = 1$. And, for the operators $\alpha^{\dagger}_k$ and α_k, we have the following anti-commutator relations:

$$\left[\alpha_k, \alpha^{\dagger}_{k'}\right]_+ = u_k u_{k'}\left[a_k, a^{\dagger}_{k'}\right]_+ + v_k v_{k'}\left[a^{\dagger}_{-k}, a_{-k'}\right]_+ = \delta_{kk'}\left(u_k^2 + v_k^2\right)$$

and

$$[\alpha_k, \alpha_{-k}]_+ = u_k v_k \left[a_k, a_k^\dagger\right]_+ - v_k u_k \left[a_{-k}^\dagger, a_{-k}\right]_+ = u_k v_k - v_k u_k = 0.$$

Differentiating (i), we obtain $\lambda_k u_k = \varepsilon_k u_k + \Delta_k v_k$. Combining this result with (ii), we derive the relation

$$\Delta_k \left(u_k^2 - v_k^2\right) = 2\varepsilon_k u_k v_k.$$

It is noted that this expression is identical to $\tan\theta_k = \frac{\Delta_k}{\varepsilon_k}$ in Sect. 12.3.2, if we write

$$u_k = \cos\frac{\theta_k}{2} \quad \text{and} \quad v_k = \sin\frac{\theta_k}{2}. \tag{iv}$$

Using the operator α_k, the ground state of the system may be represented by a wavefunction

$$\alpha_{-k}\alpha_k \Phi_{\text{vac}} = (-v_k)\left(u_k + v_k a_k^\dagger a_{-k}^\dagger\right)\Phi_{\text{vac}}.$$

Although is not normalized, omitting the factor $-v_k$, we can confirm that

$$\langle \Phi_{\text{vac}}|\left(u_k^* + v_k^* a_{-k} a_k\right)\left(u_k + v_k a_k^\dagger a_{-k}^\dagger\right)|\Phi_{\text{vac}}\rangle = \left(u_k^2 + v_k^2\right)\langle \Phi_{\text{vac}}|\Phi_{\text{vac}}\rangle;$$

hence, the function

$$\Phi_o = \prod_k \left(u_k + v_k a_k^\dagger a_{-k}^\dagger\right)\Phi_{\text{vac}} \tag{14.11}$$

is considered for the normalized wavefunction for the ground state. This wavefunction (14.11) was originally postulated in the BCS theory, implying that Cooper pairs and unpaired condensates are determined by probabilities v_k^2 and u_k^2, respectively. With Bogoliubov operators, creation and annihilation of a Cooper pair can be described conveniently; the normalization $u_k^2 + v_k^2 = 1$ assumed in the BCS theory is consistent with Bogoliubov transformation.

We can further verify that

$$\alpha_{k'}\Phi_o \propto \alpha_{k'}\left(\alpha_{-k'}\alpha_{k'}\right)\prod_{k \neq k'} \alpha_{-k}\alpha_k \Phi_{\text{vac}} = 0, \quad \text{for} \quad \alpha_{k'}\alpha_{k'} = 0,$$

signifying annihilation of a pseudo-particle. Also, from the relation

$$\alpha_{k'}^{\dagger}\Phi_o = \left(u_{k'}a_{k'}^{\dagger} - v_{k'}a_{-k'}\right)\left(u_{k'} + v_{k'}a_{k'}^{\dagger}a_{-k'}^{\dagger}\right)\prod_{k\neq k'}\left(u_k + v_k a_k^{\dagger}a_{-k}^{\dagger}\right)\Phi_{\text{vac}}$$

$$= a_{k'}^{\dagger}\prod_{k\neq k'}\left(u_k + v_k a_k^{\dagger}a_{-k}^{\dagger}\right)\Phi_{\text{vac}},$$

we see that the operator $\alpha_k^{\dagger}$ creates a single particle; $\alpha_k^{\dagger}\alpha_{-k'}^{\dagger}$ is for the pair creation.

The number operator in the $\boldsymbol{k}$-state can be expressed in terms of Bogoliubov operators as

$$n_k = a_k^{\dagger}a_k = \left(u_k\alpha_k^{\dagger} + v_k\alpha_{-k}\right)\left(u_k\alpha_k + v_k\alpha_{-k}^{\dagger}\right).$$

Using θ_k given by (iv), u_k and v_k in (14.11) are related to

$$u_k^2 = \cos^2\frac{\theta_k}{2} = \frac{1}{2}\left(1 + \frac{\varepsilon_k}{\lambda_k}\right) \quad \text{and} \quad v_k^2 = \sin^2\frac{\theta_k}{2} = \frac{1}{2}\left(1 - \frac{\varepsilon_k}{\lambda_k}\right), \tag{14.12}$$

allowing for these probabilities to be interpreted in terms of the ratio $\frac{\varepsilon_k}{\lambda_k}$.

Finally, the previous result of the energy gap E_g can be verified with Φ_o in (14.11). Namely, the expectation value of the kinetic energy is expressed as

$$\langle\Phi_o|a_{k'}^{\dagger}a_{k'}|\Phi_o\rangle = \langle\Phi_o|v_{k'}^2\alpha_{-k'}\alpha_{-k'}^{\dagger}|\Phi_o\rangle = v_{k'}^2,$$

and the potential energy term is

$$-\langle\Phi_o|a_{k'}^{\dagger}a_{-k'}^{\dagger}a_{-k''}a_{k''}|\Phi_o\rangle = \langle\Phi_o|u_{k'}v_{k'}u_{k''}v_{k''}\alpha_{-k'}\alpha_{-k'}^{\dagger}\alpha_{-k''}\alpha_{-k''}^{\dagger}|\Phi_o\rangle$$

$$= u_{k'}v_{k'}u_{k''}v_{k''}.$$

Hence,

$$\langle\Phi_o|\mathsf{H}_{\text{BCS}}|\Phi_o\rangle = 2\sum_{k}\varepsilon_k v_k^2 - V\sum_{k,k'}u_k v_k u_k{}' v_k{}'$$

$$= \sum_{k}\varepsilon_k(1-\cos\theta_{\boldsymbol{k}}) - \frac{V}{4}\sum_{k,k'}\sin\theta_k\sin\theta_{\boldsymbol{k}}{}'$$

$$= -\sum \varepsilon_k\cos\theta_k - \frac{\Delta^2}{V}.$$

The energy gap is then given as $E_g = \langle\Phi_o|\mathsf{H}_{\text{BCS}}|\Phi_o\rangle + \sum_k|\varepsilon_k|$.

14.3.4 Superconducting States at Finite Temperatures

Thermodynamically, the BCS result in Sect. 14.3.3 can be applied to absolute zero of temperature. At a finite temperature, the ground state Φ_o should be modified by adiabatic excitations of condensates, as described by Ginzburg–Landau theory. In the following, we discuss the BCS theory modified in thermodynamic environment.

The number of quasi-particles of Cooper pairs can be assumed to be temperature dependent, for which we consider the complete set of temperature-dependent states

$$|\Phi_o\rangle,\quad \alpha_k^\dagger|\Phi_o\rangle,\quad \alpha_k^\dagger\alpha_{k'}^\dagger|\Phi_o\rangle,\ldots$$

forming the basis. Here $|\ldots\rangle$ expresses temperature-dependent ket-states.

To study thermodynamic properties of a superconductor, we need to introduce the statistical average number of quasi-particles as given by the function

$$f_k = \left\langle \alpha_k^\dagger \alpha_k \right\rangle.$$

Assuming statistical independence of excitations, the entropy of the system can be given by

$$S = \mathrm{k_B}\sum_{\boldsymbol{k}}\{f_k \ln f_k + (1-f_k)\ln(1-f_k)\}.$$

We reexpress the Hamiltonian of interacting pairs $\mathrm{H}_k = \varepsilon_k\left(a_k^\dagger a_k + a_{-k}^\dagger a_{-k}\right) - V\left(a_k^\dagger a_{-k}^\dagger a_{-k} a_k\right)$ by the Bogoliubov transformation. The kinetic energy term can be converted to

$$\sum_{\boldsymbol{k}}\left\langle \varepsilon_k\left(u_k\alpha_k^\dagger + v_k\alpha_{-k}\right)\left(u_k\alpha_k + v_k\alpha_{-k}^\dagger\right)\right\rangle = \sum_{\boldsymbol{k}}\varepsilon_k\left\{v_k^2(1-f_k) + u_k^2 f_k\right\}, \quad \text{(i)}$$

in which we noted that $\alpha_k^\dagger\alpha_{-k}^\dagger = 0$. The first term on the right side of (i) can be interpreted as representing the condensate system that has no quasi-particle at the state of wavevector $\boldsymbol{k}$, with thermal probability v_k^2. The second term, on the other hand, u_k^2 is the probability for the state to be occupied by quasi-particles. The factors $1-f_k$ and f_k in these terms are the statistical weights due to average numbers of quasi-particles.

Similarly, the interaction term can be calculated as

$$\sum_{\boldsymbol{k},\boldsymbol{k}'} V(\boldsymbol{k},\boldsymbol{k}';\boldsymbol{q})(1-2f_k)(1-2f_{k'}).$$

It is noted from the definition (14.8c) that $V(\boldsymbol{k},\boldsymbol{k}';\boldsymbol{q})$ can be factorized as $V_k(\boldsymbol{q})$ $V_{-k}(-\boldsymbol{q})$. Therefore, the thermodynamic potential can be expressed as

$$g(T,p) = \langle \mathrm{H_{BCS}} \rangle - TS$$
$$= \sum_k \varepsilon_k \{u_k{}^2 f_k + v_k^2 (1 - f_k)\} - \sum_{k,k'} V_k V_{k'} u_k u_{k'} v_k v_{k'} (1 - 2f_k)(1 - 2f_{k'})$$
$$- \mathrm{k_B} T \sum_k \{f_k \ln f_k + (1 - f_k) \ln(1 - f_k)\}. \quad \text{(ii)}$$

We first minimize this Gibbs function with respect to v_k. This procedure is the same as in the zero-temperature case of Sect. 14.3.3, so that we can arrive at (14.12) without repeating calculation, that is,

$$u_k^2 = \frac{1}{2}\left(1 + \frac{\varepsilon_k}{\sqrt{\varepsilon_k^2 + \Delta_k(T)^2}}\right) \quad \text{and} \quad v_k^2 = \frac{1}{2}\left(1 - \frac{\varepsilon_k}{\sqrt{\varepsilon_k^2 + \Delta_k(T)^2}}\right),$$

where $\Delta_k(T) = \sum_k V_k u_k v_k (1 - 2f_k)$, representing one-half of the gap at the superconducting transition, which is now verified as temperature dependent.

Minimizing the Gibbs function (ii) with respect to f_k, we obtain

$$\mathrm{k_B} T\{\ln f_k + \ln(1 - f_k)\} + \sqrt{\varepsilon_k^2 + \Delta_k(T)^2} = 0,$$

that is,

$$f_k = \frac{1}{1 + \exp\dfrac{E_k}{\mathrm{k_B} T}} \quad \text{and} \quad E_k = \sqrt{\varepsilon_k^2 + \Delta_k(T)^2}.$$

This is the Fermi–Dirac distribution function for temperature-dependent E_k.

In fact, as seen from Fig. 13.2, the Gibbs potential is a function of temperature T as well as an applied magnetic field H, that is, $G(H,T)$. In the absence of H, the superconducting transition at T_c is second order characterized by no latent heat, signifying that $E_k(T_c) = 0$. On the other hand, in the presence of H, the transition is discontinuous, as signified by a finite energy gap $E_k(T)$ that is temperature dependent. With the BCS theory, the value of E_k is thus calculable, as shown in Fig. 14.3, where the calculated result is compared with experimental values from some superconducting metals, showing a reasonable agreement.

Exercise 14

1. We can define creation and annihilation operators for Cooper pairs by

$$a_k^\dagger = a_k^\dagger a_{-k}^\dagger \quad \text{and} \quad a_k = a_k a_{-k},$$

respectively, where $a_k, a^\dagger_{-k}$, etc. are one-particle Fermion operators. Show that the pair operators satisfy the commutation relations

$$[a_k, a^\dagger_{k'}] = (1 - n_k - n_{-k})\delta_{k,k'},$$

$$[a_k, a_{k'}] = 0,$$

$$[a_k, a_{k'}]_+ = 2a_k a_{k'}(1 - \delta_{k,k'}).$$

2. Show that the total number of particles is conserved with the equations of motion (i) in Sect. 14.3.3.
3. In this chapter, we defined the coherence length ξ with respect to the critical wavevector of modulation q_0. Justify the reason why it is referred to as *coherence* length.

References

1. H. Fröhlich, Phys. Rev. **79**, 845 (1950); Proc. R. Soc. **A215**, 291 (1952)
2. L. N. Cooper, Phys. Rev. **104**, 1189 (1956); Phys. Rev. Lett. **6**, 689 (1961)
3. J. Bardeen, L.N. Cooper, J.R. Schrieffer, Phys. Rev. **108**, 1175 (1957)
4. C. Kittel, *Quantum Theory of Solids* (Wiley, New York, 1963)
5. N. N. Bogoliubov, J. Phys. Moscow, 11, 23 (1947); Nuovo Cim. Ser. X, 7, 794 (1958)

Appendix

Elliptic Integrals

Elliptic integrals of the first kind:

$$F(\kappa, \varphi) = \int_0^{\varphi} \frac{d\varphi}{\sqrt{1-\kappa^2 \sin^2\varphi}} = \int_0^{\sin\varphi} \frac{dz}{\sqrt{(1-z^2)(1-\kappa^2 z^2)}},$$

and the complete elliptic integral is

$$F\left(\kappa, \frac{\pi}{2}\right) = \int_0^{\pi/2} \frac{d\varphi}{\sqrt{1-\kappa^2 \sin^2\varphi}} = \int_0^{1} \frac{dz}{\sqrt{(1-z^2)(1-\kappa^2 z^2)}} = K(\kappa).$$

Elliptic integral of the second kind:

$$E(\kappa, \varphi) = \int_0^{\varphi} \sqrt{1-\kappa^2 \sin^2\varphi}\, d\varphi = \int_0^{\sin\varphi} \sqrt{\frac{1-\kappa^2 z^2}{1-z^2}}\, dz,$$

and the complete elliptic integral is

$$E\left(\kappa, \frac{\pi}{2}\right) = \int_0^{\pi/2} \sqrt{1-\kappa^2 \sin^2\varphi}\, d\varphi = \int_0^{1} \sqrt{\frac{1-\kappa^2 z^2}{1-z^2}}\, dz = E(\kappa),$$

$$= \int_0^{K(\kappa)} dn^2(u,\kappa)\, du. \quad dn(u,\kappa) \text{ is Jacobi's dn} - \text{function.}$$

Defining $\kappa' = \sqrt{1-\kappa^2}$, $K(\kappa') = K'$, and $E(\kappa') = E'$, we have the relation $EK' + E'K - KK' = \frac{\pi}{2}$. (Legendre's relation.)

M. Fujimoto, *Thermodynamics of Crystalline States*,
DOI 10.1007/978-1-4614-5085-6, © Springer Science+Business Media New York 2013

Jacobi's Elliptic Function

From the elliptic integral $u(x) = \int_0^x \frac{dz}{\sqrt{(1-z^2)(1-\kappa^2 z^2)}}$, we write the reverse function as

$$z = \mathrm{sn}u = \mathrm{sn}(u, \kappa),$$

which is Jacobi's sn-function. Considering the relations with trigonometric and hyperbolic functions, we also define the corresponding cn- and dn-functions by

$\mathrm{cn}^2\, u - 1 - \mathrm{sn}^2\, u$ and $\mathrm{dn}^2\, u - 1 - \kappa^2 \mathrm{sn}^2\, u$.

In the limit of $\kappa \to 0$,

$\mathrm{sn}\, u \to \sin\, u$, $\mathrm{cn}\, u \to \cos\, u$, and $\mathrm{dn}\, u \to 1$.

On the other hand, if $\kappa \to 1$, we have

$\mathrm{sn}\, u \to \tanh\, u$, $\mathrm{cn}\, u \to \mathrm{sech}\, u$, and $\mathrm{dn}\, u \to \mathrm{sech}\, u$.

Differential formula:

$$\frac{d\,\mathrm{sn}\, u}{du} = \mathrm{cn}\, u\; \mathrm{dn}\, u, \quad \frac{d\,\mathrm{cn}\, u}{du} = -\mathrm{sn}\, u\, \mathrm{dn}\, u, \quad \frac{d\,\mathrm{dn}\, u}{du} = -\kappa^2\, \mathrm{sn}\, u\, \mathrm{cn}\, u.$$

Expansion formula:

$$\mathrm{sn}\ u = u - \frac{1+\kappa^2}{3!}u^3 + \frac{1+14\kappa+\kappa^4}{5!}u^5 + \cdots,$$

$$\mathrm{cn}\ u = 1 - \frac{1}{2}u^2 + \frac{1+4\kappa^2}{4!}u^4 - \frac{1+44\kappa^2+16\kappa^4}{6!}u^6 + \cdots,$$

$$\mathrm{dn}\, u = 1 - \frac{\kappa^2}{2}u^2 + \frac{(4+\kappa^2)\kappa^2}{4!}u^4 - \frac{(16+44\kappa^2+\kappa^4)\kappa^2}{6!}u^6 + \cdots$$

Reference Books

Elliptic functions and integrals are not quite familiar mathematics with physicists and engineers today. I myself learned them from Professor Morikazu Toda's book *Introduction to Elliptic Functions* in Japanese (Nippyo, Tokyo, 2001), where I found the following references in available literature.

F. Bowman, *Introduction to Elliptic Functions with Applications*, (Dover, 1961),
A.G. Greenhill, *The Applications of Elliptic Functions*, (Dover 1959),
H. Hancock, *Lecture on the Theory of Elliptic Functions*, (Dover 1958),
A. Cayley, *An Elementary Treatise on Elliptic Functions* (Dover 1961),
E.T. Whittaker and G.N. Watson, *A Course of Modern Analysis* (Cambridge Univ. Press 1958), and
L.M. Milne-Thomson, *Jacobian Elliptic Function Tables* (Dover 1950).

Index

A
Acoustic mode, 35
Action–reaction principle, 4
Adiabatic, 8
  approximation, 44
  fluctuations, 87–88
  potentials, 3, 8, 44, 198
  probability, 217
α-hematite, 208
Analyzer, 160
Annihilation, 15, 223
Antiferromagnetic, 200, 208–210
  phases, 197
  resonance, 212–214
Antiferromagnetism, 197
Atomic form factor, 157

B
Bardeen–Cooper–Schrieffer (BCS), 257
  ground states, 262–265
  Hamiltonian, 262
  theory, 255
Becker's interpretation, 60–62
β-brass CuZn, 55
Binary alloys, 55–57
Binary transition, 5
Biot–Savart's law, 190
Bipyramidal, 37
Bloch theorem, 45–49, 221–222
Bloch wave, 221
Bogoliubov transformation, 263
Bohr's magneton, 175, 192
Boltzmann constant, 3
Boltzmann factor, 27
Born-Huang relaxation, 52–53
Born-Huang's principle, 35, 88
Born-Huang's transitions, 126–129
Born-Oppenheimer's approximation, 39–45
Bose-Einstein condensation, 218
Bose-Einstein statistics, 28
Bragg-Williams theory, 57–60
Brillouin function, 194, 202
Brillouin lines, 168
Brillouin scatterings, 165–169
Brillouin zone, 49–51
Brillouin-zone boundaries, 221

C
Canonical transformation, 252
Center-of-mass coordinate, 35
Chemical potential, 217
Chemical thermodynamics, 2
Classical particles, 219
Classical pseudospins, 73–74
Classical vectors, 195
Clusters, 74–77, 200
Cnoidal potential, 101–103
Cnoidal theorem, 120–122
Coherence length, 256, 268
Complete elliptic integral, 100
Complete order, 56
Complex susceptibility, 143
Compressibility, 28–30
Conduction electrons, 226–227
Continuous fields, 137
Continuum field, 7
Cooper pair, 252–255, 264
Correlation function, 56
Correlations, 3
Coulomb interaction, 220
Creation, 15, 223
Critical anomalies, 88–90

M. Fujimoto, *Thermodynamics of Crystalline States*,
DOI 10.1007/978-1-4614-5085-6, © Springer Science+Business Media New York 2013

Critical magnetic field, 230
Critical temperatures, 208
Crystalline phases, 1–4
Crystalline potential, 176
Crystalline symmetry, 199
Crystal planes, 155, 221
Cubic to tetragonal transition, 78–80
Curie law, 194
Curie's constant, 63
Curie-Weiss law, 63, 202

D
Debye function, 28
Debye's cut-off frequency, 23
Debye's frequency, 259
Debye's-law, 25
Debye's model, 23–25
Debye's relaxation, 144
Debye's relaxator, 144
Debye's temperature, 24
Demagnetization factors, 211
Developing equation, 116
Developing operator, 116
Dielectric response function, 138
Diffuse diffraction, 37, 155
Diffuse X-ray diffraction, 67, 155–160
Discommensuration lines, 108, 131
Disorder, 4
Disordered state, 56
Dispersion relation, 112, 204
Dispersive, 8
Displacive, 70
Displacive variables, 35–39
Dissipative, 2, 8
Domains, 1
Domain-wall energy, 120
Doped crystals, 171
Doping, 173
Dulong-Petit law, 23
Dyadic tensor, 167
Dzialoshinsky–Moriya's interaction, 208

E
Easy magnetization, 198, 204
Eckart potential, 119–122, 201
Effective Bohr magneton, 194
Effective mass, 222
Ehrenfest's classification, 5
Einstein's model, 22–23
Einstein temperature, 23
Elastically scattered beam, 156
Elastic collisions, 221
Elastic scatterings, 157
Electric displacement, 139
Electron exchange, 195
Electron–phonon interaction, 8
Electron scatterings, 249
Elementary particles, 219
Elliptic function, 88
Elliptic integral of the first kind, 99
Energy gap, 237
Entropy, 3
Equilibrium, 2
Equilibrium crystals, 45–51
Ergodic hypothesis, 19
Even parity, 28
Exchange integral, 62, 197, 199, 201
Exclusive event, 25
Extensive variables, 5
External work, 3
Extrinsic pinning, 92–94

F
Fabry–Pérot interferometer, 167
Fermi–Dirac distribution, 226
Fermi–Dirac statistics, 226–227
Fermi energy, 220
Fermi level, 219
Ferrimagnetic, 208–210
Ferroelastic domains, 131
Ferroelectric phase transition, 39
Ferromagnetic, 200
Ferromagnetic order, 62–64
Ferromagnetic resonance, 211–212
Fields, 7, 8
The first law, 3
First-order, 5
First-order hyperfine energy, 180
Fluctuations, 5
Fourier amplitude, 199
Fourier series, 46, 199
Fourier transform, 97
Free electron model, 220
Free spectral range (FSR), 168
Fröhlich's condensates, 249–252
FSR. *See* Free spectral range (FSR)

G
Gaseous state, 2
Gibbs factor, 27
Gibbs potential, 4
Gibbs sum, 226

Ginzburg–Landau's theory, 256
Gorter–Casimir theory, 236
Group velocity, 21
Grüneisen's constant, 29, 146

H
Harmonic approximation, 43
Heat quantity, 2
Helmholtz free energy, 28
Hermitian operators, 14
Holes, 192
Hund's rule, 192
Hydrodynamic approach, 115
Hyperfine interactions, 180
Hypergeometric equation, 123
Hypergeometric function, 123
Hypergeometric series, 123

I
Impedance, 176
Incommensurability, 105
Incommensurate, 103
Incommensurate parameters, 80
Induced transitions, 175
Inelastic scatterings, 158, 164
Integrating denominator, 2
Interferometer, 168
Internal energy, 2
Internal field, 3
Intrinsic pinning, 88
Inversion, 37
Inversion symmetry, 194
Ionic polarization, 137
Irreversible, 2
Irrotational, 138
Isothermal, 8
Isothermal change, 128
Isotopes, 171

J
Jacobi's elliptic functions, 99, 120
Jacobi's sn-function, 119
J-multiplet, 190

K
KDP. *See* Potassium dihydrogen phosphate (KDP)
Korteweg-deVries equation, 114–117
Kronecker's delta, 13

L
Lagrange's multipliers, 61
Laminar, 138
Landau theory of binary transitions, 83–87
Landé factor, 175, 192
Laplace equation, 176
Larmor frequency, 175, 212
Larmor's precession, 200, 203
Laser oscillator, 164
Lattice displacements, 4
Lattice excitation, 46
Lattice translations, 45
Lifshitz condition for incommensurability, 103–105
Light scattering, 164–171
Logarithmic anomaly, 237
Longitudinal and transverse modes, 138
Long-range interaction, 58
Long-range order, 64, 111–135
Lyddane–Sachs-Teller (LST)
  formula, 140
  relation, 137–140

M
Magnetic anisotropy, 204–208
Magnetic dipole–dipole interactions, 195
Magnetic resonance, 171–173, 210
Magnetic symmetry, 198
Magnetization curve, 206
Magnetostriction, 206
Many-electron systems, 222–226
Matthiessen's rule, 229
Mean-field approximation, 56
Mean-field average, 3, 57
Meissner effect, 8, 231–233
Mesoscopic fields, 129–132
Mesoscopic fluctuations, 7
Mesoscopic state, 7
Microcanonical ensemble, 19
Mie–Grüneisen's equation, 29
Mirror plane, 182
Modulated structure, 2, 3
Modulus, 99
Molecular field, 63
Monochrometer, goniometer, 160
Multiplet states, 192
Mutual correlations, 40

N
Neél temperature, 164
Neutron diffraction, 204

Neutron inelastic scattering, 160–164, 204
Non-ergodic, 3
Non-linear dynamics, 8
Non-linear propagation, 88
Normal coordinates, 13
Normal modes, 12
Nuclear magnetic resonance and relaxation, 173–175

O
Odd parity, 28
One-particle state, 219
Optic and acoustic modes, 35
Optic mode, 35
Orbital, 195
  angular momentum, 190
  quenching, 177, 190
Order, 4
  parameter, 4
  variables, 4
Overdamped, 143

P
Paramagnetic resonance, 39
Paramagnetic resonance of impurity probes, 175–176
Parity, 219
Partition function, 19, 193
Pauli principle, 218–220
Pauli's exclusion principle, 192
Pauli's matrix, 258
Perfect diamagnetism, 232
Permutation, 15
Perovskites, 37, 197
Persistent current, 230
Phase, 7
Phase variables, 199
Phonon, 15
Phonon gas, 215
Phonon–phonon correlations, 217
Phonon scatterings, 51–53
Phonon spectrum, 2
Photo-elastic, 165
Physical contact, 2
Pinning by an electric field, 93–94
Pinning by point defects, 93
Pinning energy, 88
Point defects, 92
Point group, 49, 67
Polarizabilities, 165
Polarizability tensor, 170
Potassium dihydrogen phosphate (KDP), 39
Precession, 191
Pressure, 28
Principal quantum number, 192
Probabilities, 55
Probability amplitudes, 17
Pseudopotentials, 105–108
Pseudospin, 70
  clusters, 73–77
  correlations, 71–73
Pseudosymmetry, 92, 105

R
Raman scatterings, 169–171
Raman shift, 169
Rayleigh radiation, 165
Reciprocal lattice, 8, 45–47
Reduced mass, 35
Relaxation time, 144
Renormalization group, 7
Renormalized coordinates, 7, 45–47
Renormalized phase, 7
Reversible, 3
Riccati's equation, 127
Russel-Saunder's coupling, 190
Rutger's formula, 234

S
Sampling, 9
Saturated magnetization, 194
Screening length, 220
Second-law, 2
Second-order, 5
Second quantization, 251
Seed, 74, 200
Short-and long-range order, 55
Short-range correlations, 66, 78
Sine-Gordon equation, 106
Singlet, 197
Slow passage, 174
Soft modes, 52–53, 88
Soliton, 7, 115
Soliton theory, 8, 111–135
Sommerfeld model, 219
Space group, 67
Space–time uncertainties, 199
Spatial phase, 7
Specific heat, 20–22
Spin angular momentum, 189
Spin correlations, 201
Spin Hamiltonian, 176–179

Spin–lattice relaxation time, 173
Spin–orbit coupling, 190
Spin–orbit coupling constant, 177
Spin probes, 171
Spin–spin correlations, 198
Spin–spin interactions, 66
Spin–spin relaxation time, 173
Spin waves, 203–204
Spontaneous emission, 175
Standing-waves, 7
Strains, 208
Structural changes, 1
Structural form factor, 158
Structural transition, 1
Sublattices, 208
Superconducting ground state, 255–268
Superconducting states, 229–235
Superconducting transition, 8
Supercurrents, 8
Superexchange, 198
Surface pinning, 94
Susceptibility, 194
Symmetric spatial fluctuations, 159
Symmetry, 1
Symmetry change, 150–152

T
TEM. *See* Transmission electron microscopy (TEM)
Thermal contact, 18
Thermal relaxation, 88, 200
Thermal relaxation time, 69
Thermodynamic probability, 3
The third law, 19
Time inversion, 49
Timescale, 9
Total differential, 2
Transition
  anomalies, 5
  probability, 53
  temperature, 63
Transmission electron microscopy (TEM), 108
Transverse and longitudinal modes, 139
Transverse components, 101–103
Triple axis spectrometer, 160
Triplet, 197
Tris-sarcosine calcium chloride (TSCC), 39, 80–81
Tunneling model, 70–71
Type 2 superconductor, 233

U
Underdamped, 143
Unidentifiable particles, 25, 216
Uniform magnetization, 232
Unpaired electrons, 197

V
Vector field, 97
Vibrating field, 16

W
Weiss field, 203

Z
Zeeman energy, 179
Zero flux density, 231
Zero-point energy, 19
Zone boundary, 141
Zone center, 141

Printed in the United States
By Bookmasters